Science and Technology of the Undercooled Melt

NATO ASI Series

Advanced Science Institutes Series

A Series presenting the results of activities sponsored by the NATO Science Committee, which aims at the dissemination of advanced scientific and technological knowledge, with a view to strengthening links between scientific communities.

The Series is published by an international board of publishers in conjunction with the NATO Scientific Affairs Division

A	Life Sciences	Plenum Publishing Corporation
B	Physics	London and New York
C	Mathematical and Physical Sciences	D. Reidel Publishing Company Dordrecht and Boston
D	Behavioural and Social Sciences	Martinus Nijhoff Publishers Dordrecht/Boston/Lancaster
E	Applied Sciences	
F	Computer and Systems Sciences	Springer-Verlag Berlin/Heidelberg/New York
G	Ecological Sciences	

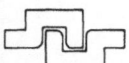

Series E: Applied Sciences – No. 114

Science and Technology of the Undercooled Melt

Rapid Solidification Materials and Technologies

edited by

P.R. Sahm
Foundry-Institute
Aachen Institute of Technology
Aachen
Federal Republic of Germany

H. Jones
Department of Metallurgy
University of Sheffield
Sheffield
UK

C.M. Adam
Materials Laboratory
Allied Corporation
Morristown, New Jersey
USA

1986 **Martinus Nijhoff Publishers**
Dordrecht / Boston / Lancaster
Published in cooperation with NATO Scientific Affairs Division

Proceedings of the NATO Advanced Research Workshop on "Rapid Solidification Technologies: Science and Technology of the Undercooled Melt", Theuern, Federal Republic of Germany, March 18-22, 1985

Library of Congress Cataloging in Publication Data

NATO Advanced Research Workshop on "Rapid Solidification
    Technologies: Science and Technology of the Under-
    cooled Melt" (1985 : Theuern, Germany)
    Science and technology of the undercooled melt.

    (NATO ASI series. Series E: Applied sciences ;
no. 114)
    "Proceedings of the NATO Advanced Research Workshop
on "Rapid Solidification Technologies: Science and
Technology of the Undercooled Melt", Theuern, Federal
Republic of Germany, March 18-22, 1985"--
    Includes index.
    1. Metals--Rapid solidification processing--
Congresses. I. Sahm, P. R. II. North Atlantic
Treaty Organization. Scientific Affairs Division.
III. Title. IV. Series: NATO ASI series. Series E,
Applied sciences ; v. 114.
TS247.N37 1985   671.3   86-14173

ISBN-13: 978-94-010-8483-3   e-ISBN-13: 978-94-009-4456-5
DOI: 10.1007/978-94-009-4456-5

Distributors for the United States and Canada: Kluwer Academic Publishers, 101 Philip Drive, Assinippi Park, Norwell, MA 02061, USA

Distributors for the UK and Ireland: Kluwer Academic Publishers, MTP Press Ltd, Falcon House, Queen Square, Lancaster LA1 1RN, UK

Distributors for all other countries: Kluwer Academic Publishers Group, Distribution Center, P.O. Box 322, 3300 AH Dordrecht, The Netherlands

TABLE OF CONTENTS

PREFACE

"SCIENCE AND TECHNOLOGY OF THE UNDERCOOLED MELT"

This title was chosen as the topical headline of the Advanced Research
Workshop (ARW) from March 17 to 22 1985, held at the Castle of Theuern.
The usual term "Rapid Solidification" is an overlapping description. Due to
the fact that nucleation is so eminently important for the undercooling of
a melt and this, in turn, is an important characteristic of rapid solidification, undercooling plays an essential role in "rapid solidification."
    The undercooled melt has caused an "accelerated evolution" (if not a
revolution) in materials science during the last decade. Several rather
exciting concepts with interesting potential for novel applications are
being pursued presently in various laboratories and companies. They concern
not only new processes and hardware developments, but also present chal-
lenging perspectives for ventures, including the founding of new companies;
or they promise growth possibilities with established larger and smaller
industrial establishments.
    The special challenge in the science and technology of the undercooled
melt lies in the fact that a very close coupling between fundamental research and pilot plant process development is not only possible but appears
to be a "conditio sine qua non" here for effective science-technology
transfer such as, for example, in controlled melt spinning processes, melt
particularization, or surface remelting techniques all of which were dealt
with during the workshop. Both materials scientists and engineers are
challenged equally in an attempt to provide, first the post-industrial
societies and then, following closely, the industrialising nations with
more energy and new materials-effective processes and technologies. This
interdisciplinary requirement was, in fact, reflected by both the invited
papers presented and extensively discussed, see Table 1, as well as the
posters displayed throughout the duration of the workshop. It indicates
that basic thermodynamics of phase changes, solidification kinetics, cry-
stallography and physical metallurgy of microstructure, experimental physics providing the necessary measurement, electronics and monitoring devices, to name but the most essential ones, are prerequisites for mastering
the posed challenges.
    These proceedings may thus serve to introduce scientists and engineers
to the field of the science and the technology of the undercooled melt,
i.e. "rapid solidification" as an overview of the current status of the
field and to a limited extent serving simultaneously also as a text-book of
sorts, be they already specialized or in the state of becoming acquainted
with the underlying issues.
    The ARW was sponsored by the NATO Scientific Affairs Division. In an
attempt to find a suitable secluded place for conducting an intensive
workshop and simultaneously realize part of the goals expressed above, e.g.
stimulate thinking or even activities into new realms of endeavor, it was
possible to gain some moral and material support from local industry (Linn
Elektronik GmbH) and government (Landkreis Amberg-Sulzbach). They not only
met some of the expenses but also took delight in offering a recreational

program, Table 2, including a reception by the district governor Dr. H. Wagner. The local newspaper reported the events at the castle, Figure 1, at some length even reflecting in detail on the involved technical significance of the subject, Figure 2.

Dr. H. Wolf, the museum director led a tour through the Museum for Mining and Industry attached to the Castle of Theuern, Figure 1, and participants visited the last underground iron ore mine operated in Germany "Leoni" and the steel works "Maxhütte". The region's highlights were thus displayed to the workshop participants.

The workshop participants came from 12 European countries and the United States of America, as shown in the listing of participants. Naturally, many German scientists and engineers used the chance to acquaint themselves with the potentials presently opened by "rapid solidification".

September 1985

P.R. Sahm, H. Jones

RAPID SOLIDIFICATION TECHNOLOGIES: SCIENCE AND TECHNOLOGY OF THE UNDERCOOLED MELT

ADVANCED RESEARCH WORKSHOP
under the auspices of the NATO Scientific Affairs Division

Monday 18-3-85	Tuesday 19-3-85	Wednesday 20-3-85	Thursday 21-3-85	Friday 22-3-85
	Topical Theme: Rapid Solidification, Undercooling, Nucleation, Thermodynamics and Kinetics. starting 8.30	Topical Theme: Microcrystalline and Metastable Phases incl. Glasses: Thermodynamics and Physical Metallurgy starting 8.30	Topical Theme: Rapid Solidification Technologies starting 8.30	Topical Theme: Properties and Applications for RSM (I), Impact on Materials Technology, "Strategic Materials", National Economics starting 8.30
	Keynote Lectures: - B. Cantor: Fundamentals in Rapid Solidification - J.H.Perepezko: Role of Nucleation in Rapid Solidification - B. Predel: Thermodynamics in Rapid Solidification - R.W. Cahn: Formation and Phase Stability in Rapid Solidification	Keynote Lectures: - W.J.Boettinger: Microstructure Formation in Rapidly Solidifying Materials - H.J. Güntherodt: Electronic Structure in Rapid Solidification	Keynote Lectures: - J. Durand: Magnetic and Superconducting Materials - G. Sepold: Surface Modification by Rapid Solidification - L. Katgerman: Continuous Products in Rapid Solidification - H. Fischmeister: Compaction of Rapidly Solidified Materials	Keynote Lectures: - H. Jones: Particulate Rapid Solidification - C.M. Adam: The Potential of RSt for Continuous Solidification Products - N.J. Grant: Engineering Properties and Applications of Rapidly Solidified Materials closing 16.00
Starting 16.00 Welcoming Party Poster Session				

March 18 - 22, 1985; Castle of Theuern, (Bergbau- und Industriemuseum Ostbayern), Portnerstraße 1, D 8451 Theuern (FRG), Tel. 09624 - 832

Table 1: A multiplicity of scientific and engineering disciplines as provided by the keynote lectures was to initiate an in-depth and extensive discussion of the particular topics.

RAPID SOLIDIFICATION TECHNOLOGIES:
SCIENCE AND TECHNOLOGY OF THE UNDERCOOLED MELT

ADVANCED RESEARCH WORKSHOP
under the auspices of the NATO Scientific Affairs Division

Time	Monday 18-3-85	Tuesday 19-3-85	Wednesday 20-3	Thursday 21-3-85	Friday 22-3-85
9–10		paper N° 1: B. Cantor	paper N° 5: W.J. Boettinger	paper N° 7: J. Durand	paper N° 9: H. Jones
11		coffee	coffee	coffee	coffee
12–13		paper N° 2: J. H. Perepzeko	paper N° 6: H.J. Güntherodt	paper N° 8: G. Sepold	paper N° 12: C. M. Adam
14–15	registration	lunch	visits to the continuous casting plant at Maxhütte, Sulzbach-Rosenberg and the iron ore mine Leoni, Auerbach	lunch reception by Linn Electronik GmbH on the premises	paper N° 13: H.J. Grant
16–17		paper N° 3: B. Predel		paper N° 10: L. Katgerman	snack lunch on the premises / end of workshop
18	"welcoming" party including a guided tour through the castle's museum and introducing posters	paper N° 4: R.W. Cahn		paper N° 11: H. Fischmeister	
20–21		"Bavarian Reception" by the District Governor on the premises			

March 18 - 22, 1985   Castle of Theuern, (Bergbau- und Industriemuseum Ostbayern),
Portnerstraße 1, D-8451 Theuern (FRG), Tel. 09624 - 832

Table 2: The course of the workshop was supported by recreational activities which included several features ranging from guided tours through the castle's museum and local industries plants to a reception offered by the local governor.

Figure 1: The Castle of Theuern

Bergbau- und Industriemuseum Ostbayern, Portnerstraße 1 Telefon: (09624) 832 8452 Theuern (Landkreis Amberg - Sulzbach).

X

„Eine weltweite Umwälzung steht bevor"

Professor Sahm spricht von Milliardenumsätzen – Tagung in Theuern von der NATO unterstützt

Von unserem Redaktionsmitglied Monika Beer

Theuern. Durch Lieder, Tänze und zünftige Blasmusik lernten die rund 70 Teilnehmer der derzeit in Schloß Theuern laufenden wissenschaftlichen Tagung etwas von der Oberpfälzer Mentalität kennen. Die Hausherren Landrat Dr. Hans Wagner und Museumsleiter Dr. Helmut Wolf sowie Unternehmer Horst Linn hatten zu einem „Bayerischen Abend" geladen, bei dem die aus Europa und Übersee stammenden Wissenschaftler und Techniker sich glänzend unterhielten.

Jugendblaskapelle und Mädchensinggruppe der Egerlandjugend, die Tanzgruppe des Trachtenvereins Stamm und dessen Jugend-Klarinettenmusik unter der Gesamtleitung von Seff Heil sorgten im Theuerner Festsaal für die richtige Einstimmung der Gäste. Prof. Dr. Peter Sahm, wissenschaftlicher Direktor der Tagung, überreichte den beiden Gastgebern gußeiserne Reliefkafeln, die im Gießerei-Institut der Rheinisch-Westfälischen Technischen Hochschule (RWTH) Aachen hergestellt wurden.

Daß an diesem Abend auch dem Small Talk gefrönt werden konnte, ist allerdings für die Tagung zum Thema „Technologie und Wissenschaft der unterkühlten Schmelze" eher die Ausnahme. Wie Prof. Sahm in einem Pressegespräch mitteilte, dient das fünftägige Seminar vor allem ausführlichen Diskussionen über diesen Spezialbereich der Werkstoffwissenschaft. „Hier findet", so Sahm, „eine weltweite Umwälzung statt. Sowohl neue Werkstoffe wie neue Technologien werden die Wir—" nachhaltig beeinflussen." Bereits für An——|ahre prognostiziere er durch di———rung" Umsätze in Milliarde—

Die durch diese Method— metalle, so Sahm weiter Eigenschaften aus, die e— ben hat: die sehr hohe F— stoffe, die besondere— schaften und die hoh— besser als bei rostfre— ren diese neuartige— dung der Katalyse— quenzen, Abschir— Elemente, Bleche schleißfeste Tor— technik.

Inzwischen — Gußeisen du— strahlen ein— haltene kor— Ein so beh— wellen in ——

werden. In den USA wurde in den letzten Jahren der Markt mit Lötfolien revolutioniert, die nach dem Prinzip der schnellen Erstarrung hergestellt werden, Japan hat nach Angaben von Prof. Sahm inzwischen die Spitzenposition bei elektromagnetischen Blechen und Bändern erreicht. „Um so wichtiger diese Tagung, denn Europa hinkt in dieser Hinsicht hinterher."

Die renommierten Wissenschaftler wollen bei dieser durch die NATO ermöglichten Tagung jedoch nicht nur aufholen: Im Ideenaustausch wird auch über die Möglichkeit nachgedacht, durch die Verbrennung von Wasserstoff eine neue, vor allem saubere Energieform zu entwickeln. „Das ist allerdings noch eine utopische

Vorstellung." Als konkretes Ergebnis des Brainstormings erwartet Sahm, daß von 100 Vorschlägen und Konzepten schließlich zehn tatsächlich realisiert und neue Wege aufzeigen können.

Daß die NATO Scientific Affais Division in Brüssel dieses Treffen der Wissenschaftler in Theuern durch finanzielle Unterstützung ermöglicht hat, sieht der Tagungsleiter als sehr positiv an. „Da die NATO", sagte Sahm, „in puncto Thematik und Zusammensetzung der Teilnehmer keinerlei Vorgaben gemacht hat, sehe ich darin auch eine Möglichkeit, deren Gelder für friedliche Forschungszwecke zu nutzen. In diesem Sinne erfüllen wir vielleicht sogar eine Mission."

Mit Streichholz und dünnem Draht versuchte Prof. Dr. Peter Sahm beim „Bayerischen Abend" Landrat Dr. Hans Wagner (rechts) das Prinzip einer technischen Neuerung zu veranschaulichen, die auch Thema der Tagung in Theuern ist. Bild: gf

„Crème de la crème" in der Oberpfalz

Theuern: Experten aus Europa und Übersee reden über unterkühlte Schmelze

Theuern. (bee) Wer weltweit in der modernen Werkstoffwissenschaft etwas zu sagen hat, gibt sich derzeit auf Schloß Theuern zum Erfahrungsaustausch ein Stelldichein: „Technologie und Wissenschaft der unterkühlten Schmelze" heißt das Thema eines 5-Tage-Seminars, zu dem 70 Fachleute aus zwölf europäischen Ländern und Übersee in die Oberpfalz gekommen sind.

Prof. Dr. Peter R. Sahm, unter anderem auch Projektleiter der Spacelab-Mission D 1, ist wissenschaftlicher Direktor dieser Tagung, die das Gießerei-Institut der Rheinisch-Westfälischen Technischen Hochschule (RWTH) Aachen unter der Schirmherrschaft der NATO Scientific Affairs Division veranstaltet. Daß die Crème de la crème der Werkstoffwissenschaft in der Oberpfalz zum Brainstorming zusammentrifft, ist dem Engagement der Firma Linn Elektronik (Eschenfelden) zu verdanken, deren Leiter Dipl.-Ing. Horst Linn auch für den Bereich der Anwendungstechnik im Seminar verantwortlich zeichnet.

Die in Fachkreisen bekannte Eschenfeldener Firma hat unter anderem für die RWTH eine Pilotanlage gebaut, bei der eine Metallschmelze

im elektromagnetischen Feld frei schwebend schockartig eingefroren werden kann. Was die Wissenschaftler unter „schneller Erstarrung" verstehen, kann Laien nur verblüffen: der Übergang einer Schmelze von der flüssigen in die feste Phase erfolgt hier binnen einer Millionstel Sekunde. Bei dieser schnellen Erstarrung – sie ist ein Hauptthema der Theuerner Tagung – können Metalle, die normalerweise feinkristallin fest werden, Strukturen erhalten, die denen von Glas entsprechen.

Die Auswirkungen dieser hochkomplizierten neuen Erstarrungstechnologie sind bedeutsam: durch sie wird es möglich, einfachen und billigen Werkstoffen, wie zum Beispiel Metallen, Eigenschaften zu geben, die sonst nur bei hochwertigem, teuren Material vorhanden sind. Im Bereich der Pulvermetallurgie soll es darüber hinaus über die schnelle Erstarrung möglich werden, den Werkstoffen gänzlich neue Eigenschaften zu geben. Fachleute sehen hier die Möglichkeit voraus, nicht nur bessere Werkstoffe auf rationellere Weise herstellen, sondern dabei auch besonders rohstoff- und energiesparend arbeiten zu können.

Innovative Erstarrungstechnologien stehen im Mittelpunkt der Tagung in Theuern, zu der rund 70 Fachleute aus Europa und Übersee gekommen sind. Unser Bild zeigt (v. L) Prof. Dr. R. W. Cahn (England), Dipl.-Ing. Horst Linn (Eschenfelden), Dr. W. J. Boettinger (USA), Prof. Dr. Marty E. Glicksman (USA) und Tagungsleiter Prof. Dr. Peter R. Sahm (RWTH Aachen).

Figure 2: The local newspaper took interest in the ARW at Theuern and reported in some detail on the future potential of "rapid solidification".

Participants List in Alphabetic Order and Role Played at ARW

L = Lecturer   C = Chairman

P = Poster   H = Organizational Help

                    W = Workshop Participant

Name and Address	Role	Name and Address	Role
C.M. Adam Materials Laboratory Allied Corporation P.O. Box 1021 R Morristown NJ 07960 USA	L	W.J. Boettinger Metallurgy Division National Bureau of Standards Gaithersburg, MD 20899 USA	L
J. Ågren Div. of Physical Metallurgy Royal Institute of Technology S-10044 Stockholm	P	G.J. Bunk DFVLR Institut für Werkstoff- Forschung Linder Höhe D-5000 Köln 90	W
L. Ajran Ardal og Sunndal Verk N-6601 Sunndalsora	W	R.W. Cahn Clare Hall Cambridge CB ODT U.K.	L
O. Arkens Katholieke Universiteit Leuven Department Metaallkunde de Croylaan 2 B-3030 Leuven	W	W. McCallum Senior Research Physicist Energy Conversion Devis. 1675 W. Maple Troy, MI 48084 USA	W
L. Arnberg Swedish Institute for Metals Research Drottning Kristians Väg 48 S-11428 Stockholm	C	B. Cantor Dept. of Metallurgy and Science of Materials University of Oxford Parks Road Oxford OX1 3 PH	L
M.D. Baro Facultad Ciencias Universidad Autonoma E-Bellaterra (Barcelona)	P	U.K. A. Chamberod D.R.F. Physique du Solide CEN, 85 X	W
H.W. Bergmann Institut für Werkstoffkunde TU Clausthal Agricolastraße 2 D-3392 Clausthal-Zellerfeld	P/C	F-38041 Grenoble Cedex G. Champier Physique des Solides E.N.S.M.I.M. Parc de Saurupt F-54042 Nancy Cedex	W

Name and Address	Role	Name and Address	Role
N. Clavaguera Facultad de Fisica Diagonal 647 E-08028 Barcelona	P	H. Fraser Dept. of Metallurgy University of Illinois Urbana, IL 61801 USA	W
M.T. Clavaguera-Mora Termologia, Fac. Ciencias Universidad Autonoma E-Bellaterra (Barcelona)	P	M.E. Glicksman Materials Eng. Dept. Rensselaer Polytechnic Institute Troy, NY 12181 USA	P
M.J. Couper Brown Boveri Research Center CH-5405 Baden	P	N.J. Grant Massachusetts Institute of Technology Dept. of Materials Science and Engineering 77 Massachusetts Avenue Cambridge, MA 02139 USA	L
J.M. Dubois Université de Nancy 1 Laboratoire de Physique du Solide B.P. 239 F-54042 Nancy Cedex	P		
J. Durand Université de Nancy 1 Laboratoire de Physique du Solide B.P. 239 F-54506 Vandoeuvre-les-Nancy Cedex	L	A.L. Greer University of Cambridge Dept. of Metallurgy Pembroke Street Cambridge, CB2 3QZ U.K.	W
K. Emmerich Vacuumschmelze GmbH Grüner Weg 37 D-6450 Hanau 1	W	H.J. Güntherodt Institut für Physik Universität Basel Klingelbergstraße 82 CH-4056 Basel	L
E.A. Feest Aere Harwell Building 35, Aere Didcot OXII U.K.	W	P.N. Hansen Termal Insulation Laboratory Technical University of Denmark Building 118 DK-2800 Lyngby	W
H. Fischmeister MPI für Metallforschung Institut für Werkstoffwissenschaften Seestraße 92 D-7000 Stuttgart 1	L	S. Hock Vacuumschmelze GmbH Grüner Weg 37 D-6450 Hanau 1	W

Name and Address	Role	Name and Address	Role
W. Hug Gießerei-Institut RWTH Aachen Intzestraße 5 D-5100 Aachen	P/H	U. Köster AG Werkstoffe u. Korrosion Abt. Chemietechnik Universität Dortmund D-4600 Dortmund	P/C
H. Jones The University of Sheffield Dept. of Metallurgy Mappin Street Sheffield S1 3 JD U.K.	L	G. Lapasset ONERA 29, Ave. de la Civision le clerc F-92320 Chautillon	W
L. Kallien Gießerei-Institut RWTH Aachen Intzestraße 5 D-5100 Aachen	P/H	F.J. Lenze Stahlwerke Bochum Castroper Straße 228 D-4630 Bochum 1	W
L. Katgerman Alcan International Ltd. Banbury Laboratories Southam Road Banbury OXI6 7 SP U.K.	L	R.E. Lewis Lockheed Palo Alto Research Laboratories 3251 Hannover Street Palo Alto, CA 94304 USA	W
T. Kaup Gießerei-Institut RWTH Aachen Intzestraße 5 D-5100 Aachen	H	H. Linn Linn-Elektronik GmbH Heinrich-Hertz-Platz 1 D-8459 Hirschbach 1	H
W.A. Kaysser MPI Metallforschung Heisenbergstraße 5 D-7000 Stuttgart	W	H. Litterscheidt Thyssen Stahl AG Franz-Lenze-Straße 1 D-4100 Duisburg	W
P.O. Kettunen Tampere Univ. of Technology Inst. of Materials Science SF-33101 Tampere 10	P	M.H. Loretto Dept. of Metallurgy and Materials University of Birmingham Birmingham B 152 TT U.K.	P/C
C.S. Kiminami Gießerei-Institut RWTH Aachen Intzestraße 5 D-5100 Aachen	P/H	R.E. Maringer Batelle Laboratories 505 King Avenue Columbus, OH 43201 USA	C

Name and Address	Role	Name and Address	Role
F. Moret Centre d'Etudes Nucleaires de Grenoble DMG-SEM-LES, 85X F-38041 Grenoble Cedex	P	G. Sepold Bremer Inst. f. angewandte Strahltechnik Ermlandstraße 59 D-2820 Bremen 71	L
J.S. Muñoz Dept. of Electricity Facultad Ciencias Universidad Autonoma E-Bellaterra (Barcelona)	W	T. Sheppard Dept. of Metallurgy and Materials Science Imperial College London SW7 2BP U.K.	C
J.H. Perepezko University of Wisconsin-Madison Dept. of Metallurgical and Mineral Engineering 1509 University Ave. Madison, WI 53706 USA	L	F. Sommer MPI Metallforschung Institut für Werkstoffwissenschaften Seestraße 92 D-7000 Stuttgart	W
K. Peters Stahlwerke Bochum Castroper Straße 228 D-4630 Bochum 1	W	G. Staniek DFVLR-Institut für Werk- stoff-Forschung Linder Höhe D-5000 Köln 90	W
H. von Philipsborn Universität Regensburg Postfach 397 D-8400 Regensburg	W	A. Thorvaldsen Norsk Hydro Box 110 N-3901 Porsgrunn	W
B. Predel Universität Stuttgart Inst. f. Metallkunde Seestraße 75 D-7000 Stuttgart 1	L	R. Trivedi Iowa State University 100 Metallurgy Building Ames, IA 50011 USA	P
P.R. Sahm Gießerei-Institut RWTH Aachen Instzestraße 5 D-5100 Aachen	P	E. Vogt MPI für Eisenforschung Max-Planck-Straße 1 D-4000 Düsseldorf	P
R. Schulz Vacuumschmelze GmbH Grüner Weg 37 D-6450 Hanau	W	H. Warlimont Vacuumschmelze GmbH Grüner Weg 37 D-6450 Hanau 1	C

Name and Address	Role	Name and Address	Role
H. Wolf Bergbau- und Industrie- museum Ostbayern/Schloß Theuern Portnerstraße 1 D-8451 Theuern	H		
J.V. Wood Faculty of Technology The Open University Milton Keynes MK 7 6 AA U.K.	W		

A. INVITED PAPERS AND SUMMARIZED DISCUSSIONS

FUNDAMENTALS OF RAPID SOLIDIFICATION

B. CANTOR
Department of Metallurgy and Science Materials
University of Oxford
Oxford OX1 3 PH, U.K.

ABSTRACT
    This paper reviews several fundamental aspects of the rapid solidification of metals and alloys. Different methods of measuring cooling rates during rapid solidification are described, and the observed cooling curves are analysed according to various cooling models. The experimental effects of rapid solidification conditions on heat transfer coefficient and cooling rate in the vicinity of the solidification point are also described, and the efficiencies of different rapid solidification processes are compared. Measured undercoolings and the shape of the solidification arrest are analysed according to various models of nucleation and growth during solidification, and the effects of cooling rate and undercooling on important microstructural features such as grain size, dendrite arm spacing, and extent of segregation are discussed.

1. INTRODUCTION

    Rapid solidification processing can be used to manufacture many new materials with improved properties (1, 2). A wide variety of different rapid solidification processes have therefore been developed, including gun and piston splat quenching (3, 4), melt spinning and planar flow casting (5), melt extraction (6), inert gas and water atomisation (7), plasma and atomised spray deposition (8), and laser and electron beam surface melting (9). Although the detailed technology varies very greatly from one rapid solidification process to another, the fundamental principles are virtually the same. In all rapid solidification processes, heat is removed as fast as possible from a mass of liquid metal, and this is achieved by manipulating the liquid mass so as to be thin in at least one dimension and at the same time in good thermal contact with an efficient heat sink. With a high rate of heat removal, the liquid cools rapidly, often to a temperature well below the equilibrium freezing point. Solidification then takes place under non-equilibrium conditions of high liquid undercooling and high solidification front velocity, which lead to a variety of desirable microstructural effects, including small grain size, fine scale or zero microsegregation, extended solubility of alloying elements, and formation of metastable crystalline and amorphous phases. The objective of this paper is to examine the interrelationships between (a) imposed thermal conditions during rapid solidification, (b) resulting solidification conditions of cooling rate, liquid undercooling, and solidification front velocity, and (c) final microstructural effects, in particular grain size, segregation, solubility and post-solidification phase transformations.

2. COOLING RATES

    In most rapid solidification processes, the thin dimension of the liquid is typically ≤ 100 µm with a corresponding cooling rate of $\geq 10^5$ Ks^{-1}, so that cooling and solidification are completed within a matter of a few milliseconds. Recording temperature variations in such a short space of

time is experimentally difficult and very few cooling rate measurements
have been reported (10, 11). A number of cooling curves have been obtained
from liquid metal droplets piston-quenched between two plates, by embedding
a temperature sensor in the surface of one of the plates (12 - 15). The
temperature sensor can be either a photo-electric cell (12), a thin junction
conventional thermocouple (13, 14), or an intrinsic thermocouple in which the
junction is completed by the liquid metal (15). An embedded intrinsic
thermocouple has also been used with gun-quenched liquid metal droplets
(16). Figure 1 shows a typical series of cooling curves obtained in this
way for liquid Fe droplets quenched between two Cu plates using an embedded
thin junction chromel-alumel thermocouple (13). This efficiency of the
cooling process can be seen to increase with increasing plate velocity. A
number of cooling curves have also been obtained from melt-spun ribbons, by
taking black and white (17, 18) or colour (19 - 21) photographs of the
unchilled ribbon surface during melt spinning, under exposure and development conditions for which a correlation between film density and surface
temperature has previously been established.

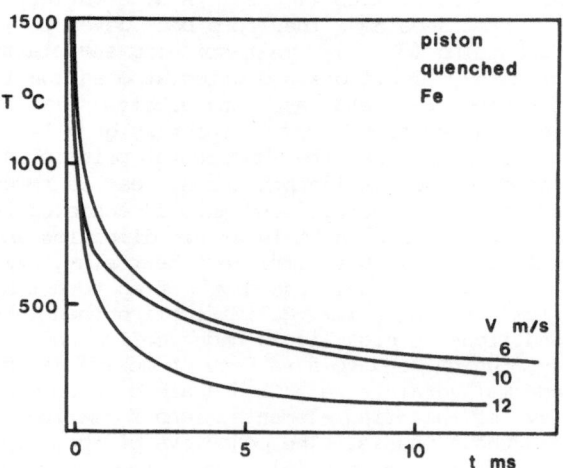

Figure 1: Cooling curves for liquid Fe droplets piston splat -quenched between two Cu
plates, using an embedded chromel-alumel thin
junction thermocouple, at plate velocities of
6, 10 and 12ms^{-1}.(13)

Figure 2 shows a typical series of cooling curves obtained in this way for Ni-5wt%Al ribbons melt-spun on a Cu wheel using colour photography (19). The efficiency of the cooling process can again be seen to increase with increasing wheel velocity.

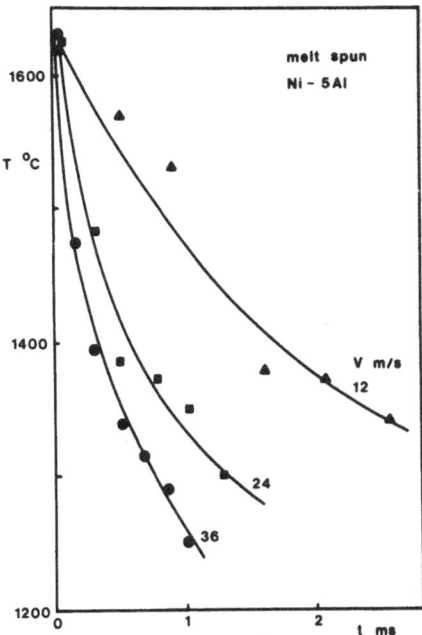

Figure 2: Cooling curves for Ni-5wt%Al ribbons melt-spun on a Cu wheel, using temperature-calibrated colour photography, at a superheat of 200 K, ejection pressure of 42kPa, and wheel speeds of 12, 24 and 36ms^{-1}.(19)

Figure 3 shows all available measurements of mean cooling rate in the vicinity of the soldification point for piston-quenched liquid droplets and melt-spun ribbons, taken from cooling curves such as in Figures 1 and 2 and plotted as a function of plate or wheel velocity. Figure 4 shows the same data plotted as a function of specimen thickness. The cooling rate measurements in Figures 3 and 4 cover a wide variety of different metals and alloys being rapidly solidified under a wide variety of different conditions of melt superheat, heat sink material, heat sink surface condition, surrounding atmosphere etc. The measured cooling rates range from 5.10^4 – $2.10^7 Ks^{-1}$ for plate and wheel velocities in the range 1-30ms^{-1} and specimen thicknesses in the range 20-130 μm. Within a fair amount of scatter, there is a tendency for the cooling rate to increase with increasing plate or wheel velocity and decreasing specimen thickness. The most remarkable feature of the data in Figures 3 and 4 is that measured cooling rates during melt spinning show excellent agreement with a linear variation with wheel velocity, virtually independent of melt-spun ribbon material or other melt spinning variables. This is shown more clearly in Figure 5, and can be represented by:

$$\dot{T} = aV \qquad (1)$$

where $\dot{T}$ is the mean cooling rate during melt spinning in the vicinity of the solidification point, V is wheel velocity, and the constant a has a value of $1.2 \cdot 10^4$ K m^{-1}. The reason for such a straightforward dependence of cooling rate on wheel speed is difficult to explain. Nevertheless, Figure 5 and equation (1) are extremely useful, since cooling rates can be predicted for a wide range of melt spinning conditions and melt-spun materials.

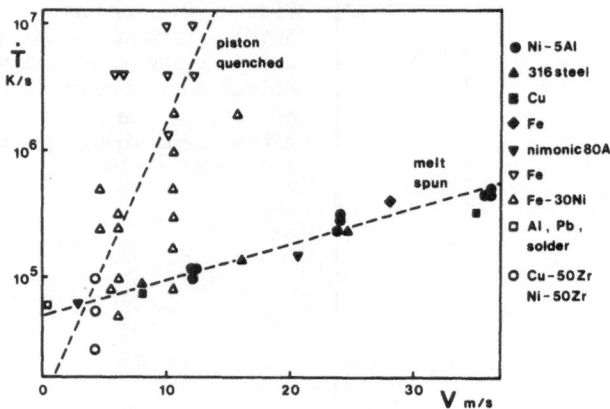

Figure 3: Mean cooling rate $\dot{T}$ in the vicinity of the solidification point as a function of plate or wheel velocity V for a variety of different piston-quenched and melt-spun metals and alloys (12 - 21). Open symbols, piston-quenched; full symbols, melt-spun.

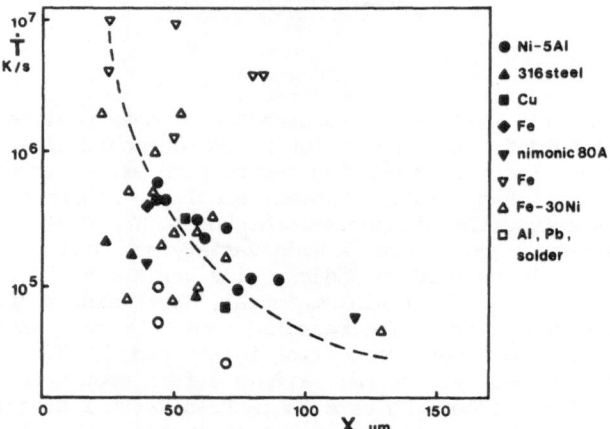

Figure 4: Mean cooling rate $\dot{T}$ in the vicinity of the solidification point as a function of thickness X for a variety of different piston-quenched and melt-spun metals and alloys (12-21). Open symbols, piston-quenched; full symbols, melt-spun. (X = half-thickness for piston-quenched material).

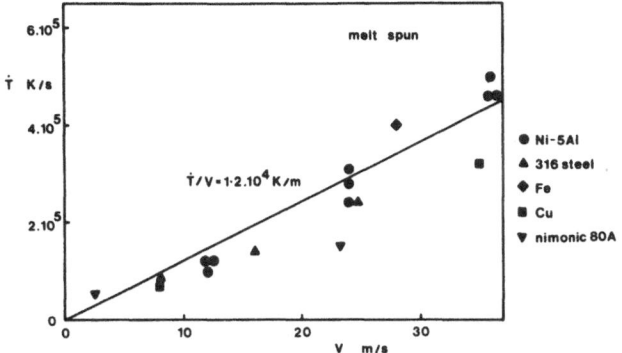

Figure 5: Linear relationship between mean cooling rate $\dot{T}$ in the vicinity of the solidification point and wheel speed V for a variety of different melt-spun metals and alloys and a variety of other melt spinning process parameters (17 - 21).

3. 1-D HEAT FLOW

To analyse in detail cooling curves such as in Figures 1 and 2, consider the situation shown schematically in Figure 6. A slab of material of thickness X in the x direction and of infinite extent in the y and z directions is initially at a uniform temperature T_I. One face of the slab is brought at time t = 0 into contact with a heat sink which is initially at a lower uniform temperature T_s. The slab cools by 1-D heat flow in the x direction normal to the slab/heat sink interface at x = X. Heat flow in the slab is by conduction, with the rate of heat flow per unit area q given by Fourier's law:

$$q = - k \partial T/\partial x \qquad (2)$$

where k is the thermal conductivity of the slab material and T(x,t) is the temperature. The heat sink is taken to be perfectly efficient so that its temperature remains constant at T_s. Thermal contact between slab and heat sink is described by Newton's law of cooling, with the rate of heat flow through unit area of interface given by:

$$q = h(T_X - T_s) \qquad (3)$$

where h is the heat transfer coefficient and T_X is the slab temperature adjacent to the interface i.e. at x = X.

The resulting temperature profiles in the slab at different times can be found (22) by solving the 1-D thermal diffusion equation:

$$\partial T/\partial t = \alpha \partial^2 T/\partial x^2 \qquad (4)$$

where $\alpha = k/\rho C$ is the thermal diffusivity of the slab material, ρ is its density and C its specific heat.

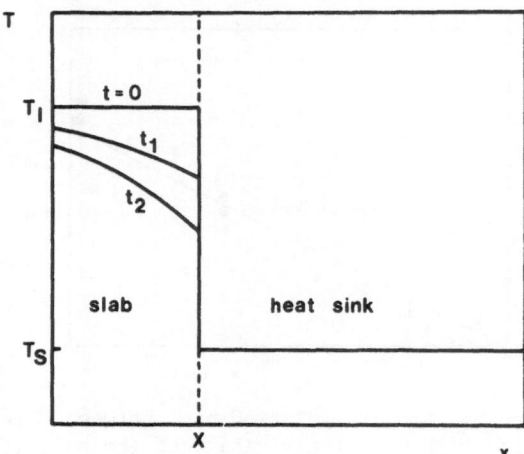

Figure 6:   Schematic temperature profiles
through a slab of thickness X and initial
temperature T_I, cooling one-dimensionally
in contact with a perfectly efficient heat
sink at initial temperature T_s.

The temperature profiles are most conveniently expressed in terms of dimensionless temperature $\theta = (T - T_s)/(T_1 - T_s)$, dimensionless distance $X = x/X$, and dimensionless time $\tau = \alpha T/X^2$:

$$\theta = \sum_{n=1}^{\infty} \frac{2N\cos(A_n\chi)\sec(A_n)\exp(-A_n^2\tau)}{N(N+1) + A_n^2} \tag{5}$$

where $N = hX/k$ is the Nusselt number and A_n is an infinite series of constants given by the positive roots of the equation $A \tan A = N$. The slab temperature at the slab/heat sink interface $T_X = T(X,t)$, the slab temperature at the unchilled surface $T_0 = T(0,t)$, and the average slab temperature $T_{AV} = X^{-1} \int_0^X T(x,t)dx$ are given by:

$$\theta_X = \sum_{n=1}^{\infty} \frac{2N\exp(-A_n^2\tau)}{N(N+1) + A_n^2} \tag{6}$$

$$\theta_0 = \sum_{n=1}^{\infty} \frac{2N\sec(A_n)\exp(-A_n^2\tau)}{N(N+1) + A_n^2} \tag{7}$$

$$\theta_{AV} = \sum_{n=1}^{\infty} \frac{2N^2\exp(-A_n^2\tau)}{A_n^2 N(N+1) + A_n^4} \tag{8}$$

Identical results are obtained for a slab of thickness 2X cooling with both faces in contact with a heat sink, except that T_χ now gives the temperatures of both faces of the slab, and T_0 gives the temperature at the centre. Perfect Newtonian cooling conditions correspond to the extreme case of N > 1, with negligible temperature gradients in the slab, and heat transfer across the slab/heat sink interface providing the only resistance to heat flow. Equation (5) then becomes

$$\theta = \exp(-N\tau) \tag{9}$$

Ideal cooling conditions correspond to the other extreme case of N >> 1, with perfect thermal contact between slab and heat sink, and conduction in the slab providing the only resistance to heat flow. Equation (5) then becomes:

$$\theta = \sum_{n=1}^{\infty} \frac{2\cos(B_n\chi)\exp(-B_n^2\tau)}{B_n^2} \tag{10}$$

where $B_n = (n+1/2)\pi$.

Modified analytical expressions equivalent to equations (5) – (10) can be obtained for more complex initial temperature distributions in the slab (22). Such effects are unlikely to be significant when analysing cooling curves such as in Figures 1 and 2 from piston-quenched liquid droplets and melt-spun ribbons. Finite difference computer heat flow calculations (23, 24) indicate relatively little effect of an imperfectly efficient heat sink during piston quenching or melt spinning, as long as the heat sink thermal conductivity is greater than the slab conductivity.

4. HEAT TRANSFER COEFFICIENTS

Cooling curves obtained during rapid solidification often show little or no evidence of a solidification arrest (12 – 21), as can be seen on Figures 1 and 2, because of insufficient resolution in the temperature measurements. In these circumstances, the most convenient way to allow for evolution of the latent heat of solidification L in equations (5) – (10) is to replace the specific heat C by $C + L/\Delta T_s$ where ΔT_s is the effective temperature range over which solidification takes places. Figure 7 shows equations (6) and (7) for the chilled and unchilled surface temperatures T_χ and T_0 over a range of values of Nusselt number N plotted in the form $\ln_e\theta$ and $\ln_e\theta$ versus $\tau = \alpha t/x^2$ where $\alpha = k/\rho(C+L/\Delta T_s)$. The parameter $L/\Delta T_s$ is taken to be $1.5\cdot10^3$ J kg^{-1} K^{-1} to allow for liquid undercoolings of ~ 200 K before the onset of solidification, as discussed further in a later section. Also included on Figure 7 are data points from a variety of cooling curves obtained during rapid solidification such as in Figures 1 and 2. Piston - quenched data correspond to measurements of the chilled surface temperature T_χ, and melt-spun ribbon data correspond to measurements of the unchilled surface temperature T_0. As shown in Figure 7, Nusselt numbers are typically 0.5-1 during piston quenching and 0.05-0.1 during melt spinning, significantly lower because of poorer thermal contact between slab material and heat sink.

Heat transfer coefficients calculated from the Nusselt numbers obtained from Figure 7 are shown as a function of plate or wheel velocity in Figure 8. Heat transfer coefficients are in the range $3-6.10^5$ W m^{-2} K^{-1} for piston quenching and $10^{-4} - 10^{-5}$ W m^{-2} K^{-1} for melt spinning, again significantly lower because of poorer thermal contact between slab material and heat sink. In addition, increasing plate or wheel velocity leads to improved thermal contact and a correspondingly higher value of heat transfer coefficient. In melt-spun ribbons, this effect is associated with a clearer replication of the wheel surface topography at the higher wheel velocities (18, 19, 25). Relative variations of Nusselt number and heat transfer coefficient are shown correctly in Figures 7 and 8, but the absolute values should be treated with some caution because of uncertainty in the value of ΔT_s.

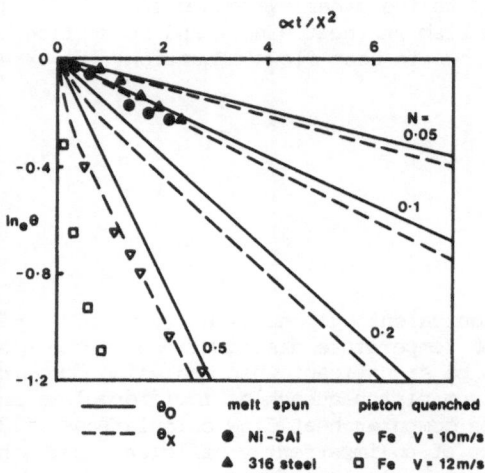

Figure 7: Non-Newtonian heat flow analysis (22) of chilled surface temperature T and unchilled surface temperature T_0 as a function of time t for 1-D slab in Figure 6, plotted as $\ln_e \theta_X = \ln_e (T_X-T_s)/(T_I-T_s)$ and $\ln_e \theta_0 = \ln_e (T_0-T_s)/(T_I-T_s)$ versus dimensionless time $\tau = \alpha t/X^2$, where α is thermal diffusivity and X is slab thickness. The calculated lines are shown for a variety of values of Nusselt number N, together with experimental data points from melt-spun Ni-5wt%Al and 316 stainless steel (21) and piston-quenched Fe (13).

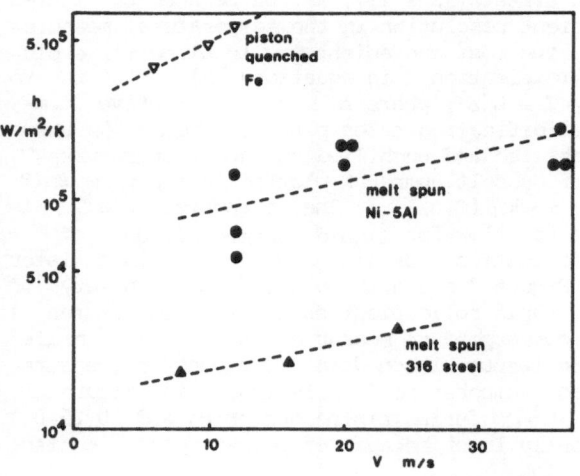

Figure 8: Heat transfer coefficient h as a function of plate or wheel velocity V for melt-spun Ni-5wt%Al (19) and 316 stainless steel (21), and piston-quenched Fe (13).

5. 2-D AND 3-D HEAT FLOW

For 2-D cooling of a circular-cross section cylinder as in melt extraction, equation (5) is replaced by (22):

$$\theta = \sum_{n=1}^{\infty} \frac{2N J_0(C_n \chi) \exp(-C_n^2 \tau)}{(C_n^2 + N^2) J_0(C_n)} \tag{11}$$

where x is now the radial dimension, X is the radius of the circular cross section, C_n are roots of $C J_1(C) = N J_0(C)$, and J_0 and J_1 are Bessel functions of the first kind of order zero and one respectively. For 3-D cooling of a sphere as for an atomised liquid droplet, equation (5) is replaced by (22):

$$\theta = \sum_{n=1}^{\infty} \frac{2N\{D_n^2 + (N-1)^2\} \sin(D_n \chi) \sin(D_n) \exp(-D_n^2 \tau)}{\chi\{D_n^2 N(N+1) + D_n^4\}} \tag{12}$$

where D_n are the roots of $D \cot D = 1-N$. Under perfect Newtonian cooling conditions, equations (11) and (12) become:

$$\theta = \exp(-2N\tau) \tag{13}$$

$$\theta = \exp(-3N\tau) \tag{14}$$

for cylinder and sphere respectively. Under ideal cooling conditions, equations (11) and (12) become:

$$\theta = \sum_{n=1}^{\infty} \frac{2 J_0(E_n \chi) \exp(-E_n^2 \tau)}{E_n J_1(E_n)} \tag{15}$$

$$\theta = \sum_{n=1}^{\infty} \frac{2(-1)^n \sin(F_n \chi) \exp(-F_n^2 \tau)}{F_n \chi} \tag{16}$$

for cylinder and sphere respectively, where E_n are the roots of $J_0(E) = 0$ and $F_n = n\pi$. Inspection of equations (5) - (16) shows that cooling increases in efficiency in going from 1-D to 2-D to 3-D under equivalent conditions of heat transfer coefficient, slab material, slab thickness etc. Unfortunately, no experimental cooling curves are available to test the validity of the 2-D and 3-D equations (11)-(16).

6. LIQUID UNDERCOOLING AND SOLIDIFICATION RATE

Solidification arrests have been detected on one or two rapid solidification cooling curves (14, 15, 18, 26). The cooling rates in the liquid before the onset of solidification are found to be approximately one order of magnitude higher than the mean cooling rates shown in Figures 3 and 4 over the whole solidification range (15, 18). Measured undercoolings in the liquid before the onset of solidification are found to be rather variable, and range from 50 - 250 K (12, 14, 15, 26).

Figure 9 shows a schematic solidification arrest during rapid solidification. The liquid undercools to a temperature ΔT_N below the equilibrium freezing temperature T_f before solidification is nucleated, followed by recalescence to a steady state growth plateau at a temperature T_G below the equilibrium freezing point. Assume that perfect Newtonian conditions persist into the solidification region for the 1-D slab in Figure 6, i.e. that no thermal gradients develop in the slab. The latent heat liberated by solidification is dissipated partly by heat transfer across the slab/heat sink interface and partly by a rise in slab temperature:

$$L\rho v = h(T-T_s) + X\rho CdT/dt \qquad (17)$$

where $v = dy/dt$ is the solidification rate, y is the thickness solidified, and dT/dt is the rate of increase in slab temperature. Assume that solidification kinetics are linear so that the solidification rate is proportional to undercooling:

$$v = m\Delta T \qquad (18)$$

where m is the solid/liquid interface mobility and $T = T_f-T$ is the undercooling. Equations (17) and (18) can be used in different ways to describe both the recalescence region and the steady growth plateau on the cooling curve in Figure 9.

On the steady state growth plateau in Figure 9, $dT/dt = 0$ so that the last term in equation (17) can be ignored. The steady state undercooling ΔT_G and solidification rate v_G on the plateau in Figure 9 are then given by:

$$\Delta T_G = h(T_f-T_s)/(h+L\rho m) \qquad (19)$$

$$v_G = hm(T_f-T_s)/(h+L\rho m) \qquad (20)$$

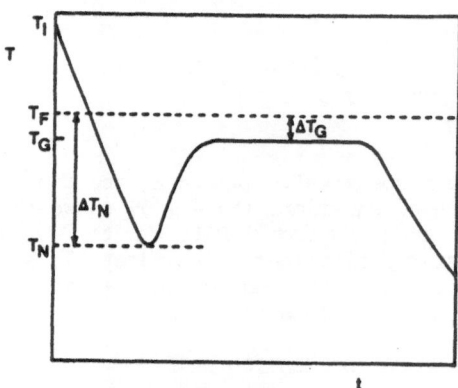

Figure 9: Schematic solidification arrest on cooling curve during rapid solidification of slab in Figure 6, showing nucleation undercooling ΔT_N below the equilibrium freezing temperature T_f, followed by recalescence to a steady state growth plateau at an undercooling ΔT_G below T_f.

Taking $h \sim 10^5$ W m^{-2} K^{-1} from Figure 8 and $m \sim 0.1$ms^{-1} K^{-1}, (27) values of ΔT_G and v_G are typically 0.5K and 5mm/s for most metals. In the recalescence region of Figure 9, dT/dt is large so that the second term in equation (17) can be ignored. Using dT/dt = vdT/dy, equation (17) can then be integrated from a nucleation undercooling ΔT_N at y = 0, to show that the fractional thickness solidified during recalescence η = y/X increases linearly with decreasing undercooling:

$$\eta = C(\Delta T_N - \Delta T)/L \qquad (21)$$

Recalescence is complete and the steady state growth plateau is reached in Figure 9 when $\Delta T = \Delta T_G$, which is sufficiently small to be ignored in equation (21). The fractional thickness solidified at the transition from recalescence to steady state growth η_G is therefore given by:

$$\eta_G = C\Delta T_N/L \qquad (22)$$

For most metals L/C is in the range 250–450K. (28) When the nucleation undercooling ΔT_N is greater than L/C, solidification is completed during recalescence and there is no steady state growth plateau on the cooling curve. Half of the thickness is solidified during recalescence when $\Delta T_N = \frac{1}{2}$L/C, a quarter of the thickness when $\Delta T_N = \frac{1}{4}$L/C, and so on.

Equation (17) can be integrated with all terms included to describe the whole of the solidification arrest on the cooling curve in Figure 9 (11).

$$\eta = A(\Delta T_N - \Delta T) + AB \ln \{(\Delta T_N - B)/(\Delta T - B)\} \qquad (23)$$

where A = Cρm/(h+Lρm) and B = h($T_f - T_s$)/(h+Lρm). When Newtonian conditions are not maintained during solidification, latent heat remains concentrated in the vicinity of the solid/liquid interface. A temperature inversion builds up ahead of the solid/liquid interface, and equations (17) – (23) underestimate the rate of recalescence. Finite difference computer heat flow calculations confirm that this can happen at the higher values of heat transfer coefficient (29).

7. GRAIN SIZE

The microstructure of melt-spun ribbons typically consists of columnar grains through the ribbon thickness (19–21). Figure 10 shows a typical example of this type of microstructure in a melt-spun ribbon of Ni-5at%NbC (30).

Figure 10: Optical micrograph of longitudinal through thickness section of a melt-spun ribbon of Ni-5at%NbC, showing columnar as-solidified grains through the ribbon thickness (30). Wheel surface at bottom of micrograph.

14

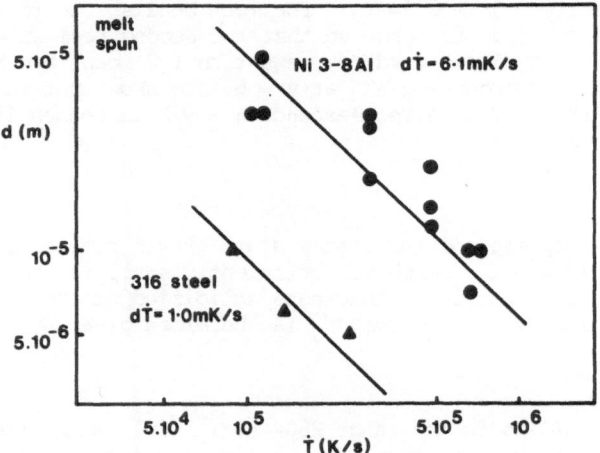

Figure 11: Grain size d as a function of mean
cooling rate $\dot{T}$ in the vicinity of the solidification point for melt-spun Ni-Al alloys   (19)
and   316 stainless steel (21).

Figure 11 shows the variation of columnar grain diameter d with mean
cooling rate $\dot{T}$ during solidification for melt spun ribbons of   Ni3-8wt%Al
(19) and 316 stainless steel (21). The grain size is inversely proportional
to cooling rate and can be expressed by:

$$d = b/\dot{T} \tag{24}$$

where the proportionality constant b has a value of 6.1m K s^{-1} for the Ni-
Al alloys and 1.0m K s^{-1} for 316 stainless steel. To explain these microstructural observations, consider the nucleation of the solidification
process   at an undercooling ΔT_N for the 1-D slab in Figure 6. Nucleation
takes place at the slab/heat sink interface, partly because the temperature
is lowest at this point and partly because the heat sink surface can act as
a nucleation catalyst. Nucleated grains grow laterally until the slab/heat
sink interface is completely covered. A stable set of grains then continue
to grow in a columnar way through the slab thickness. Although some competitive columnar grain growth may take place, the columnar grain diameter d
is primarily determined by a balance between the initial rates of nucleation and lateral growth on the slab/heat sink interface.

Under Newtonian cooling conditions, the time taken to cool from the
equilibrium freezing temperature T_f to a nucleation undercooling ΔT_N can be
obtained approximately from equation (9):

$$t = \Delta T_N/\dot{T} \tag{25}$$

when $\Delta T_N \ll T_f-T_s$. Let t_s be the time taken to cover completely the slab/-
heat sink interface by nucleation and lateral growth, and assume that t_s is
some small fraction f of the total cooling time t:

$$t_s = ft \tag{26}$$

The mean area of the slab/heat sink interface covered by each grain can be expressed in terms of either the lateral grain diameter d, the lateral solidification rate v, or the nucleation rate per unit area of slab/heat sink interface N_A:

$$\tfrac{1}{4}\pi d^2 = \tfrac{1}{4}\pi v^2 t_s^2 = 1/N_A t_s \qquad (27)$$

The lateral solidification rate is given by the linear kinetic equation (18) $v = m\Delta T_N$, and the nucleation rate can be written approximately as(31);

$$N = E\exp(-\epsilon/\Delta T_N^2) \qquad (28)$$

where $E = n_A(k_B T_F/h_p)\exp(-\Delta G_D/k_B T_F)$, n_A is the number of available nucleation sites per unit area of slab/heat sink interface, k_B and h_p are Boltzmann's and Planc's constants, ΔG_D is the activation energy for addition of an atom to a critical nucleus, $\epsilon = 16\pi\gamma_{SL}^3 T_F g(\theta)/3L^2\rho^2 k_B$, γ_{SL} is the solid/liquid surface energy, $g(\theta) = \tfrac{1}{4}(2-3\cos\theta + \cos^3\theta)$ is a factor to allow for catalytic nucleation, and θ is the contact angle for a solid nucleus on the heat sink surface.

Equations (18) and (25) - (28) can be solved to give independently the variation of grain size and nucleation undercooling with cooling rate:

$$2mf\epsilon/d\dot{T} = \ln_e(\pi^2 f E^2 d^5/16m\dot{T}) \qquad (29)$$

$$\epsilon/\Delta T_N^2 = \ln_e(\pi m^2 f^3 E\Delta T_N^5/4\dot{T}^3) \qquad (30)$$

A simpler expression can also be obtained which includes all three variables:

$$d = mf\Delta T_N^2/\dot{T} \qquad (31)$$

If nucleation undercooling is approximately constant in equation (31), then grain size varies inversely with cooling rate as observed experimentally in Figure 11 and equation (24). Finite difference computer heat flow calculations also indicate an inverse dependence of grain size on cooling rate (32). Equations (24), (29) and (31) can be used to predict grain size after rapid solidification by splat quenching or melt spinning, but not atomization or surface melting. Nucleation during gas or water atomization depends critically on the presence or absence of any heterogeneous nucleation catalyst in each individual atomised liquid droplet. Widely different nucleation undercoolings are therefore possible in droplets of similar size and cooling rate(33). There is no nucleation barrier during laser or electron beam surface melting, and the rapidly-solidified grain size is strongly influenced by the initial grain size of the underlying self substrate (34).

8. SEGREGATION

In melt-spun alloy ribbons, the microstructure frequently exhibits a segregation - free zone in the initial stages of solidification adjacent to the chilled surface, with a cellular or dendritic segregation pattern which develops within the columnar grains in the later stages of solidification near the unchilled surface. Figure 12 shows a typical example of this type of microstructure in a melt-spun ribbon of Ni-11wt%Al (19). The thickness of the segregation—free zone is often variable as shown in Figure 12. Similar microstructures are also seen in alloys rapidly solidified by splat quenching (35), melt extraction (36), and atomization (37).

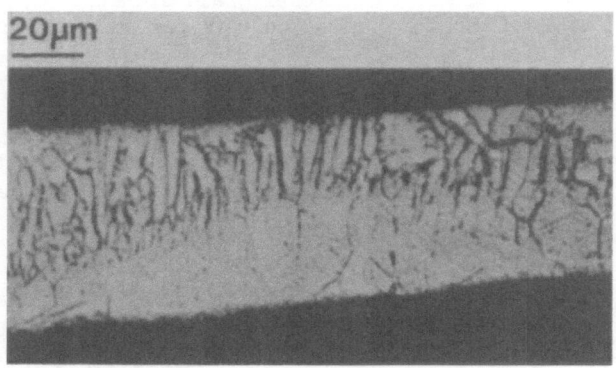

Figure 12: Optical micrograph of longitudinal
through-thickness section of a melt–spun ribbon
of Ni-11wt%Al, showing transition from segregation-free to cellular solidification (19). Wheel
surface at bottom of micrograph.

Consider the schematic alloy phase diagram shown in Figure 13. A rapidly-
solidified alloy of composition c_0 nucleates at an undercooling ΔT_N below
the solidus temperature, followed by recalescence to a steady state growth
plateau at an undercooling ΔT_G below the solidus. At the nucleation temperature T_N, the maximum driving force for solidification can be found from a
parallel tangent construction (11), as shown on the molar free energy/com-
position curve in Figure 14. Since other factors in the nucleation process
are unlikely to depend strongly on composition, solid nuclei are therefore
formed with a composition c_N, significantly lower than the alloy composition c_0. As the solid nuclei begin to grow, the temperature recalesces to
the steady state growth plateau on the cooling curve. For a planar solidification front, the solid composition changes from c_N to c_0, the only solid
composition for which planar, steady state growth is possible. Figure 15
shows the molar free energy/composition curve at the steady state growth
temperature T_G, and Figure 16 shows corresponding profiles of composition
and chemical potential through the solidifying interface. The steady state
solidification rate v_G and undercooling T_G are given by the heat balance
equations (19) and (20) with T_f taken as a solidus temperature. The liquid
composition and the solid/liquid composition at the solid/liquid interface
c_L adjust to provide a suitable driving force for the two alloy components:

$$v_G = m_A(\mu_{AL} - \mu_{AS}) = m_B(\mu_{BL} - \mu_{BS}) \qquad (32)$$

where m_A and m_B are solid/liquid interface mobilities, μ_{AL} and μ_{BL} are
chemical potentials for a liquid composition c_L, μ_{AS} and μ_{BS} are chemical
potentials for a solid composition c_0, and solidification kinetics are
assumed to be linear.

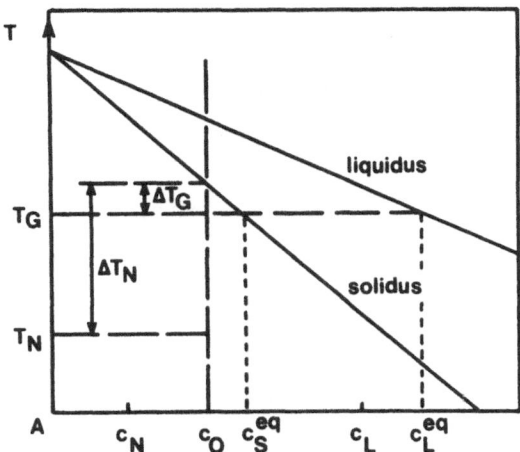

Figure 13: Schematic A-B alloy phase diagram for rapid solidification of alloy composition c , which undercools by ΔT_N below the solidus before the onset of nucleation, followed by recalescence to a steady
state growth plateau at an undercooling ΔT_G
below the solidus. Composition of solid
nuclei = c_N ; composition of liquid adjacent to steady state solid/liquid interface
= c_L.

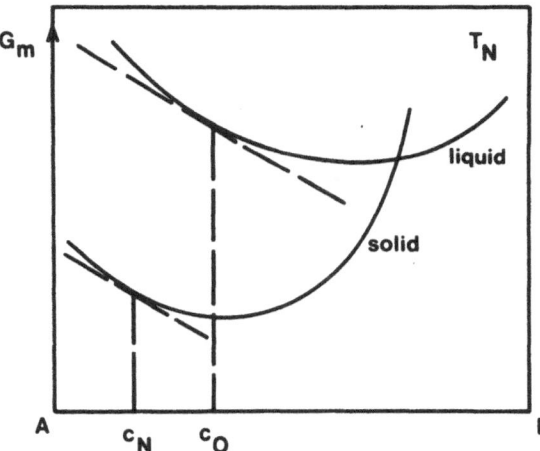

Figure 14: Molar free energy G_m versus alloy
composition c at nucleation temperature T_N, in
rapidly-solidified alloy   of composition c_0
with schematic A-B alloy phase diagram in Fig.
13. Composition of solid nuclei = c_N given by
parallel tangent construction.

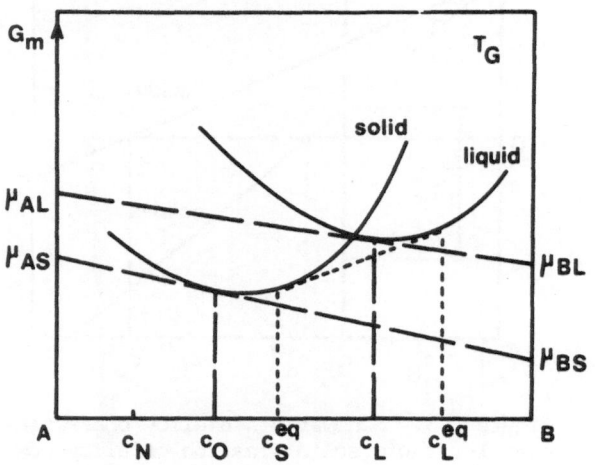

Figure 15: Molar free energy G_m versus alloy composition c at steady state rowth temperature T_G, in rapidly solidified alloy of composition c_0 with schematic A-B alloy phase diagram in Figure 13. Composition of liquid at steady state solid/liquid interface = c_L determined by relative chemical potentials of A and B in liquid and solid μ_{AL}, μ_{AS}, μ_{BL} and μ_{BS}.

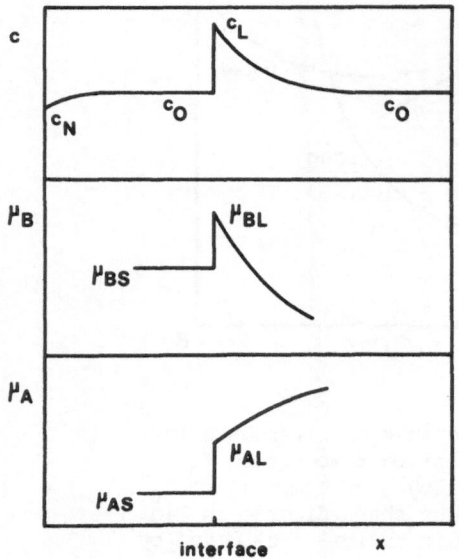

Figure 16: Profiles of alloy composition c and chemical potentials μ_A and μ_B through steady state solid/liquid interface, during rapid solidification of an alloy of composition c_0 with schematic A-B alloy phase diagram in Figure 13.

Equations (17), (18) and (32) are sufficient to fix values of v_G, ΔT_G and c_L for given conditions of heat transfer coefficient and variation of chemical potential with temperature and composition. At sufficiently high solidification rate, solute trapping can cause deviations from equation (32) (38, 39). The critical solidification rate for the onset of solute trapping v_s is given approximately by:

$$v_s = D_L/a_o \tag{33}$$

where D_L is the liquid diffusion coefficient and a_o is the atomic separation. A typical value of v_s is $1ms^{-1}$ for most metals, considerably greater than v_G from equation (20). Solute trapping is therefore only significant in the initial stages of recalescence, and equation (32) is valid during steady state growth.

As recalescence takes place, the solidification rate and undercooling decrease towards v_G and T_G. Cellular breakdown and segregation at the solid/liquid interface is initiated when the solidification rate falls below the absolute stability velocity v_A, which corresponds to an undercoolings ΔT_A (40):

$$v_A = m_L D_L c_L \rho(1-k_o)/k_o^2 T_F \gamma_{SL} \tag{34}$$

$$\Delta T_A = v_A/m \tag{35}$$

where m_L is the liquidus slope and k_o is the equilibrium partition coefficient. Different segregation patterns result from different nucleation and growth conditions:
(1) For $\Delta T_N > \Delta T_A + L/C$, solidification is completed during recalescence, the solidification rate and undercooling never fall below v_A and ΔT_A, and the resulting microstructure is completely segregation-free.
(2) For $\Delta T_A + L/C > \Delta T_N > \Delta T_A$ and $\Delta T_G > \Delta T_A$, recalescence gives way to a steady state growth plateau, but the solidification rate and undercooling still never fall below v_A and ΔT_A, so the resulting microstructure is again completely segregation-free.
(3) For $\Delta T_A + L/C > \Delta T_N > \Delta T_A$ and $\Delta T_G < \Delta T_A$, the microstructure exhibits a transition from segregation-free to cellular solidification, as in Figure 12.
(4) For $\Delta T_N < \Delta T_A$ and $\Delta T_G < \Delta T_A$, the solidification rate and undercooling are always smaller than v_A and ΔT_A, so the resulting microstructure is completely segregated.
(5) For $\Delta T_N < \Delta T_A$ and $\Delta T_G > \Delta T_A$, there is no recalescence, the solid/liquid interface accelerates towards steady state growth, and the resulting microstructure exhibits a transition from cellular to segregation-free solidification. This happens when there is no barrier to nucleation, as in alloys rapidly solidified by surface melting (40).

This description of segregation during rapid solidification is only valid for Newtonian cooling and planar front solidification in an alloy with a small freezing range. Non-Newtonian cooling leads to faster recalescence with a temperature inversion at the solid/liquid interface, and therefore an earlier onset of cellular segregation. The solidification of a dendritic array is not fully understood even at low solidification rates and attempts to allow for dendritic solidification indicate that the temperature at the solid/liquid interface rises above the solidus temperature (57). In alloys with a large freezing range, nucleation can take place between the solidus and liquidus temperatures (40).

When a microstructure such as Figure 12 shows a transition from segrega-

tion-free to cellular solidification, measurements of the fractional thickness of the segregation-free zone η_T can be used together with equation
(21) to estimate the nucleation undercooling ΔT_N:

$$\eta_T = C(\Delta T_N - \Delta T_A)/L \tag{36}$$

Relative variations of nucleation undercooling can be determined in this
way, but absolute values should be treated with caution, since, with a high
heat transfer coefficient, Newtonian conditions are not maintained during
recalescence and equation (36) underestimates ΔT_N. As shown in Figure 12,
the fractional segregation-free thickness varies along the length of a melt-
spun ribbon, indicating a variation in nucleation undercooling from place
to place on the wheel surface. Figure 17 shows maximum nucleation
undercoolings obtained from equation (36) by measuring maximum values of η_T
in melt-spun Ni-11wt%Al (19), Fe-35wt%Ni (20) and 316 stainless steel (21).
Nucleation undercooling increases linearly with increasing cooling rate,
but the increase is not sufficiently steep to prevent the inverse dependence of grain size on cooling rate, predicted in equation (31) and observed experimentally in Figure 11 and equation (24).

Unlike melt-spun ribbons of Ni-11wt%Al, which show a transition from
segregation-free to cellular solidification in Figure 12, Ni-Al alloys
containing 3-8wt%Al show no evidence of segregation when melt-spun under
identical conditions (19). From equation (36), this seems to indicate a
sharp increase in nucleation undercooling with decreasing Al-content in the
alloy. A more likely explanation however, is given by the variation of ΔT_G
and ΔT_A with alloy composition shown on a schematic phase diagram in Figure
18. The microstructural transition in Ni-11wt%Al implies $\Delta T_G < \Delta T_A$ with a
nucleation undercooling from Figure 17 of 300-350K. Figure 18 and equations
(19) and (35) show that ΔT_A decreases linearly with decreasing alloy
content while ΔT_G remains constant. Below a critical alloy content in the
range 8-11wt%Al, ΔT_A becomes smaller than ΔT_G, and solidification is
completely segregation-free without any change in nucleation undercooling.

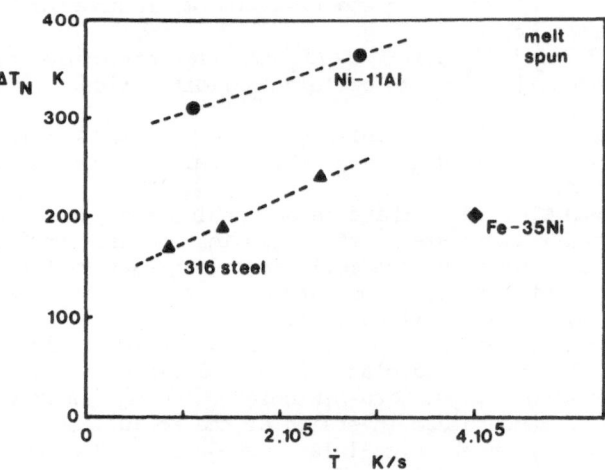

Figure 17: Maximum nucleation undercooling in liquid before onset
of solidification ΔT_N, measured from maximum segregation-free thickness η_T, as a function of measured cooling rate T in melt-spun ribbons of Ni-11wt%Al, (19)Fe-35wt%Ni (20) and 316 stainless steel (21).

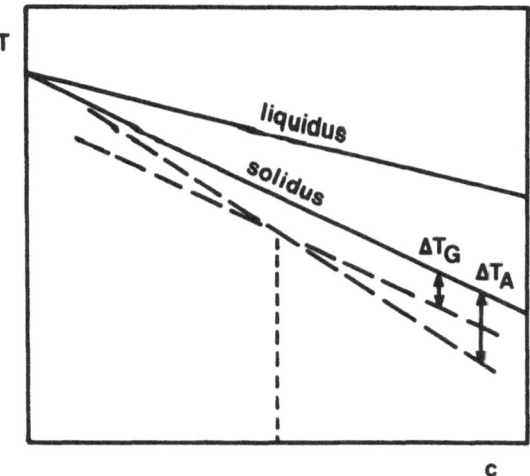

Figure 18: Schematic phase diagram showing variation of steady state growth undercooling at a solid/liquid interface
ΔT_G, and undercooling ΔT_A corresponding to absolute stability velocity
v_A, as functions of alloy composition c.

9. DENDRITE ARM SPACING

Cooling rates are difficult to measure directly during rapid
solidification, and are often estimated indirectly (2). At low cooling
rates, secondary dendrite arm spacing λ exhibits a well—established
variation with cooling rate (2, 41):

$$\lambda = k_1 \dot{T}^{-n} \tag{37}$$

where k_1 and n are constants for a given material. By assuming that
equation (37) remains valid when extrapolated to high cooling rates,
dendrite arm or cell spacing in a segregated rapidly-solidified
microstructure can be used to estimate cooling rate (2, 41, 42). Figure 19
shows cell spacing as a function of directly measured cooling rate in melt-
spun ribbons of Ni-11wt%Al (19), Fe-35wt%Ni (20) and 316 stainless steel
(21), together with extrapolations of equation (37) established at low
cooling rates in other Fe-and Ni-based alloys (2, 41). The melt-spinning data
do not agree particularly well with the extrapolations of equation (37)
This probably reflects differences in cell, primary dendrite arm and
secondary dendrite arm spacing, as well as differences in alloy
composition. Some care must therefore be exercised when using equation (37)
to estimate cooling rates during rapid solidification from measured dendrite
arm and cell spacings.

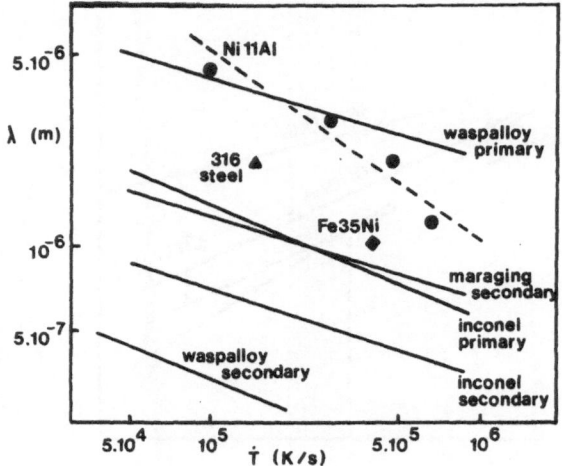

Figure 19: Cell spacing λ as a function of mean cooling rate $\dot{T}$ in the vicinity of the solidification point for melt spun Ni-11wt%Al, (19), Fe-35wt%Ni (20) and 316 stainless steel (21) together with extrapolated lines for secondary dendrite arm spacing at low cooling rates in other Fe- and Ni-alloys (2, 41).

10. SOLUBILITY

The cooling rate, grain size and segregation measurements described above are for alloys with a narrow equilibrium freezing range, typically $\sim$ 10K. Estimates such as in Figure 17 of the nucleation undercooling ΔT_N below the solidus are virtually equivalent to the undercooling, $\Delta T_N'$, below the liquidus. With increasing freezing range, segregation patterns can become more complex since solidification is more likely to be nucleated between the liquidus and solidus (40). In many alloy systems, solubility is limited by a eutectic or peritectic reaction, and rapid solidification often extends the solubility beyond the equilibrium solubility limit (2, 43). At the maximum extended solubility, the nucleation undercooling below the liquidus $\Delta T_N'$, can be estimated as approximately equal to the temperature difference between liquidus and extrapolated solidus (11). Table 1 shows some values of $\Delta T_N'$, obtained in this way for Fe-based alloys. The data in Table 1 are very approximate because of uncertainties in both the measured maximum solubility and the extrapolation of the solidus.

Table 1: Solid solubility extension and estimated nucleation undercooling below the liquidus for rapidly-solidified Fe-alloys.

Alloy	Maximum Equilibrium Solubility c_s^{eq} (at%)	Extended Solubility c_s^{ex} (at%)	Nucleation Undercooling $\Delta T_N'$ (K)	Ref.
Fe-Cu	7.2	15.0	50	52
Fe-Ga	47.0	50.0	130	53
Fe-Mo	26.0	40.6	80	54
Fe-Ti	9.8	16.0	140	55
Fe-W	13.0	18.0	80	56

11. POST-SOLIDIFICATION PHASE TRANSFORMATIONS

A small rapidly-solidified grain size suppresses post-solidification martensite transformations. Figure 20 shows a typical rapid solidification microstructure of fine-scale columnar austenite grains in melt-spun Fe-26wt%Ni-0.4wt%C, and Figure 21 shows the resulting plate martensite microstructure obtained after cooling below room temperature (44, 45). By various combinations of deformation and heat treatment, the austenite grain size and dislocation density can be varied independently, and Figure 22 shows their effect on M_s temperature during subsequent sub-zero cooling. The small rapidly-solidified grain size is clearly effective in depressing M_s by ~50K compared with conventional large grain size material. In splat quenched Fe-25wt%Ni, austenite is retained to room temperature with the conventional lath martensite transformation completely suppressed (20, 46). Similar effects are seen in a range of other alloys (47, 48) and can also be produced by extended solubility of alloying elements (49, 50).

Post-solidification diffusion-controlled transformations can be stimulated by a small rapidly-solidified grain size, because of efficient nucleation on the large area of grain boundary. In melt-spun Fe-Ni alloys, the as-solidified austenitic microstructure is similar to that shown in Figure 20 for melt-spun Fe-26wt%Ni-0.4%C (20). On cooling to room temperature, the small as-solidified austenite grain size in melt-spun Fe-15wt%Ni stimulates a diffusion-controlled ferrite transformation instead of the lath martensite transformation which is seen at large grain sizes (44). Figure 23 shows the resulting ferritic microstructure in melt-spun Fe-5wt%Ni. In a similar way, a small rapidly-solidified grain size often promotes grain-boundary-nucleated discontinuous precipitation rather than continuous matrix precipitation (51).

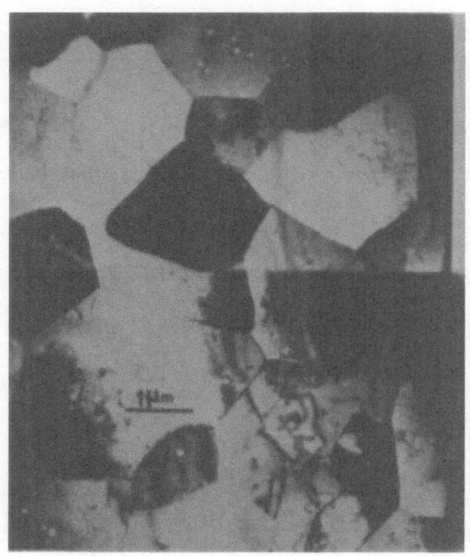

Figure 20:   Transmission electron micrograph showing transverse section of columnar austenite grains in melt-spun Fe-26wt%Ni-0.4wt% (44).

Figure 21. Transmission electron micrograph showing plate martensite formed in melt-spun Fe-26wt%Ni-0.4wt%C after cooling below room temperature (44).

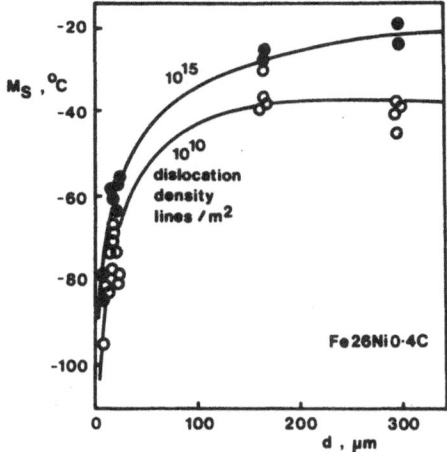

Figure 22: Start temperature of sub-
zero plate martensite transformation
M$_s$ as a function of prior austenite grain size in recrystallised,
low dislocation density and deformed,
high dislocation density Fe-26wt%Ni-
0.4wt%C(44).

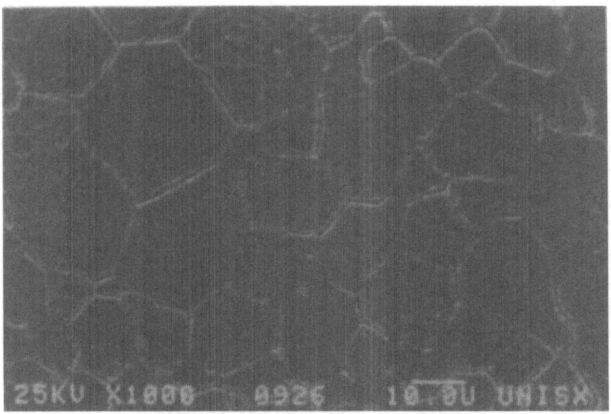

Figure 23: Scanning electron micrograph showing massive ferrite grains, together with
internal cellular segregation structure inheri-
ted from the as solidified austenite, in melt-
spun Fe-15wt%Ni.

ACKNOWLEDGEMENTS
    I would like to thank Professor Sir Peter Hirsch and Professor J.W. Christian for provision of laboratory facilities, and the UK Science and Engineering Research Council and General Electric Corporate R and D Labs for financial support.

REFERENCES
1. Masumoto T, Suzuki K (eds), "Rapidly Quenched Metals IV" (Japan Institute of Metals, Sendai, 1982).
2. Jones H, "Rapid Solidification of Metals and Alloys" (Institution of Metallurgists, London 1982).
3. Duwez P, Willens R H and Klement W, J. Appl. Phys., 31 (1960) 1136.
4. Cahn R W, Krishnanand K D, Laridjani M, Greenholz M and Hill R, Mat.Sci. Eng., 23 (1976) 83.
5. Pond R and Maddin R, TMS-AIME, 245 (1969) 2475.
6. Maringer, R E, Mobley C E and Collings E W, in "Rapidly Quenched Metals II" ed. Grant N J and Giessen B C (MIT Press, Cambridge 1976) vol. 1 p29.
7. Martin P L, Lipsitt H A and Williams J C, in "Rapid Solidification Processing II" ed. Mehrabian R, Kear B H and Cohen M (Claitors, Baton Rouge, 1980) p123.
8. Apelian D, Paliwal M, Smith R W and Schilling W F, Int.Met.Rev., 28 (1983) 271.
9. Appleton B R and Celler G K (eds), "Laser and Electron Beam Processing of Solids" (North Holland, New York, 1982).
10. Jones H, Rep. Progr. Phys., 36 (1973) 1425.
11. Cantor B, in "Rapidly Solidified Amorphous and Crystalline Alloys", ed. Kear B H and Giessen B C (North Holland, New York 1982) p317.
12. Strachan R W, PhD Thesis, MIT (1967).
13. Duflos F and Cantor B, Acta Met., 30 (1982) 323.
14. Nishi Y, Morohoshi T, Kawakami M, Suzuki K and Masumoto T, in ref. (1) vol.1 p111.
15. Harbur D R, Anderson J W and Maraman W J, TMS-AIME, 245 (1969) 1055.
16. Predecki P, Mullendore A W and Grant N J, TMS-AIME, 233 (1965) 1581.
17. Warrington D H, Davies H A and Shohoji N, in ref. (1), vol.1 p69.
18. Tenwick M J and Davies H A in "Rapidly Quenched Metals V", ed. S Steeb and h Warlimont (North Holland, 1985), p. 67
19. Gillen A G and Cantor B, Acta Met., in press
20. Hayzelden C, Rayment J J and Cantor B, Acta Met., 31 (1983) 379.
21. Bewlay B P and Cantor B, to be published.
22. Carslaw H S and Jaeger J C, "Conduction of Heat in Solids" (OUP, Oxford, 1959).
23. Ruhl R C, Mat. Sci. Eng., 1 (1967) 313.
24. Hayzelden C, D Phil Thesis, Sussex University, 1984.
25. Huang S C and Fiedler H C, Mat. Sci. Eng., 51 (1981) 39.
26. Lohberg K und Muller H, Z. Metallk., 60 (1969) 231.
27. Jones D R H and Chadwick G A, Phil. Mag., 24 (1971) 1327.
28. Smithells C J and Brandes E A, "Metals Reference Book" (Butterworths, London 1976).
29. Clyne T W, Met. Trans., 15B (1984) 369.
30. Vincent A J B and Cantor B, to be published.
31. Cantor B and Doherty R D, Acta Met., 27 (1979) 33.
32. Boswell P G and Chadwick G A, Scripta Met., 11 (1977) 459.
33. Duflos F and Stohr J F, in "Rapidly Solidified Amorphous and Crystalline Alloys", ed. Kear B H and Giessen B C (North Holland, New York 1982) p167.

34. Kadalbal R, Montoya-Cruz J and Kattamis T Z, in "Rapid Solidification Processing II, ed. Mehrabian R, Kear B H and Cohen M'(Claitors, Baton Rouge, 1980) p195.
35. Wood J V and Honeycombe R W K, J. Mat. Sci., 9 (1974) 1183.
36. Robertson S R, Gorsuch T J and Adler R P I, in "Rapid Solidification Processing I", ed. Mehrabian R, Kear B H and Cohen M (Claitors, Baton Rouge 1978) p188.
37. Levi C G and Mehrabian R, Met Trans., 13A (1982) 13.
38. Hillert M and Sundman B, Acta Met., 24 (1976) 731.
39. Baker J C and Cahn J W, Acta Met., 17 (1969) 575.
40. Boettinger W J, Shechtman D, Schaefer R J and Biancaniello F S, Met. Trans., 15A (1984) 55.
41. Mehrabian R, in "Rapid Solidification Processing I", ed. Mehrabian R, Kear B H and Cohen M (Claitors, Baton Rouge, 1978) p9.
42. Matyja H, Giessen B C and Grant N J, J. Inst. Met., 96 (1968) 30.
43. Anantharaman T R and Suyanarayana C, J. Mat. Sci., 6 (1971) 1111.
44. Hayzelden C and Cantor B, Acta Met., in press.
45. Hayzelden C and Cantor B, in "Solid-Solid Phase Transformations", ed. Aaronson H I, Laughlin D E, Sekerka R F and Wayman C M (AIME, New York 1982) p1397.
46. Inokuti Y and Cantor B, Acta Met., 30 (1982) 343.
47. Inokuti Y and Cantor B, J. Mat. Sci., 12 (1977) 946.
48. Banerjee S and Cantor B, in "ICOMAT-79", ed. Owen W S (MIT Press, Cambridge 1979) p195.
49. Ruhl R C and Cohen M, TMS-AIME, 245 (1969) 253.
50. Bee J V and Wood J V, in "Rapid Solidification Processing II", ed. Mehrabian R, Kear B H and Cohen M (Claitors, Baton Rouge, 1980) p172.
51. Sahin E and Jones H, in "Rapidly Quenched Metals III", ed. Cantor B (Metals Society, London 1978) p138.
52. Klement W, TMS-AIME, 233 (1965) 1180.
53. Luo H L, TMS-AIME, 239 (1969) 119.
54. Polesya A F, Slipchenka L P, Burov L M, Gudzenko V N and Demeshkin V I: Izv. VUZ Chernaya Met, 9 (1971) 114.
55. Giessen B C, in "Developments in Structural Chemistry of Alloy Phases" (Plenum, New York 1969) p227.
56. Mirkin L I, Izv.VUZ Chernaya Met., 14 (1971) 114.
57. Kurz W, Giovanola B and Trivedi R, Acta Metallurgica in press

ABSTRACTED DISCUSSION OF THE PAPER BY B. CANTOR

Participants: W.J. Boettinger, R.W. Cahn, H. Fischmeister, M.E. Glicksman
H. Jones, R.E. Lewis, J.H. Perepezko and J.V. Wood

The discussion raised a number of issues concerning realism of assumptions
necessarily made in modelling solidification at high undercooling including
the regime of heat transfer and type of nucleation operative and the role
of growth rate in relation to solute redistributions and recalescence. The
speaker emphasized that he had restricted himself to only the simplest
possible models with particular reference to alloys with small freezing
ranges for which the complication of competing metastable phases could be
left out of account. He believed that maximum thermodynamic driving force
determined the composition of the single phase solid forming in such circumstances. It was suggested that the nature and activity of heterogeneous
sites actually present could be the determining factor in situations where
metastable phases form. Concern over issues such as whether or not nucleation behaved as a single valued-function of temperature (as assumed in the
model) or not, had to be seen in that context and comparisons with soldification of atomized droplets had to made cautiously since two droplets,
identical in size, could solidify with quite different results in the same
spray because of different levels and activities of nucleant particles
within them. Attention was drawn to the possibility that segregation-free
solidification might occur in small enough particles because the wavelength
of standard interfacial stabilities was larger than the particle size, as
discussed by Coriell et al. at one of the Reston Conferences. The possible
role of gas bubbles trapped between the solidifying ribbon and dull-surface
had likewise still to be clearly established, not only in relation to the
heterogeneous nucleation process. Some apparent inconsistencies between
parametric relationships and measurements could be the result of heat
transfer changing from near-ideal to near-newtonian as the liquid is drawn
out of the melt pool to solidify in melt spinning. Transient effects, for
example on dislocation density, along the length of melt-spun ribbons had
also to be taken into account in any fully realistic model.

ROLE OF NUCLEATION IN RAPID SOLIDIFICATION

J.H. PEREPEZKO
Dept. of Metallurgical and Mineral Engineering
University of Wisconsin
Madison, WI 53706,
USA

ABSTRACT
    During rapid solidification processing (RSP) the amount of liquid under-
cooling is an important factor in determining microstructural development
by controlling phase selection during nucleation and morphological evolution during crystal growth. When a liquid is dispersed into fine droplets,
an effective isolation of the most potent nucleants allows for a deep
undercooling in the range of 0.3-0.4 T_m before the onset of nucleation.
Liquid undercooling behavior can be influenced by powder size refinement,
size distribution characteristics and coating treatments. At high undercooling the nucleation of an equilibrium phase may be superseded by metastable
product structures. In this case the use of metastable phase diagrams is
important for the interpretation and prediction of solidification products
such as supersaturated solutions, metastable intermediate phases and glasses. With known heterogeneous sites in droplets it has been possible to
control nucleation and to identify specific pathways for metastable phase
formation and microstructural development. The operative reaction is determined by a competitive nucleation kinetics which can be modified at different undercooling levels by the effect of cooling rate. These advances have
allowed for a clearer assessment of the interplay between undercooling,
cooling rate and particle size statistics in controlling nucleation during
RSP structure formation.

1. INTRODUCTION
    One of the most attractive features of RSP (Rapid Solidification Pro-
cessing) is directly related to the unique microstructural morphologies
that have been developed in a fairly wide variety of alloy systems (1-3). A
first level of microstructural evolution is the formation of highly refined
microstructural constituents, such as cells and dendrites of equilibrium
phases which are directly associated with the reduction in local solidification time during a rapid solidification. While a refined microstructural
scale is certainly a valuable product of RSP, there is a further level of
modification associated with alternative metastable phases and very fine
grain sizes that can develop under RSP conditions. These products yield
novel microstructures. Most often, it has been difficult to identify the
specific details of the solidification kinetics that favor novel, metastable phase microstructures. However, it is clear on the basis of thermodynamic considerations of relative stability that metastable phases form
from a liquid which is undercooled below the stable phase melting temperature. Indeed, the formation of metastable phases is evidence that an
undercooled liquid existed prior to crystal nucleation at high cooling
rate even if the actual temperature was not measured.
    Although RSP can be viewed in a broad sense to encompass a variety of
materials processing procedures and the associated nonequilibrium structures, for the focus of the discussion the emphasis will be placed upon

melt processing. Even with this restriction there are two general approaches to melt processing of importance involving powders with cooling mainly by convective exchange with a surrounding gas environment and thin melt streams in melt spinning or surface melting with conductive cooling to a perfect or imperfectly wetted substrate (4). Inspite of the appearance of seemingly different sample configurations and conditions, there is a reasonable justification for treating some of the nucleation characteristics involved in these treatments on a common basis. Such a normalization of nucleation behavior in terms of metastable phase formation is possible if the kinetics of phase selection are viewed in terms of melt undercooling (5). At least from a thermodynamic basis it is known that the formation of a given phase can only occur at or below its melting point. Similarly, metastable phases have melting points below that for the equilibrium stable phase. Since metastable phases have been reported in numerous substrate cooling and powder studies, it must be admitted that undercooling conditions can pervail for both sample types. For powder samples, the melt subdivision that accompanies atomization provides an effective nucleant isolation which in turn promotes liquid undercooling as clearly demonstrated in droplet solidification studies (6,7). With substrate cooled samples a simple, physical subdivision of the melt is not active, but instead it appears that a transient isolation is possible perhaps due to a localization of nucleant influence so that the required undercooling is obtained for a sufficient time period to allow metastable phases the opportunity to develop (8). Within this common basis, the examination of nucleation during RSP may be developed along a unified theme. Similarly, many of the crystallization reactions during devitrification of metallic glass may be viewed in terms of metastable phase domains and relative undercooling ranges (9,10).

There has been recent progress in developing descriptions for nucleation kinetics appropriate to high melt undercooling conditions and including cooling rate affects (5-7), distribution and effectiveness of heterogeneous nucleation sites (11,12) and the proper representation of transient or non steady state nucleation conditions (13,14). Similar progress in growth kinetics analysis for morphological stability, limits to growth with solute segregation and the onset of solute trapping in partitionless growth has also been reported, but will be discussed in depth elsewhere in the proceedings (15).

A key feature in understanding the microstructural evolution and product phase selection during RSP is the operation of competitive kinetics (16). The competitive kinetics is important both in terms of alternate nucleation paths as well as in the subsequent growth kinetic competition. Once this feature is recognized and identified with key processing parameters some measure of structural control is possible. In developing these points the present state of nucleation kinetics analysis will be highlighted and further illustrated with a number of examples of structural evolution and control.

2. NUCLEATION KINETICS IN UNDERCOOLED LIQUIDS

In the present discussion the main interest in nucleation kinetics will be focussed on the possible analytical relationships between undercooling, melt cooling rate and alloy composition during the solidification of a distribution of liquid droplets. Similar rate expressions can be applied for the formation of equilibrium and metastable phases during the rapid quenching of melt streams. These kinetic relationships may be examined by nucleation rate calculations, but the computed results are highly dependent on the specific values assigned to the kinetic rate parameters. Reliable information is available for some parameters; however, in most cases the kinetic parameters are assigned values based upon theoretical models which will require experimental verification. In order to minimize the uncertainty associated with calculations based on model dependent parameter values, it is of interest to examine the range of nucleation kinetics in undercooled liquids in a general manner. With this approach, it is possible to deduce several useful guiding relationships for rapid solidification behavior.

Nucleation of a crystalline phase in an undercooled liquid phase may be classified into the mechanistic types. When nucleation occurs without an association with any catalyic nucleation sites, it can be classified as homogeneous nucleation. While this represents a limiting case, it has received the most theoretical attention. Based upon simple classical theory (17) an expression for the steady state volume nucleation rate, J_v, can be represented as

$$J_v = \Omega_v \, exp \, [-\Delta G^*/kT] \tag{1}$$

where $\Omega_v = 10^{35}/cm^3$-sec and is proportional to the number of liquid atoms as well as the jump frequency of liquid atoms, ΔG^* is the activation barrier for nucleation and kT has the usual meaning. For nuclei with a spherical geometry

$$\Delta G^* = \frac{16\pi\,\sigma^3}{3\Delta G_v^2} \tag{2}$$

where σ is the interfacial energy between the nucleus and the liquid, and ΔG_v is the driving free energy for nucleation of unit volume of the product phase.

In the vast majority of experimental conditions, nucleation is initiated at some catalytic nucleation site as a heterogeneous nucleation. For this case an expression for the heterogeneous nucleation rate J_a can be developed on the basis of homogeneous nucleation theory (17) as

$$J_a = \Omega_a \, exp \, [-\Delta G^* f(\theta)/kT] \tag{3}$$

where Ω_a is of the order of $10^{27}/cm^2$ sec and is proportional to the number of liquid atoms in contact with unit area of the catalytic surface as well as the jump frequency of atoms and $f(\theta)$ is a contact angle factor. For nuclei having the form of a spherical sector on a planar catalytic site this factor can be represented as

$$f(\theta) = (2-3cos\theta + cos^3\theta)/4 \tag{4}$$

with θ as the contact angle between the nucleus and the catalytic site. During rapid melt quenching and especially at high undercooling near the glass transition temperature, atomic transport may not be adequate to maintain diffusional equilibrium. The classical nucleation model is based upon the formation of critical clusters by the addition or dissolution of individual atoms or molecules. At steady state the distribution of cluster sizes is stationary and the isothermal nucleation rate is a constant as represented by equations (1) or (3). Under conditions of relatively sluggish diffusional kinetics the initial cluster distribution can be different from the steady state value and the time interval required to reach steady state represents an effective delay time (13). Under these conditions the time dependent nucleation rate J(t) rises continuously towards the steady state value where the number of nucleated crystals, N can be represented as

$$N = J_v \ (t - \tau) \tag{5}$$

where τ is an effective delay time. The influence of transient nucleation is likely to be most important for glass formation during rapid quenching (14) or during devitrification (18,19) where the product phase with the shortest delay time can dominate the phase selection.

3. NUCLEATION RATE PARAMETERS

The nucleation rate is a relatively steep function of temperature with a steady state magnitude determined principally by the value of the exponential terms in ΔG^* at the nucleation temperature T_n and to a lesser amount by the prefactor term. In evaluating the temperature dependence of the nucleation rate, usually a constant value based upon theory is taken for Ω_v, but for Ω_a it is necessary to consider that the catalytic site density and active surface area may vary for different conditions. Similarly, little information is available to judge the catalytic potency so that a range of values for $f(\theta)$ is used in calculations. However, the most important parameters in determining the nucleation rate and hence the maximum undercooling level are ΔG_v and σ which have received continued experimental and theoretical study.

In pure metals the value of ΔG_v can be calculated directly from

$$\Delta G_v = \frac{\Delta H_v \ \Delta T}{T_m} - \int_T^{T_m} \Delta Cp \ dT + T \int_T^{T_m} \frac{\Delta Cp}{T} \ dT \tag{6}$$

where ΔH_v is the heat of crystallization for unit volume of solid, ΔCp is the heat capacity difference between undercooled liquid and crystal and $\Delta T = T_m - T$ is the undercooling. An example of the experimental determination ΔCp is shown in figure 1 for undercooled In. Although the magnitude of the effect varies between metals, the measurements available to date indicate that there is a continuous rise in the liquid heat capacity C_p^l with the decreasing temperature and a concommitant growing divergence in ΔCp with increasing undercooling (20,21). Most often measurements are not available and the temperature dependence of ΔG_v is approximated by assuming various forms of ΔCp in equation (6). In general, the experimental values are represented to within a few percent by the relation proposed by Turnbull (17) based upon $\Delta Cp = 0$ for pure metals giving:

$$\Delta G_v = \frac{\Delta H_v \ \Delta T}{T_m} = \Delta H_v \ (1 - T_r) \tag{7}$$

where $T_r = T/T_m$ is the reduced temperature.

The approximations have also been considered by Thompson and Spaepen (22) who propose that a reasonable expression for easy glass forming alloys is

$$\Delta C_p = \Delta H_v (T_m)/T_m \tag{8}$$

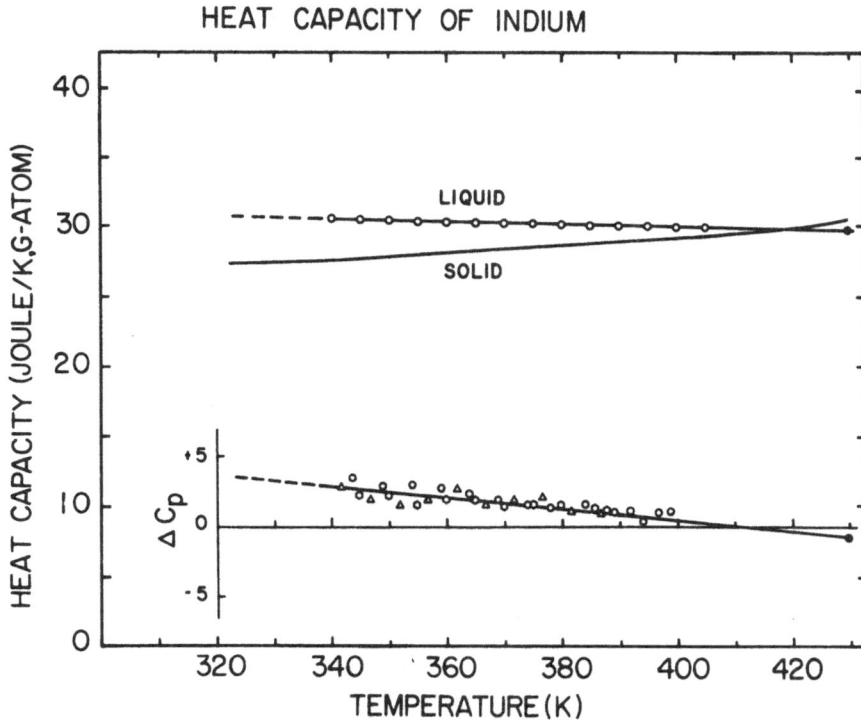

Figure 1: Temperature dependence of the heat capacity for undercooled In.

which in turn can be shown to yield

$$\Delta G_v = \frac{\Delta H_v \ \Delta T}{T_m} \left(\frac{2T}{T_m + T}\right) \tag{9}$$

One consequence of the correct evaluation of ΔG_v relates to the proper analysis of crystallization kinetics. For example, the use of equation (7) for ΔG_v at maximum undercooling would not change the value of the pre-factor in the nucleation rate expression by more than a factor of 10^2. However, the use of other approximations which involve an error in ΔG_v in excess of 10% could alter the value of the prefactor or more than a factor of 10^6 which is near the magnitude of the ratio of prefactors for hetero-geneous and homogeneous nucleation (21).

For alloys the value of ΔG_v is determined from the composition of the nucleating phase at T_n. The composition dependence of ΔG_v can be expressed as

$$\Delta G_v = [X_A \ \Delta\mu^A + (1-X_A) \ \Delta\mu^B]/\bar{V} \tag{10}$$

where X_A is the mole fraction of A, $\Delta\mu^A$ and $\Delta\mu^B$ are the chemical potential differences between the undercooled liquid and crystal and $\overline{V}$ is the molar volume for the crystalline phase at composition X_A. It is apparent from equation (10) that ΔG_v is maximized when $\Delta\mu^A = \Delta\mu^B$ so that $\Delta G_v = \Delta\mu^A/\overline{V}$. This condition has been proposed by Hillert (23) to assess the nucleus composition. However, the favored nucleus composition should yield a minimum value of ΔG^*. Only if $\overline{V}$ and σ are composition independent can the maximum ΔG_v value occur at the same composition as that for a minimum in ΔG^*. For alloys ΔG^* is a function of both composition and nucleus size. The critical values of nucleus composition, X_A^* and size, r^*, therefore correspond to a saddle point location on a contour map representing the free energy of nucleus formation as a function of size and composition. Based upon calculations (24) it appears that the maximum undercooling nucleation behavior in some alloy systems may correspond to a maximum driving force situation. From this condition and with a specific model for σ (25) the composition dependence of the nucleation temperature that is calculated can be shown to reproduce the observed experimental trends which reflect the composition dependence of the liquidus, but the generality of the calculations and the assumptions used require further evaluation. For example, in a number of binary alloy systems with nucleation temperature trends that mirror the liquidus in composition dependence the product phase is either an equilibrium two-phase mixture or a metastable phase. In Sn-Bi alloys for Sn-rich compositions a supersaturated Sn solid solution forms while for Bi-rich compositions that product changes from a two-phase mixture of metastable Bi(II) phase + Sn to a metastable (BiSn) phase. Regardless of the product identity in Sn-Bi alloys, the nucleation temperature trend is found to parallel the equilibrium liquidus. Other similar observations have been made in Pb-Sn and Pb-Bi alloys. Another type of behavior occurs in a system such as Pb-Sb where the nucleation temperature follows the liquidus trend with composition for different undercooling levels due to different droplet coatings. In this case careful x-ray diffraction examination has established that at low undercooling equilibrium products form while at high undercooling metastable solid solutions are generated. Furthermore, the reported alloy nucleation temperature trends do not refer to homogeneous nucleation behavior. Based on these observations it appears that the changes in the nucleation process and the associated product structure do not reflect the application of the maximum ΔG_v condition in all cases. A further examination of the composition dependence of the nucleation temperature in alloys is needed in order to resolve the key kinetic factors that control the observed behavior.

In the study of nucleation, one of the most important parameters is the solid-liquid interfacial energy, σ. Since the nucleation rate is proportional to $\exp(-\sigma^3)$ as shown in equations (1) and (3), it is very sensitive to even slight errors in the σ value. A direct measurement of σ even with a large uncertainty has been very difficult. Usually values of σ are calculated from classical nucleation theory by assuming that the experimentally obtained maximum undercooling values are associated with homogeneous nucleation (26). Unless the maximum undercooling value is proven to refer to homogeneous nucleation, the estimated σ value is expected to underestimate the true value. Furthermore, the nucleation derived values for σ refer to the undercooling temperature T_n and not the melting temperature where an equilibrium measurement is possible. This raises an important question relating to the temperature dependence of σ. There have been a number of theoretical approaches to study the structural properties of the solid-liquid interface on the basis of various model constructions. The models range from limiting cases in which σ values are calculated for enthalpic bond breaking arguments (27) or derived solely from a basis of entropic considerations (25) to intermediate approaches where a weighted contribution of enthalpic and entropic terms is incorporated into the evaluation (28,29). At the present time, the theoretical values appear to underestimate σ compared to values calculated from nucleation undercooling results. Certainly, an improvement in the independent experimental determination of σ and the temperature dependence, dσ/dT would be invaluable in allowing for a clearer evaluation of theoretical interface models and more reliable nucleation calculations.

4. NUCLEATION KINETICS DURING CONTINUOUS COOLING

Since the onset of nucleation in a highly undercooled liquid is expected to be a sharp function of temperature (26), it is useful to consider a limiting case for the effect of cooling rate on the nucleation temperature. In general, for a given nucleation kinetics the critical condition to observe nucleation experimentally during isothermal holding over a time, t, in a liquid volume v is given by $J_v vt \simeq 1$ for homogeneous nucleation. For heterogeneous nucleation in a liquid volume under the influence of an active catalytic surface area, a, the nucleation condition is $J_a at \simeq 1$. For continuous cooling (30) these conditions require some modification to account for the total number of nuclei, n formed during cooling from T_m to T_n. At a constant cooling rate, $\dot{T}$ and for heterogeneous nucleation, n is represented by

$$ n = \frac{1}{\dot{T}} \int_{T_n}^{T_m} J_a(T) \, dT \tag{11} $$

While this expression may be evaluated numerically, a useful estimate is possible by expanding $J_a(T)$ about the rapid nucleation region where T is near T_n and retaining the first order term to yield

$$ \frac{K J_a a \Delta T}{\dot{T}} - 1 \tag{12} $$

where K is about 0.02. From this basis the heterogeneous nucleation time for the appearance of crystals can be represented as a function of temperature as

$$ \ln t = - \ln [a \, \Omega_a \, K] + 16\pi\sigma^3 f(\theta)/[3k\Delta H_v^2 T_m (1-T_r)^2 T_r] \tag{13} $$

where Ω_a represents the heterogeneous nucleation site population associated with a liquid volume v and catalytic potency $f(\theta)$ and $t \simeq \Delta T/\dot{T}$. Furthermore, the rate dependence of the nucleation temperature which represents the shape of the transformation diagram associated with a specific nucleation kinetics that is active is represented in general as

$$ \frac{d T_r}{d(\ln t)} = \frac{3 k \Delta H_v^2 T_m}{16 \pi \sigma^3 f(\theta)} [\frac{(1-T_r)^3 T_r^2}{(3T_r-1)}] \tag{14} $$

where the possible temperature dependence of Ω_a, ΔH_v, σ and $f(\theta)$ has been neglected in order to examine a limiting value and to minimize the uncertainty of model dependent values. For homogeneous nucleation $f(\theta) = 1$, but

for heterogeneous nucleation $0 \leq f(\theta) < 1$. Therefore, transformation diagrams for heterogeneous nucleation can have different breadths representing
a different temperature dependence of the nucleation rate for different
types of active heterogeneous sites. In equation (14), $dT_r/d(\ln t)$ becomes
infinite at T_r = 1/3 as the nucleation time is minimized during cooling.
Therefore, T_r = 1/3 is believed to estimate a lower bound to the onset of
sensible nucleation of a crystalline phase from an undercooled liquid
during cooling. For pure metals this undercooling limit has not been a-
chieved although as the summary presented in table 1 indicates, the undercooling for gallium approaches this limit.

TABLE 1: Maximum undercooling limits.

Element	Previous Studies			Current Studies	
	ΔT (°C)	$\Delta T/T_m$	Reference	ΔT (°C)	$\Delta T/T_m$
Al	130	0.14	26	160	0.17
Sb	135	0.15	26	210	0.23
Bi	90	0.16	26	227	0.41
Cd	–	–	–	110	0.19
Ga	150	0.50	31	174	0.58
In	–	–	–	110	0.26
Pb	80	0.13	26	153	0.26
Hg	80	0.34	17	88	0.38
Te	–	–	–	236	0.32
Sn	117	0.23	32	192	0.38

In alloys a similar limit may be expected although the simple expression
presented in equation (14) cannot be expected to hold without modifications. For example, near glass forming composition ranges the diffusive
frequency in the liquid changes markedly with undercooling and would affect
an undercooling limit and the nucleation temperature-cooling rate relationship significantly. In fact, inspection of the reduced glass temperatures
that have been reported in the range of about 0.4-0.65 T_m(4) indicates that
the effect of reduced atomic mobility upon approaching a glass transition
can increase the possible undercooling. A reduced atomic mobility also
increases the importance of transient nucleation which effectively reduces
n at each temperature during cooling. In comparison with steady state
conditions transient conditions allows the glass transition to be reached
at a lower cooling rate as shown for Au-Si alloys (14) or requires a larger
undercooling for the detection of sensible crystallization as reported for
Cu-Te alloys (33).   At the deepest undercooling levels the competition
between crystallization and glass formation may also require a consideration of the type of crystal structure. For example the number of atoms, n*
in a spherical nucleus is given by $n* = [(4\pi/3 \, v_a) \, (-2\sigma/\Delta G_v)^3]$ where v_a
is the atomic volume. With very deep undercooling n* can approach values
(e.g. 20 atoms) which fall below those required in unit cells of crystals
with complex structures so that nucleation of these crystals may be difficult. Formation of such crystals is also hindered by their relatively slow
growth kinetics. Another factor of uncertainty for both pure metals and
alloys relates to the temperature dependence of σ and $f(\theta)$. Little experi-

mental information is available on these factors. However, based upon the nucleation undercooling (7) the estimated value for bismuth is σ = 92 mJ/m^2 while the reported measurements by Glicksman and Vold (34) indicate that at the melting point σ = 72 mJ/m^2 implying that dσ/dT < 0. This result is contrary to several theoretical models and clearly requires further study before resolution of even the sign of dσ/dT is possible.

If the undercooling limit of a liquid sample represents the solidification of the nucleant-free volume, the crystallization temperature is determined either by the catalytic nature of the operating nucleant or by the intervention of homogeneous nucleation kinetics. Also, during cooling the operation of a given catalyst kinetics and the kinetic competition between different catalysts and different product phase structures will be influenced strongly by the active catalyst distribution at the undercooling limit. In order to reach this limiting undercooling the nucleant isolation as effected for example by the size refinement in powder samples becomes one of the most important factors. A further analysis of the distribution of active nucleants within undercooled volumes then becomes necessary in order to understand the nucleation behavior during rapid solidification of undercooled samples.

5. SIZE DISTRIBUTIONS STATISTICS IN NUCLEATION

As long as a single nucleation kinetics with a steep temperature dependence is operating, the effect of the amount of undercooled volume on the nucleation temperature is relatively small (17). In fact, for isokinetic behavior, a sharp nucleation peak over a narrow temperature range (i.e.δT< 10°C) is expected even with a sample of a relatively broad powder size distribution. However, in the case of a number of powder samples of pure metals and alloys a broad crystallization peak is often exhibited over a wide temperature range (i.e.δT >20°C). In this case the operation of a single nucleation kinetics is unlikely, but the broad crystallization range can be associated with multiple nucleation events that are continuously activated during the cooling process (6,17). Moreover, at a given cooling rate, nucleation of a crystalline phase in each droplet is associated with the catalytic potency of the most active nucleant within the undercooled volume. As a result, nucleation at different levels of undercooling due to crystallization over a wide temperature range can result in the formation of different structures in different portions of isolated undercooled volumes. Even if the same structure is nucleated at different levels of undercooling, this structure will not necessarily be retained completely in the final solidification pattern. For example, the final product structure can develop variations due to different thermal history associated with the recalescence period following nucleation at different undercooling levels in separated volume elements.

The influence of volume element size on undercooling, nucleation and crystallization behavior of liquids has been emphasized in a number of previous observations relating to droplet samples. For example, if the nucleating sites contained within a given liquid volume are distributed randomly, the arrangement of nucleants among the droplets for a high degree of dispersal maybe described by a Poisson distribution (5-7). For this case the nucleant free droplet fraction X among droplets of volume, v is represented by X = exp(-mv) where mv is the average number of nucleants per volume element. With such a nucleant distribution it is necessary for a sample to contain an active catalyst concentration of only 3 x 10^{15} m^{-3} to yield crystallization for all volume elements of 30 micron size. However, a size refinement to a 4 micron scale will yield X = 0.9. Of course, if a further size refinement yields an increase in undercooling then the active

38

catalyst concentration would be increased accordingly. The relative population of the nucleant-free fraction in an overall powder size distribution is more clearly illustrated in Figure 2 which represents a measured distribution for rapidly solidified Al-9 w/o Fe powders (35). The size

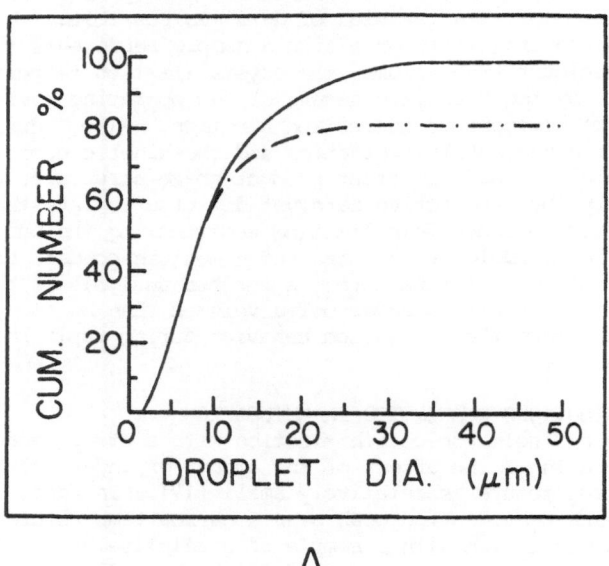

A

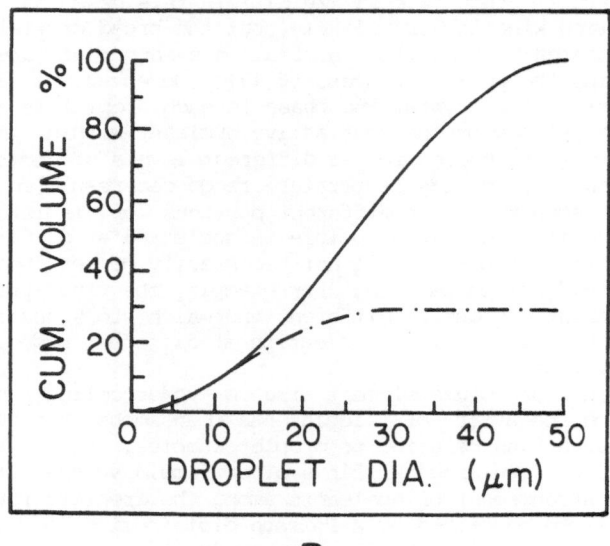

B

Figure 2: Particle size distribution for Al-9 w/o Fe powders.
A) cummulative number percent format;
B) cummulative volume percent format. The solid curve represents the measured population while the broken curve represents the calculated nucleant-free fraction.

distribution results are presented in two formats represented in terms of cummulative number percent in Figure 2a and in terms of cummulative volume percent in Figure 2b. For illustration in each format, it was assumed that for fine 10 μm particles, the internal nucleant-free fraction is 0.9 such that nucleation at this particle size is catalyzed by the powder surface coating. On a calculated cummulative population basis, about 80 percent of the number of powders are without internal nucleants and have a size below 30 μm. Almost all powders of larger size are likely to contain internal nucleants. An even more striking comparison becomes apparent when it is realized that the cummulative volume of the nucleant-free fraction represents only about 29 percent of the total volume of the powder sample. Of course, the characteristics revealed in Figure 2 are also dependent on size distribution parameters such as range and average size. However, the results do show that the contribution of the nucleant-free powder fraction to the volume of refined RSP microstructure in a powder sample can be small with typical powder size distributions even if a large percentage of the number of powders is nucleant-free. This indicates the important role of size refinement in achieving large undercooling, but also that very small active catalyst concentrations can influence the undercooling and amount of RSP product strongly. Accordingly, there is some evidence to support the fact that some impurities in specific cases can play a role in limiting the undercooling behavior. Therefore, high purity melts are not an absolute guarantee of large undercooling behavior (7).

An interesting example of the effect of size distribution in producing a kinetic transition in product phase selection due to crystallization behavior over a broad powder sample size range may be illustrated by considering some of the careful studies on Fe-Ni alloys (36). In fine droplet samples, (< 40 μm) it was demonstrated that droplets of a metastable bcc structure existed in addition to droplets of the equilibrium fcc structure. Size distribution analysis results showed that small droplets with diameters less than 13 μm were more likely nucleated at high undercooling since the major fraction had a metastable bcc structure. Since there are also droplets of a metastable structure within a large size grouping (> 20 μm), it is necessary to consider the mode of nucleant distribution among droplets. Depending on the distribution mechanism of nucleants during powder production (mv) is proportional either to the volume or to the corresponding surface area of the powder. Since a nucleant distribution proportional to the droplet volume would result in a very rapid decrease in the yield of metastable product with increasing powder size, it appears that for the Fe-Ni case where the change in metastable product amount with size was not drastic, a surface area dependent nucleant distribution was most probable.

A further application of the Poisson distribution function has been reported for the vitrification of droplet samples (11,12). In this case the droplet fraction which solidifies continuously as a glass is represented by X which can also be expressed in terms of droplet diameter, d as

$$x = \exp\left[-(d/d_0)^n\right] \qquad (15)$$

where d_0 is related to the nucleant site concentration (11). When mv is proportional to the droplet surface area, n = 2 for heterogeneous surface nucleation. For heterogeneous volume nucleation n = 3 and for homogeneous nucleation n = 4.6 for free fall cooling conditions. As in a number of other droplet studies the solidification behavior of Pd-Si powders (11) and $ZnCl_2$ droplets (12) were described most closely by a heterogeneous surface nucleation kinetics (i.e. n = 2) as illustrated in figure 3. It is important to note that due to the large difference in optical properties of glassy and crystalline $ZnCl_2$ droplets, it was possible to distinguish clearly between the glass and crystalline fractions. Partially crystalline droplets were treated to be nucleated during cooling prior to the glass transition temperature and these droplets were not included in the count shown in figure 3 although they represented a significant contribution to the population in the middle size range. It appears that because of the sluggish growth kinetics that occur in the nucleation temperature range, a significant of volume fraction of the liquid can be retained to the glass transition temperature even at slow cooling rates. In some cases this fraction appears in conjunction with crystalline portions in the same droplet. Such a mixed product structure within a given undercooled volume is not expected for crystalline products due to the usually very rapid growth rates and recalescence effects which act to preclude formation of other nucleation centers or product structures within a droplet volume at slow cooling rates.

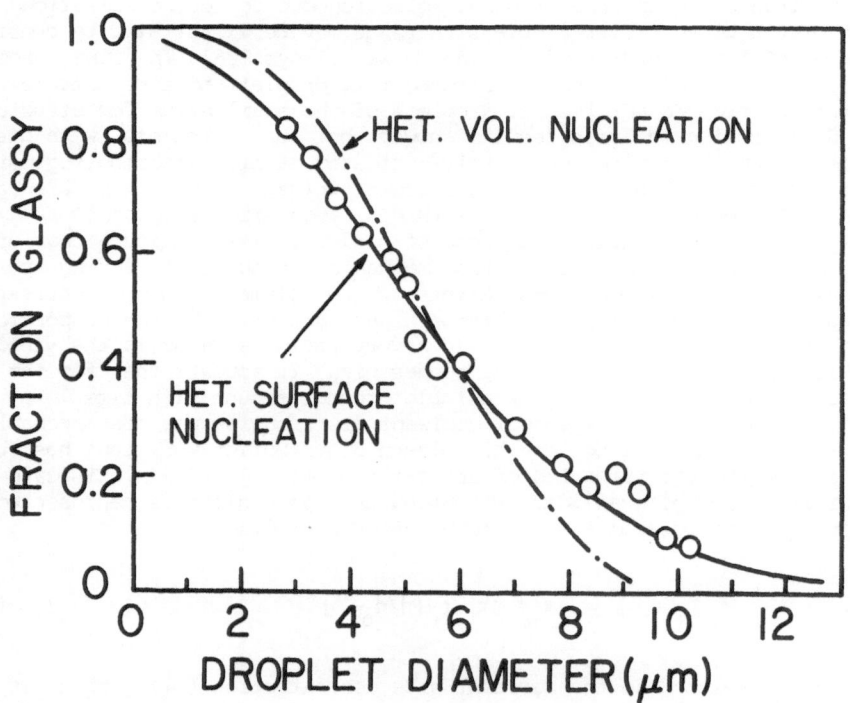

Figure 3: Size distribution of glassy $ZnCl_2$ droplets in comparison with predictions of nucleant distribution kinetic models.

Beyond the effect of size distribution statistics on nucleant isolation to allow for the operation of a kinetic transition in product phase selection, it is important to learn if the size distribution effects can also make it possible to observe a transition in the nature of a nucleation process. The transition from heterogeneous surface to homogeneous volume dependent nucleation is of interest. Of course, with sufficient undercooling, homogeneous nucleation is believed to be possible provided a glass transition does not develop. However, due to the statistical nature of nucleation and sample size distribution effects, a sharp cut off transition in kinetics is not expected with changing undercooling or volume size. One way of examining such a case involves considering when J_v and J_a become of comparable magnitude for a given powder size. This can be expressed in terms of a ratio

$$\frac{J_v}{J_a} = \frac{\Omega v}{\Omega a} \exp \left[- \frac{\Delta G^*}{kT} (1-f(\theta)) \right] \tag{16}$$

When J_v and J_a become comparable at one nucleus/drop-sec. for a 20 μm volume element, than $J_v = 2.4 \times 10^8$/sec.-cm^3 and $J_a = 7.9 \times 10^4$/sec.-cm^2. Further, if sensible nucleation is represented by a limiting value for ΔG^* of 60 kT, then the resulting value of $f(\theta)$ from equation (16) implies a contact angle of about 118° for a constant catalytic site density. For nucleants described by contact angles in excess of 118°, homogeneous nucleation is favored to dominate against a background of heterogeneous nucleation as demonstrated in figure 4. This behavior is mainly related to the larger value of the prefactor for homogeneous nucleation. Consequently, it is not necessary for a given undercooled volume to be free of all nucleants; it is only necessary for the isolation induced by subdivision or other processes to confine the influence of the most potent nucleants in order to observe the largest undercooling at the onset of homogeneous nucleation. This observation can have a number of an important implications extending for example to the understanding of grain size effects during rapid solidification processing (8).

6. METASTABLE PHASE FORMATION AND SELECTION KINETICS
6.1 Thermodynamic Features
    A large number of examples of metastable phase formation have been demonstrated during rapid quenching experiments. The nucleation and growth difficulties of some stable phases that can occur during RSP, naturally lead to an expanded utilization of phase diagram concepts; in particular these constraints require a close examination of the metastable equilibrium features of phase diagrams (16). At the onset it should be emphasized that a metastable equilibrium is a true reversible equilibrium. It is also useful to note that with stable equilibrium, there is a certainty of phase prediction. With metastable equilibrium this certainty of prediction becomes a choice of product phases depending on the particular constraints imposed on the system and the controlling phase selection kinetics. There is a single stable equilibrium phase diagram, but there are several possible metastable phase diagrams for a given alloy system. While the possible reaction paths leading to metastable phase formation in undercooled alloy systems are numerous, a consideration of thermodynamic constraints can be useful in providing guidelines and limiting bounds on the reaction sequences that are possible.
    One example of the application of thermodynamic constraints is illustrated in figure 5 where the free energy relationships are presented that are relevant to the generation of solid solutions. Two features are of interest in this case. For the formation of an equilibrium solid solution

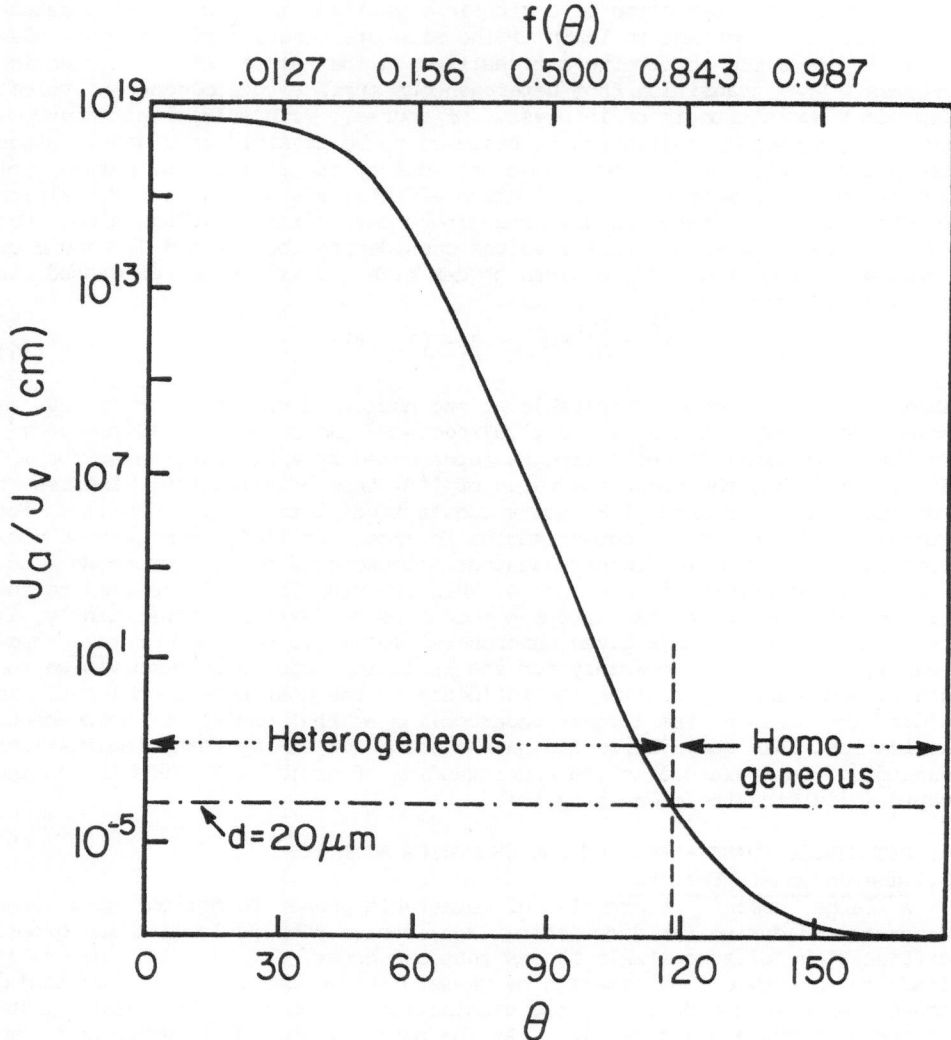

Figure 4: Relative magnitudes of heterogeneous surface to homogeneous nucleation in a 20 μm droplet as a function of contact angle.

of composition C_0 at T_n and overall free energy change of ΔG_1 is involved in reaction. However, in the early nucleation stage of reaction when a small amount of α forms from the liquid the driving free energy is maximized for an α phase with a composition of C_1 rather than C_0. When the molar volume of solution does not change appreciably with composition, the C_1 composition may be determined by constructing a tangent to the α free energy curve which is parallel to a tangent to the liquid free energy curve at the composition C_0. As noted previously, this consideration indicates that α with a solute content different from C_0 can develop initially during

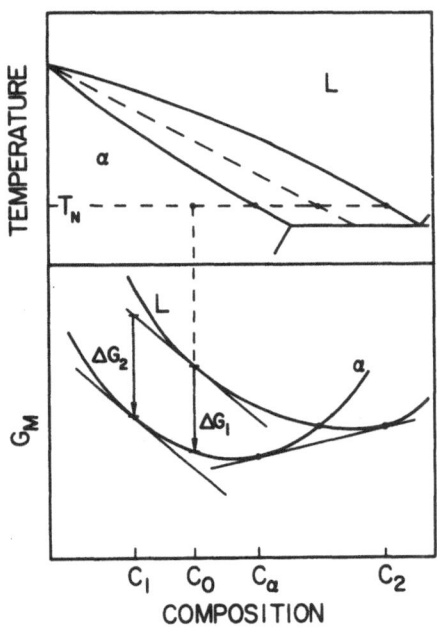

Figure 5: Schematic free energy relationship between undercooled liquid and
α phase. The T_0 curve between the α solidus and liquidus is
shown as a dashed curve.

nucleation. The range of possible solid composition within this bounding
limit makes uncertain the precise specification of nucleation kinetics in
alloys and also requires further consideration in the rapid growth of the
nucleated phases (37).

Another feature of interest in figure 5 concerns the bounds set by the
T_0 curve between the liquidus and solidus, which represents the locus of
equal free energy between liquid and solid. Formation of α product with the
same composition as the liquid is excluded at compositions beyond the T_0
limit. Composition regions beyond the T_0 curve trace require multiple phase
solidification with solute redistribution. Due to the growth kinetic difficulties associated with solute partitioning the composition regions between
T_0 curves for adjacent dissimilar phases have been considered to be favorable for glass formation during RSP (38,39).

A thermodynamic consideration is of most value when the analysis also
can provide an estimate of the potential metastable intermediate phases
that can enter in the kinetic competition. For example, in alloys the
requirement that the free energy, ΔG, exhibit an overall decrease with
reaction can lead to several interesting possibilities for phase reaction
sequences (40) as illustrated by one example in figure 6. In this case the
equilibrium solidification of liquid of composition of C_0 would yield a
two-phase mixture of α with composition C_α and β with composition C_β.
However, if the nucleation kinetics are favorable, an intermediate reaction
step can occur involving the formation of the metastable γ phase with the
composition C_0 and free energy reduction ΔG (L→γ). The presence of γ phase
affects the reaction sequence in that α phase cannot form since the free

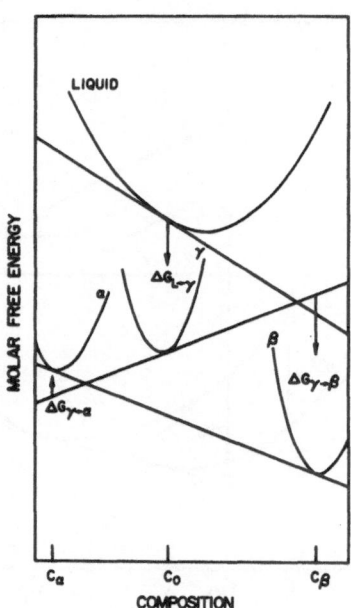

Figure 6: Free energy relationship between undercooled liquid and several crystalline phases (40).

energy change ΔG (γ-a) is positive. Only after the γ phase transforms into β phase can the α phase form with a free energy reduction. When there are several metastable phases possible, the number of intermediate reaction steps can increase and present interesting opportunities for structure control during both solidification and subsequent solid state processing. One way of summarizing the potential for structure modification in alloys is to construct metastable phase diagrams for the various combinations of phases (16).

Since the driving force of the formation of a metastable phase is directly related to the amount of undercooling below the melting temperature of the phase, a high level of undercooling can expand the product selection involved in nucleation to an increased variety of types of structures. This is illustrated by the nucleation transition from FCC to BCC in Fe-Ni alloys, stainless steels and tool steels. Other examples have been observed for Pb-Bi, Bi-Cd and Pb-Sn alloys as well as pure Ga, Bi and Sb (5). For Pb-Bi alloys a multiple phase selection transition has been observed in droplets. As the level of undercooling increases an equilibrium eutectic mixture is replaced by a subsaturated (HCP) phase which in turn is superceeded by a metastable intermediate phase at the highest undercooling. In addition, for Pb-Sn, Sn-Ge, Ge-Sb and Cd-Sb alloys, the operation of the nucleation product transition has been related to the catalytic affect of specific heterogeneous sites (5). One type of effective nucleation site is a primary solid solution which can participate in catalytic reactions such as those in figure 7. When the α phase is a potent catalyst, the continued growth of α is possible during cooling to form a supersaturated α product as in figure 7a. When α is strongly catalytic in promoting the nucleation of a metastable intermediate γ phase as indicated in figure 7b, the

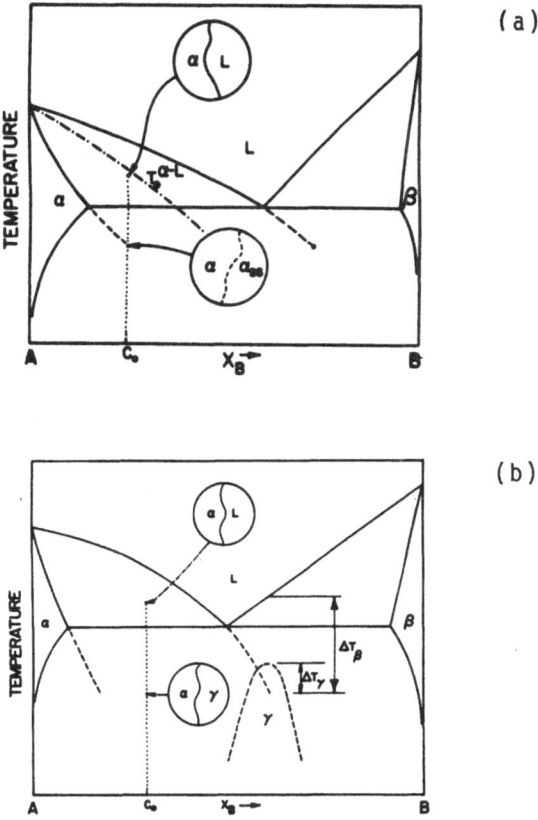

Figure 7: Schematic illustration of a possible reaction path for primary phase catlysis of a metastable solid solution (a) and a metastable γ intermediate phase (b).

formation of an equilibrium β phase may be bypassed as the liquid is undercooled below the metastable γ phase melting point. Metastable phase diagrams can be useful in such cases in providing guidance on the range of available product structure options when equilibrium phases do not exhibit favorable formation kinetics. Thus, for an alloy that may exhibit metastable products, it is possible to select a particular phase by controlling the degree of undercooling prior to solidification and by providing a high density of favorable catalytic sites. The precise kinetic perscription for the most potent site for a given product phase structure is uncertain at this time although it appears that similarity of structural arrangement and low lattice misfit disregistry along close packed planes and directions may be important in some cases; especially at relatively low undercooling levels.

6.2 Analysis of Competitive Kinetics

In order to understand the kinetic competition between different phases during nucleation, the comparison of thermodynamic stability may not be sufficient. In terms of overall nucleation kinetics, the driving free energy is not the only parameter of importance, but also the nucleation site density and the magnitude of ΔG^* associated with a given product phase are critical. As an illustration of the application of nucleation kinetics to reveal the relationships between cooling rate, undercooling and phase selection, it is useful to consider the competition between different nucleation kinetics during continuous cooling in several situations that are illustrated in figure 8. As an initial example, when two catalysts of different nucleation site density and potency are in the same droplet (B and C), the undercooling limit of the droplet at T_{r1} is determined by catalyst C at cooling rate $\dot{T}_1$. At cooling rate $\dot{T}_2$, however, the undercooling of the droplet increases to T_{r2} by circumventing the catalytic affect of C. Thus, at a continually increasing level of cooling rate, there can be a significant improvement of undercooling as the catalytic effects of highly potent nucleants are circumvented. During this process the controlling nucleation kinetics involves catalysts of decreasing potency. When two different phases are competing with each other at the same catalytic surface present in an undercooled liquid, according to nucleation theory, the magnitude of the activation free energy barrier becomes the most dominant factor. Therefore, in order to favor nucleation of a metastable phase at a given cooling rate, the activation free energy barrier for the metastable phase must be lower than that for the equilibrium phase. In terms of equation (3) this condition may be expressed as

$$\frac{\sigma_{Lm}^3 \, f(\theta_m)}{(\Delta H_v^m)^2 \, (1-T_r^1)^2} < \frac{\sigma^3 f(\theta)}{\Delta H_v^2 \, (1-T_r)^2} \qquad (17)$$

where σ_{Lm} and θ_m are the interfacial energy and the contact angle between the metastable phase the catalytic surface, ΔH_v^m is the heat of fusion per unit volume of the metastable phase and T_r^1 is the reduced nucleation temperature of the metastable phase. It has been shown in equation (14) that the breadth of the transformation diagram associated with different nucleation kinetics is proportional to $\Delta H_v^2 T_m / \sigma^3 \, f(\theta)$. When the melting temperatures of two different phases nucleated by the same catalyst are similar and the metastable phase dominates the kinetics, the condition given by equation (17) may be approximated by

$$\frac{(\Delta H_v^m)^2}{\sigma_{Lm}^3 \, f(\theta_m)} > \frac{\Delta H_v^2}{\sigma^3 \, f(\theta)} \qquad (18)$$

In this case, the breadth of transformation diagram for the nucleation of the metastable phase (A) is larger than that for the equilibrium phase (B), but the noses of the two diagrams are placed at the same position as illustrated in figure 8. Unless there are nucleants of higher catalytic potency for nucleation of the equilibrium phase, nucleation of the metastable phase is dominant regardless of the cooling rate. Alternately, if the nucleation barrier for A is higher than that for B, the equilibrium phase will nucleate at all cooling rates. In addition, it is also possible for the nucleation of the equilibrium phase to be favored by an increase of the cooling rate, if the nucleation site density for the equilibrium phase (curve D) is greater than that for the metastable phase. For these conditions an increase in cooling rate can result in a decreasing yield of a metastable product. In fact, a transition in nucleation from surface dependent to volume dependent kinetics with cooling such as C to D in figure 8 can yield a large increase in nucleation rate and may contribute to a propensity for multiple nucleation as reported in quenched Al-Si alloy droplets (41). Other factors involving the specific thermal history of droplets can also contribute to the development of multiple nucleation (8).

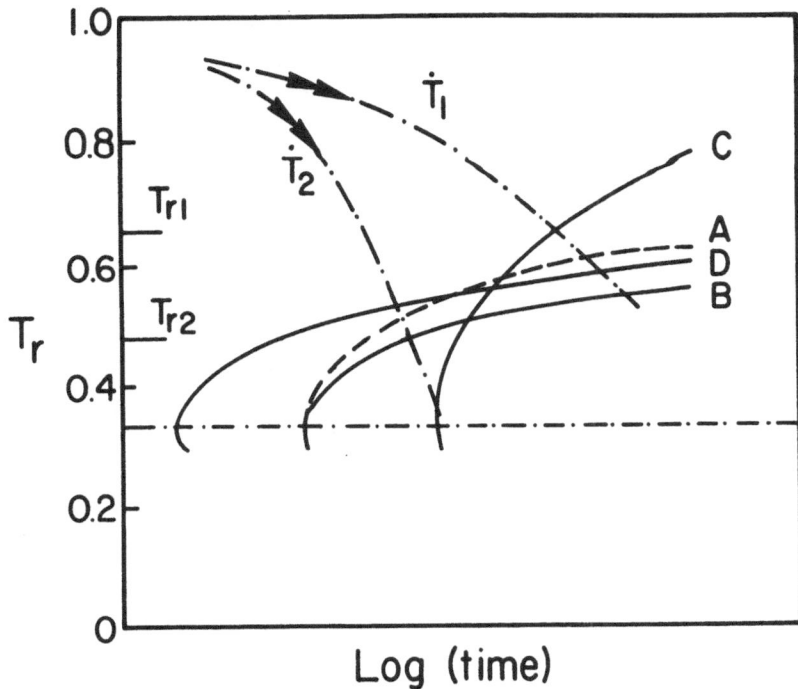

Figure 8: Time-temperature-transformation diagram representing different nucleation kinetics that may occur during continuous cooling of undercooled droplets. Transformation curves B, C and D will tend to $T_r = 1.0$ at long time, while transformation curve A will tend to $T_r = T_m^1/T_m < 1.0$ at a long time.

Furthermore, the combination of kinetic transitions related to size distribution statistics and cooling rate effects can provide a useful experimental probe of nucleation behavior and guidance for optimizing or controlling the yield during RSP.

An illustration of the application of continuous cooling analysis of the crystallization kinetics is represented by some recent measurements for eutectic Pb-56a/o Bi alloy droplets (42). In this alloy extensive undercooling to $0.3\ T_m$ occurs and yields nucleation of a metastable single phase product with a well defined crystallization onset. Over the cooling range from 10°C/min to 320°C/min, DSC measurements indicate that T_n decreases from 265 K to 255 K. Also other droplet work indicates that these nucleation temperatures do not represent homogeneous nucleation since lower T_n values are possible with different droplet preparation treatments.

The continuous cooling nucleation results are plotted in terms of equation (13) in figure 9 where t is given by $\Delta T/\dot{T}$. In this case instead of the usual approximation for ΔG_v as equation (7), the ΔG_v values used in figure 9 were derived from direct calorimetric measurements of the latent of the metastable phase, the temperature dependence of ΔC_p and reported molar volumes. Except for the results at the highest cooling rate, the kinetics measurements can be fitted to equation (13) in terms of a single nucleation frequency with a slope of 1.74×10^7 K–J^2–cm^{-6} and the value of $(\Omega_a \, a \, K)$ of 3.3×10^{22} sec^{-1}. The prefactor term, Ω_a, is in reasonable agreement with the classical theory result if only a portion of the surface area presents an active catalytic site for metastable phase nucleation.

Based upon the analysis in figure 9 it is useful to develop the complete transformation diagram that is described by equation (13). If Ω_a is

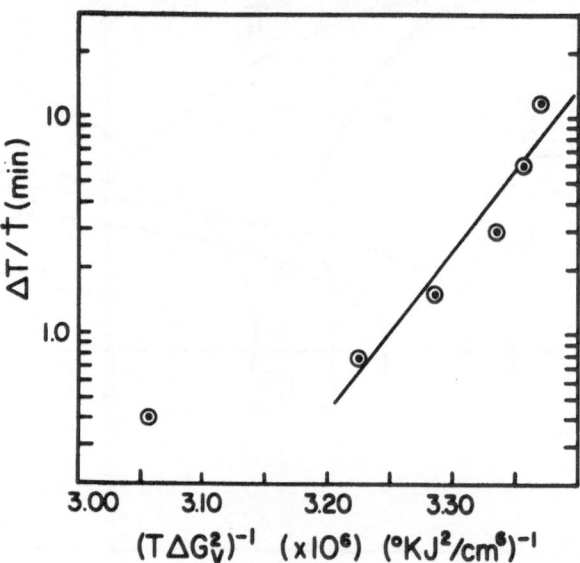

Figure 9: Nucleation kinetics plot for Pb–56 a/o Bi eutectic droplets forming a metastable x phase.

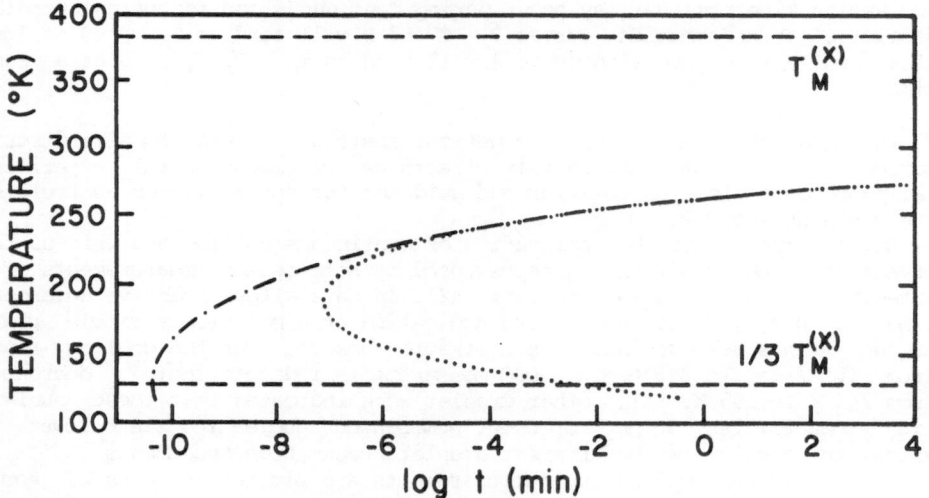

Figure 10: Calculated heterogeneous nucleation kinetics for metastable x phase in Pb–56 a/o Bi droplets. The dotted curve is based on measured ΔG_v and the broken curve is derived from the linear ΔG_v approximation. The solid curve segment gives the range of measured kinetics.

treated as approximately constant, the evaluation of equation (13) yields the transformation diagram shown in figure 10. While the calculation is approximate due to estimates of some of the parameters, the results do offer an interesting viewpoint on the kinetics. For example, if the linear approximation for ΔG_v had been applied instead of the measured values, the nose of the transformation diagram would be located at $1/3 \; T_m = 128$ K instead of 187 K. Also, the cooling rates to reach the nose differ by a factor of about 10^3 for the different ΔG_v values. This observation reveals the importance of using measured values, even if they represent a limited temperature range, for the nucleation parameters instead of estimates. It is also of interest to compare the droplet kinetics behavior with rapid quenching results. In this regard the calculated transformation diagram is consistent with the observation that suppression of the metastable phase nucleation by splat quenching is difficult (43). According to figure 10, a critical cooling rate in excess of 5×10^7 °C/sec is required to bypass metastable phase nucleation. Further, other droplet results indicate that another nucleation site for the metastable phase is active at a greater undercooling than that shown in figure 10. As a result at high cooling rate a transition from one type of nucleation site to another for formation of the same phase is possible. This calculation indicates that the use of even limited measurements can yield useful guidelines in assessing rapid solidification processing treatments.

SUMMARY

Rapid solidification has stimulated a renewed interest in understanding and controlling undercooling in liquids. As a result, more effort has been directed to the examination of nucleation in undercooled liquids. This increase of attention to nucleation kinetics has been yielding new insights on structure control during RSP from several viewpoints.

From a thermodynamic perspective the consideration of metastable phase diagrams and T_0 curve constructions has been useful in defining the range of operating kinetics and the reaction pathways for some nucleation processes. A number of important processing variables have been identified in terms of their influence on the optimization of the undercooling level. While control of these variables has allowed for the attainment of new maximum undercooling limits for a number of metals, the observed crystallization kinetics emphasizes the importance of heterogeneous nucleation in dominating the initial product selection in most cases. In addressing the examination of heterogeneous nucleation, good progress has been made in experimentally defining the proper conditions for the valid measurement of the catalytic potency of known heterogeneous nucleation sites. Based upon these experimental advances and new observations on nucleation in undercooled liquids, the operation of competitive kinetics to promote the formation of different product structures has been established and is now receiving a quantitative evaluation.

Overall, it appears that the experimental experience and methods of nucleation kinetcs analysis are evolving to a level that should allow for progress beyond the rationalization of post solidification structural observations to a realization of some predictive capability for phase selection structural control and RSP alloy design in the near future.

ACKNOWLEDGEMENT

The support of the ARO (DAAG-29-80-K-0068) and the NSF (DMR-79-15802) is gratefully acknowledged. It is a pleasure to acknowledge many stimulating discussions with Dr. W.J. Boettinger of NBS.

REFERENCES

1. Cohen M, Kear B H and Mehrabian R, in Rapid Solidification Processing: Principles and Technologies II, eds. Mehrabian R, Kear B H and Cohen M (Claitor's Pub., Baton Rouge, LA) 1980,p 1.
2. Jones H, Mat. Sci. Eng.,1984, 65 145
3. Hornbogen E, in Rapidly Quenched Metals V, eds. Steeb S and Warlimont H (Elsevier, Amsterdam) 1985, 785.
4. Jones H, Rapid Solidification of Metals and Alloys, Monograph #8, (Institution of Metallurgists, London) 1982.
5. Perepezko, J H, Shiohara Y, Paik J S and FLemings M C in Proc. 3rd Int. Conf. on Rapid Solidification Processing: Principles and Technologies, ed. Meharabian R, (Nat.Bur.Stds.,Washington, DC), 1983, 28.
6. Perepezko J H and Paik J S, in Rapidly Soldified Amorphous and Crystalline Alloys, eds. Kear B H, Giessen B C and Cohen M, (North Holland, Amsterdam) 1982, 49.
7. Perepezko J H, Mat. Sci. Eng.,1984, 65, 125.
8. Boettinger W J and Perepezko J H, in Rapidly Solidified Crystalline Alloys, ed. Das S, Kear, B H and Adam, C M (TMS-AIME, Warrendale,PA) 1986, 21.
9. Blanke H and Köster U, in Rapidly Quenched Metals V, eds. Steeb S and Warlimont H (Elsevier, Amsterdam) 1985, 227.
10. Hornbogen E and Schmidt I, in Rapidly Soldified Amorphous and Crystalline Alloys, eds. Kear B H, Giessen B C and Cohen M, (Elsevier, Amsterdam) 1982, 199.
11. Drehmann A J and Turnbull D, Scripta Met., 1981, 15, 543.
12. Paik J S and Perepezko J H, J.Non-Cryst. Solids, 1983, 56, 405.
13. Kelton K F, Greer A L and Thompson C V, J.Chem.Phys. 1983, 79, 6261.
14. Kelton K F and Greer A L, in Rapidly Quenched Metals V, eds. Steeb S and Warlimont H (Elsevier Amsterdam) 1985, 223.
15. Boettinger W J and Coriell S R, these proceedings.
16. Perepezko J H and Boettinger W J, Mat.Res. Soc. Symp.Proc.,1983,19, 223.
17. Turnbull D, J.Chem. Phys., 1952, 20, 411.
18. Thompson C V, Greer A L and Spaepen F, Acta Met., 1983,31, 1883.
19. Köster U and Herold V, in Rapidly Quenched Metals IV, eds. Masumoto T and Suzuki K (Japan Inst.Metals, Sendai) 1982, 717.
20. Chen H S and Turnbull D, Acta Met. 1968,16, 369.
21. Perepezko J H and Paik J S, J.Non-Cryst. Solids 1984,61, 113.
22. Thompson C V and Spaepen F, Acta Met. 1979,27, 1855.
23. Hillert M, Acta Met.,1953,1, 763.
24. Thompson C V and Spaepen F, Acta Met. 1983,31, 202.
25. Spaepen F, Acta Met., 1975,23, 729.
26. Turnbull D, J. Appl. Phys.,1950,21, 1022 (1950)
27. Skapski A S, Acta Met., 1956,4, 576.
28. Bonissent A, Finney J L and Mutaftschiev B, Phil.Mag. B,1985, 42, 223.
29. Waseda Y and Miller W A, Trans. JIM, 1978, 19, 547.
30. Hirth J P, Met. Trans., 1978, 9A, 401.
31. Bosio L, Defrain A and Epelboin I, J.Phys. (Paris), 1966, 27, 61.
32. Pound G M and La Mer V K, J.Amer.Chem. Soc., 1952, 74, 2323.
33. Perepezko J H and Smith J S, J.Non-Cryst. Solids,1981, 44, 65.
34. Glicksman M E and Vold C L, Scripta Met., 1971, 5, 593.
35. Perepezko J H, LeBeau S E, Miller B A and Hildeman G J, in Rapidly Solidified Powder Aluminium Alloys, ASTM 1984, (in press).
36. Cech R E, J. Metals, 1956, 206, 585.
37. Boettinger W J, Coriell S R and Sekerka R F, Mat.Sci. Eng.,1984,65, 27.

38. Massalski T B, in Proc.Fourth Int.Conf.on Rapidly Quenched Metals, eds. Masumoto T and Suzuki K (Japan Inst.Metals, Sendai), 1982, 203.
39. Boettinger W J, ibid. p. 99.
40. Baker J C and Cahn J W, in Soldification (ASM, Metals Park, OH),1971, 23.
41. Levi C G and Mehrabian R, Met. Trans. A, 1982, 13, 221.
42. Perepezko J H, Mueller B A, Richmond J J and Cooper K P, in Rapidly Quenched Metals V, eds. Steeb S and Warlimont H (Elsevier,Amsterdam), 1985, 43.
43. Borromee-Gautier C, Giessen B C and Grant N J, J.Chem.Phys., 1968,48, 1905.

ABSTRACTED DISCUSSION OF THE PAPER BY J.H. PEREPEZKO

Participants: W.J. Boettinger, R.W. Cahn, B. Cantor, M.E. Glicksman,
L. Katgerman, R.E. Lewis and P.R. Sahm

Discussion commenced with a question regarding nucleation temperature
ranges for specific droplet size fractions. J.H. Perepezko stated that
nucleation temperature exotherms measured from his DTA experiments were
frequently about thirty degrees wide, but with much finer droplets, these
were ten dregrees in width. In cases where hypercooling was achieved the
thermal effects of recalescence appear minor because the liquid medium
surrounding the droplet is a good conductor. The prospects for finding
homogeneous nucleation in specific alloy systems appears to be good, and
J.H. Perepezko stated that for a 20 μm diameter particle with competition
between various heterogeneous nucleants, homogeneous nucleation should
predominate if the contact angle is greater than about 120 degrees. This
nucleation effect of various coatings is clearly important and indicates a
promising area for future research. Questions about surface nucleation and
volume nucleation clearly indicated that much remains to be understood
about the role of catalytic surface agents. Discussion of the role of small
silicon additions in modifying the observed nucleation undercooling in
Al-Pb alloys led to general observations about the need for better sur-
face-science analytical tools as an investigative technique. In Al-5wt.%Fe
alloys produced by gas atomization substantial undercoolings ($\Delta T \doteq 260$ K)
were found in only ~10 μm size fractions, and only in those solidified
droplets was the microeutectic (Zone A) structure observed. These slow
cooling experiments therefore give rather accurate undercooling requirements for metastable microstructural phase formation. In these classes of
industrially important alloys J.H. Perepezko felt some degree of progress
had now been made toward designing undercooling requirements for specific
alloy systems.

THERMODYNAMICS IN RAPID SOLIDIFICATION

B. PREDEL
Universität Stuttgart
Institut für Metallkunde
D-7000 Stuttgart, F.R. Germany

SUMMARY
    Given suitable kinetic conditions, rapid solidification is characterized
by a large reduction of the Gibbs free energy of the system during nucleation and crystallization processes. In this context, the thermodynamic
properties of supercooled melts are discussed. Possibilities for calculating the Gibbs free energy of metastable melts are indicated, the association model being discussed in particular. Finally, the glass forming ability of liquid alloys and the crystallization of metallic glasses and metastable crystalline phases formed from strongly supercooled melts are
considered.

1. INTRODUCTION
    As is well known, crystallization of a melt only occurs if a reduction
of the Gibbs free energy results. For this thermodynamic reason, crystallization is not possible above the melting point. This crystallization may be
suppressed also at very high supercoolings, however, for kinetic reasons
even when this reduction of the Gibbs free energy would take place on
crystallization. At the glass temperature T_g, the viscosity of the melt
amounts to 10^{12} Nsm^{-2}. This means that no crystallization will occur within
finite times. Therefore, the temperature range between the melting temperature T_m and the glass temperature T_g will be of particular interest. Both
the rate of nucleus formation I and the crystal growth velocity v show
exponential dependences on the Gibbs free energy

$$\Delta G = G^L - G^S \qquad (1)$$

(L = Liquid, S = Solid) which is reduced as a result of the liquid-to-
crystalline transition. This is strictly true only if nucleation does not
involve a concentrational change. ΔG depends on the amount of supercooling
$\Delta T = T_m - T$. The interactions between I and v may strongly affect the
microstructure and thus the properties of the solid product.
    With increasing supercooling ΔT among other things dendrite arm spacings
decrease, and in alloys increasing supersaturation of the primary phase may
occur, metastable crystalline phases can be formed, and ultimately even a
noncrystalline solid - a glass - can be produced. These changes can be
utilized technologically to engineer microstructure and properties.
    In the following, the thermodynamic properties of supercooled metallic
melts will be treated, rather than the kinetics of rapid solidification.
Correlation between the thermodynamics and kinetics of solidification of
metallic melts as well as new technological capabilities have resulted from
the intensive studies of supercooled metallic liquids carried out during
the last few years. Also research on metallic glasses has been very helpful
in this respect. Unfortunately, the method of extremely rapid quenching of
suitable metallic melts does not permit direct investigations of the supercooled state. This is due to the short time interval during which the melt

goes through the temperature range between T_m and T_g. Some progress has been made by utilizing small droplet techniques in which the probability of operation of the more active heterogeneous nucleants becomes quite small (1). Thus, Perepezko (2) has been able to achieve supercooling to temperatures as low as $0.3\ T_m$.

In addition to that, there are theoretical considerations and model concepts which yield certain insights into the behaviour of supercooled metallic liquids.

2. FUNDAMENTALS OF SUPERCOOLED MELTS

At first, both G-T and enthalpy-temperature plots for a pure metal (figures 1 and 2) will be recalled, in order to identify ΔG and ΔH. Often, for simplicity, it is assumed that $\Delta C_p = 0$. Then, ΔG is easily shown for an arbitrary supercooling ΔT to be given by:

$$\Delta G = G^L - G^S = (S^L - S^S) \cdot \Delta T = \Delta S \cdot \Delta T = \frac{\Delta H_m}{T_m} \cdot \Delta T \qquad (2)$$

The assumption $\Delta C = 0$, however, is only an approximation. Precise measurements show in fact that $\Delta C_p \neq 0$ (figure 3). The deviation from $\Delta C_p(T_m)$ will steadily increase with increasing supercooling, figure 4. This, consequently, affects the enthalpy-temperature curve, figure 5.

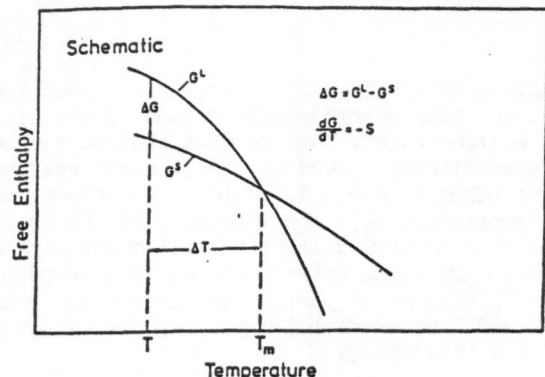

Figure 1: Gibbs free energy (free enthalpy) as a function of temperature for a pure substance near its melting point T_m (schematically).

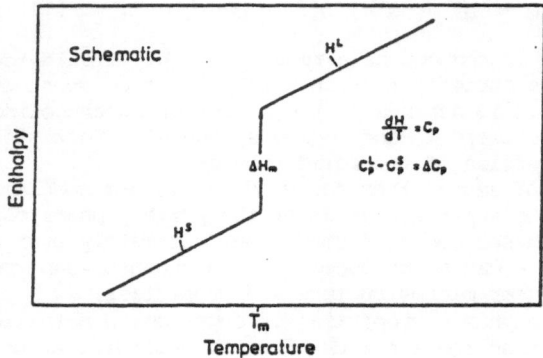

Figure 2: Enthalpy as a function of temperature for a pure substance near its melting point T_m (schematically).

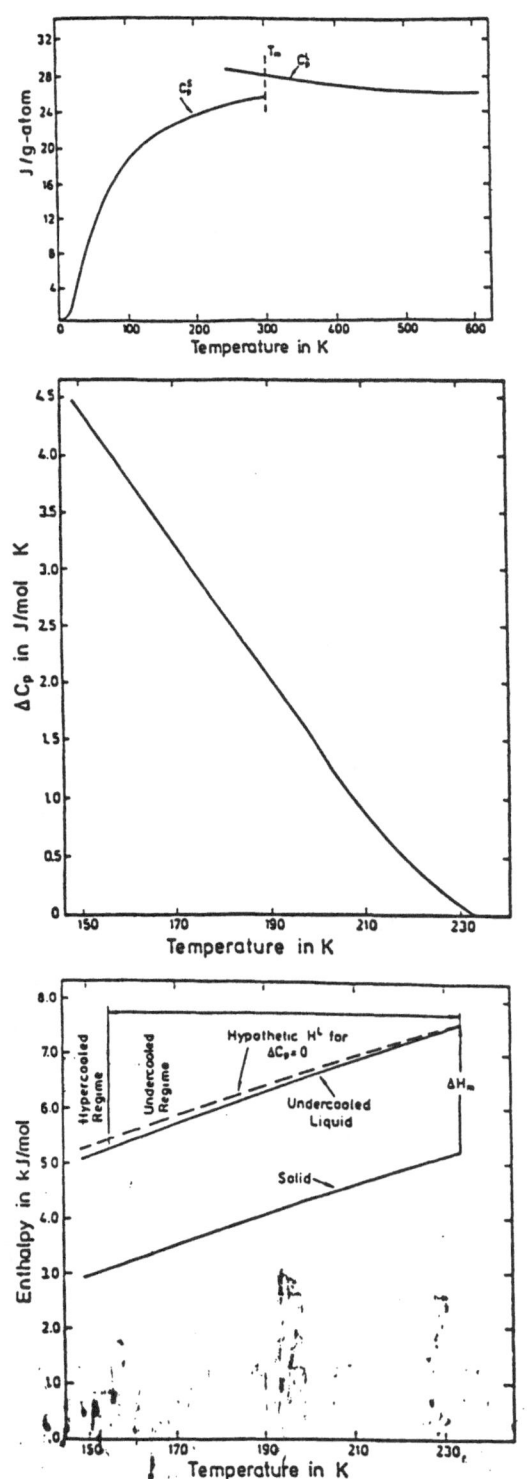

Figure 3: Molar heat C_p of pure gallium as a function of the temperature (L = liquid, S = crystalline solid). After Chen and Turnbull (3).

Figure 4: Difference ΔC_p in molar heat between the supercooled melt and the (thermodynamically stable) crystalline solid as a function of the temperature for pure mercury. After Perepezko and Rasmussen (4).

Figure 5: Enthalpy of mercury as a function of temperature below its melting point T_m. After Perepezko and Rasmussen (4).

Three regimes of temperature can be distinguished for supercooled melts:

1. The first extends from the solidification temperature T_m down to a nucleation temperature T_n, which is low enough to ensure that the enthalpy of melting just suffices to reheat the solidifying mass to T_m. This region is characterized by the validity of the following relationship (Perepezko and Rasmussen (4)):

$$\Delta H_m(T_n) \geq \int_{T_n}^{T_m} c_p^L \, dT. \tag{3}$$

2. Below regime 1, a "hypercooled melt range" may be defined. It extends down to the glass temperature T_g at which no nucleation occurs any more due to the low mobility of atoms.

3. Not much below T_m, i.e. in slightly supercooled melts, the structural units are sufficiently mobile, that they can assume positions corresponding to minimal Gibbs free energy conditions in response to any change in temperature. Then, numerous physical properties of the melt, the molar heat for example, can be unambiguously measured. In a glass, the mobility of the structural units is so low that within finite times temperature changes do not generate appropriate structural adjustments which will be in accordance with a minimal Gibbs free energy. In relation to the stable solid phase(s) the supercooled melt (as well as the glasses) are metastable. The supercooled melt, however, displays an internal equilibrium, in contrast to the glass. In the glass, the structure is frozen-in which is characteristic of the supercooled melt at the glass point T_g. The glass temperature depends on the cooling rate. Glass formation is a relaxation phenomenon.

If one could succeed in cooling down so slowly that with decreasing temperature the equilibrium obtains during each stage of cooling, glass transition would not occur. Then, a truly supercooled melt would persist exhibiting peculiar properties. By extrapolating the entropy below T_g, a situation could be visualised in which the entropy of the liquid phase could become lower than that of the crystalline phase and corresponding behavior would be found for the enthalpy and for the molar volume. Figure 6 for example, shows the dependence on temperature of the enthalpies of lactic acid and of the glass-forming alloy $Au_{81.4}Si_{18.6}$. Since the hypothesized cases, i.e., such $S^L - S^S < 0$ and $H^L - H^S < 0$, are not possible, we can conclude that such far-reaching extrapolations are inadmissible. Generally, it can be said, that on approaching the so-called Kauzmann temperature (5) T_0, at which $H^L = H^S$, the activation energy for nucleation is of the same order of magnitude as the activation energy of the movement of the structural units of the supercooled melt. Then, both nucleation and growth of nuclei can be proceed with kinetics similar to those governing movement of the structural units in the melt at this temperature.

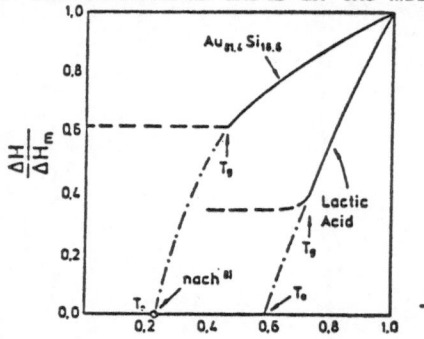

Figure 6: Fractional enthalpy $\Delta H/\Delta Hm$ of $Au_{81.4}Si_{18.6}$ and lactic acid as a function of homologous temperature T/T_m (5); data from (6), (7), (8).

Although the glass possesses the atomic structure of the melt, with regard ·to the temperature dependent enthalpy change, it is behaving more like the crystalline state. This is because, in the glass, some of the melt s degrees of freedom are frozen-in. This becomes particularly clear for the molar heat. The quenched-in degrees of freedom can be of very different kinds, depending on the sort of material. In glassy alloys this may obviously be the ability to adjust the equilibrium of association which is at a standstill in the glass. Because

$$C_p = (\frac{\partial H}{\partial T})_p \ , \tag{4}$$

on increasing the temperature of supercooled melt the enthalpy must be increased because the degree of dissociation will have been lowered. In a glass, this contribution to C_p amounts to zero. The relation between C_p^l and the dissociation process is indicated in figure 7. The average value $\overline{\Delta C_p^l}$ (see below, equation (30)) can be represented completely on the basis of the equilibrium of association using the mass action law. The effects of association equilibria in liquid alloys will be treated below.

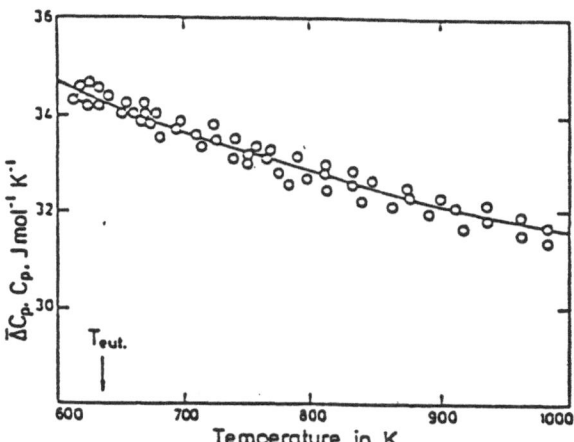

Figure 7: Molar heat as a function of temperature for liquid $Au_{72}Ge_{28}$ alloy:
o ΔC_p experimental values according to (9)
– $\overline{\Delta C_p^l}$ calculated on the basis of the
association model (10).

3. APPROXIMATIONS FOR THE CALCULATION OF ΔG

The relationship given in equation (2) often correctly describes the circumstances in the immediate neighborhood of T_m, as experiment shows, that ΔC_p is usually small at the melting point. More generally

$$\Delta G = \frac{\Delta H_m}{T_m} \cdot \Delta T - \int_T^{T_m} \Delta C_p \ dT + T \int_T^{T_m} \Delta C_p \ d \ln T \tag{5}$$

This relationship, however, requires a knowledge of ΔC_p as a function of the temperature. Therefore, various approximations have been developed which do not require an explicit knowledge of $\Delta C_p(T)$. Thus, for equation, (2) different correction terms are recommended which are more or less useful. Figure 8 comprises a survey of data for lead and indicates various ways of correction as follows:

a) ΔC_p = const. Hoffmann (11) approximates

$$\Delta G = \frac{\Delta H_m}{T_m} \cdot \Delta T \cdot \frac{T}{T_m} \qquad (6)$$

b) ΔC_p = const. Thompson and Spaepen (12) suggest:

$$\Delta G = \frac{\Delta H_m}{T_m} \cdot \Delta T \cdot (\frac{2T}{T_m + T}) \qquad (7)$$

c) ΔC_p = const. Battezzati and Garrone (13) offer:

$$\Delta G = \frac{\Delta H_m}{T_m} \cdot \Delta T - 0,8 \frac{\Delta H_m}{T_m} (\Delta T - T \cdot \ln \frac{T_m}{T}) \qquad (8)$$

d) $\Delta C_p = \Delta C_p^m$ = const. Jones and Chadwick (14) propose:

$$\Delta G = \frac{\Delta H_m}{T_m} \cdot \Delta T - \frac{\Delta C_p^m \cdot (\Delta T)^2}{(T_m + T)} \qquad (9)$$

e) An approximation by Dubey and Ramachandrarao (7) from the hole theory of liquids gives

$$\Delta G = \frac{\Delta H_m}{T_m} \cdot \Delta T - \frac{\Delta C_o^m \cdot (\Delta T)^2}{2T} \cdot (1 - \frac{\Delta T}{6T}) \qquad (10)$$

which is closely related to equation (9).

f) $\Delta C_p(T)$; ΔC_p is a linear function of T. For this case Singh and Holz (15) approximate

$$\Delta G = \frac{\Delta H_m}{T_m} \cdot \Delta T \cdot \frac{7T}{T_m + 6T} \qquad (11)$$

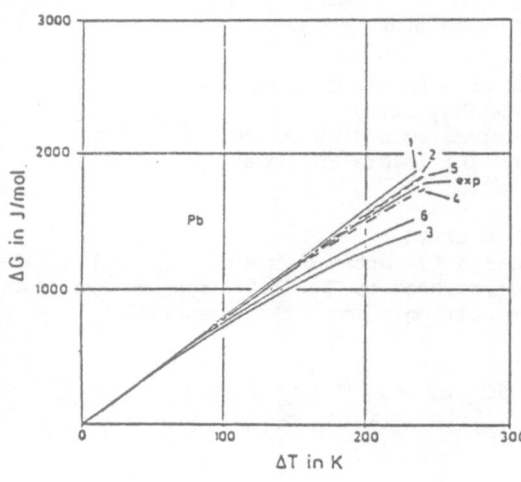

Figure 8: Free enthalpy change Δ G (Gibbs free energy) upon crystallization of pure lead as a function of supercooling Δ T according to different models:
1) eq. (2),
2) eq. (9),
3) eq. (7),
4) eq. (11),
5) eq. (10),
6) eq. (8).
Experimental data from (12) and (15).

Numerous papers by Perepezko and co-workers (e.g.(4), (16), (17))as well as by Chu et al. (1) have shown that large supercoolings can be achieved at low cooling rates (about 10 K min^{-1}). They emulsify the metal in non-metallic liquids and thus create small droplets. They thus gain the possibility to determine C_p^L experimentally in the range of supercooling. Values obtained in this experimental way are evidently preferable to those gained with the aid of the approximate formulae cited above.

Finally, Miura, Isa and Omuro (18) proposed to compute the change of the Gibbs free energy $\Delta G = G^L - G^S$. For that purpose a knowledge of T_c (crystallization temperature of the quenched-in glass), T_m and the crystallization enthalpy ΔH_c are required. They argue:

$$\Delta G = \frac{\Delta H_m}{T_m} (T_m - T) - \left[T_m - T - T\ln \frac{T_m}{T}\right] \cdot a - \frac{(T_m - T)^2}{2} \cdot b \quad (12)$$

In doing so, it is assumed that

$$\Delta C_p = a + b \cdot T; \quad \Delta C_p^m = a + b \cdot T_m \quad (13)$$

$$\Delta H_m = -\Delta H_c + \int_{T_c}^{T_m} (a + b \cdot T) \, dT \quad (14)$$

$$a = \Delta C_p^m - \frac{2T_m \cdot \Delta C_p^m}{T_m - T_c} + \frac{2T_m (\Delta H_m + \Delta H_c)}{(T_m - T_c)^2} \quad (15)$$

and

$$b = \frac{2 \Delta C_p^m}{T_m - T_c} - \frac{2 (\Delta H_m + \Delta H_c)}{(T_m - T_c)^2} \quad (16)$$

Figure 9 plots ΔG for the transition of $Au_{81.4}Si_{18.6}$ from the glassy into a metastable intermetallic state with the γ-brass structure. The approximations, equations (12) and (7), both deviate by almost equal amount but with opposite signs from the experimental data for large supercooling.

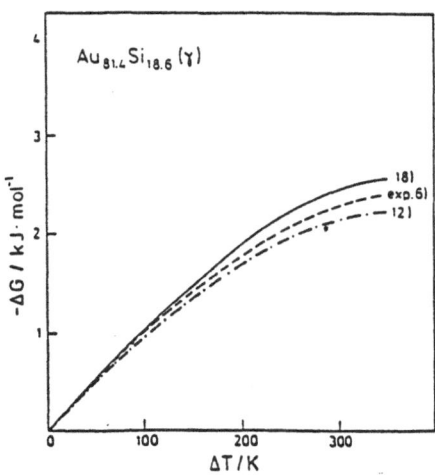

Figure 9: Dependence on supercooling ΔT of the Gibbs free energy change for the transition of the metallic glass $Au_{81.4}Si_{18.6}$ into a metallic phase with γ-brass structure
--- experimental values according to (6),
—— approximation according to (18),
-.- approximation according to (12).

4. EFFECT OF ASSOCIATION ON THE ENERGETIC PROPERTIES OF SUPERCOOLED LIQUID
    ALLOYS

Metallic glasses can be obtained by extremely rapid cooling of alloy
melts. To be sure, glass formation can only be achieved if certain
structural and energetic conditions are fulfilled. A necessary, though not
sufficient, prerequisite is that the melt shows a tendency towards compound
formation. Obviously then, the liquid alloy will exhibit a non-random
distribution of the various atoms present; a chemical short range order. As
is well known, this can be proven on the basis of X-ray or neutron diffraction experiments, e.g. (19) and (20). The degree of this chemical short
range order will increase with decreasing temperature. In the glassy state,
this will be "frozen-in" as already stated. Thus, the atomic arrangement
will be the same one as is present in the supercooled melt, at the glass
temperature T_g.

Suzuki, Fukunaga, Misawa and Masumoto (21) have shown by diffraction
experiments that the atomic arrangement in Pd-Si glasses correlates with
that of the crystalline intermetallic compound Pd_3Si. By means of nuclear
magnetic resonance Kemeny, Vincze, Fogarassy and Arajas (22) have shown
for amorphous Fe-B alloys that the atom positions are ordered correspondingly to those in the metastable crystalline compound Fe_3B.

It seems obvious to comprehend stoichiometric units such as mentioned
above as molecule-like species (often also referred to as "associates") and
assume them also to be present above the glass temperature T_g. Above T_g
the number of atoms belonging to such associates would be accordingly
smaller. It is possible to describe the degree of short range order within
the melt by a degree of association as follows: an equilibrium of association between the associates A_iB_j and the monatomic species A_1 and B_1 is
established by way of

$$iA_1 + jB_1 \rightleftharpoons A_iB_j \qquad (17)$$

This description of the chemical short range order in equation (17) is
equivalent to expressing it by the Cowley short range order parameters.
Sommer (23) has demonstrated this by a comparison of melts in the Cu-Ti
system. Bhatia and Singh (24) introduced a general correlation between
these quantities.

The existence of associates naturally influences the thermodynamic
mixing functions of the melts. For a random distribution of the different
types of atom, the enthalpy of mixing (utilizing the regular solution
model) is a parabolic function of the atomic fractions involved, see figure
10. For a very high degree of association, a nearly triangular ΔH-x curve
is prevalent the maximum of which marks the composition of the associates.
A linear ΔH-x behavior between e.g. the pure component A and the associate
indicates that here a simple mixture exists between the two species involved, namely A_iB_j and A_1.

If, however, there is a considerable interaction between these species,
a curved line between zero (for pure A) and the extreme value of ΔH at the
composition of the associates will result. Also, if associate formation is
incomplete, a curved ΔH-x curve will be the result lying between a parabola
of the second degree and a triangle. As the shape of the ΔH-x curve is
determined by the equilibrium of association it is possible to calculate
the degree of association from the shape of the curve. This depends, in an
analytically representable way, on the concentration and the temperature.
Utilizing the association model, one may thus derive thermodynamic behavior
of supercooled melts from data easily accessible by extrapolating down
below the liquidus temperature.

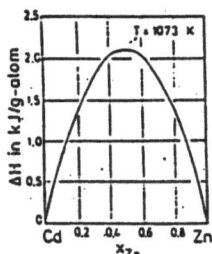 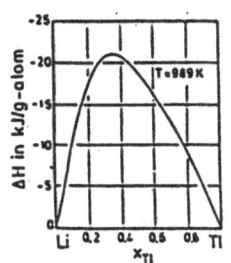 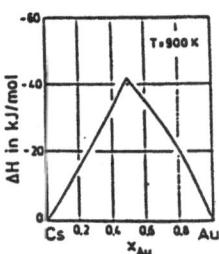

Figure 10: Showing the influence of the association in alloy melts on the concentration dependence of the mixing enthalpies ΔH for

- Cd-Zn (25): approximately random distribution of atoms;
- Li-Tl (29): stoichiometry of the associates is Li_2Tl; degree of association for Li_2Tl is 44 % (10);
- Au-Cs (26): stoichiometry of the associates is AuCs (10) and their degree of association: 100 %.

For illustrating such a procedure we follow Sommer, e.g. (28). For the association equilibrium in a homogeneous phase, equation (17), the equilibrium condition

$$\left(\frac{\partial \; \Delta G}{\partial \; n_{A_iB_j}}\right)_{T,p,n_A} = 0 \qquad (18)$$

is valid. Here, ΔG the change of the Gibbs free energy of the system upon associate formation and $n_{A_iB_j}$ is the number of moles of associate A_iB_j which are present at equilibrium. The total number of moles is then given by:

$$n = n_{A_1} + n_{B_1} + n_{A_iB_j} \qquad (19)$$

The number of moles of particular species are connected with the total numbers n_A and n_B as follows:

$$n_{A_1} = n_A - i \cdot n_{A_iB_j} \qquad (20)$$

$$n_{B_1} = n_B - j \, n_{A_iB_j} \qquad (21)$$

The partial Gibbs free energy of the species A presents itself as

$$\bar{G}_{A_1} = G^L_{A_1} + RT \ln a_{A_1}$$

$$= G^L_{A_1} + RT \cdot \ln \frac{n_{A_1}}{n} + RT \ln \gamma_{A_1} \qquad (22)$$

Here, a_{A_1} is the thermodynamic activity, γ_{A_1} is the activity coefficient of the species A_1 and $G^L_{A_1}$ is the Gibbs free energy of the pure liquid component A at temperature T.

The relationships for the partial Gibbs free energies of the other two species, namely $\bar{G}^L_{B_1}$ and $\bar{G}_{A_iB_j}$, are expressed in an analogous way. The integral Gibbs free energy ΔG of the liquid alloy, in which an equilibrium according to equation (17) is assumed, then may be expressed as:

$$
\begin{aligned}
\Delta G &= \sum_\nu n_\nu \, \bar{G}_\nu - n_A \, G^L_A - n_B \, G^L_B \\
&= n_{A_iB_j} \, (G^L_{A_iB_j} - i \cdot G^L_{A_1} - j \cdot G^L_{B_1}) + RT \sum_\nu n_\nu \cdot \ln \left(\frac{n_\nu}{n}\right) \\
&\qquad + RT \sum_\nu n_\nu \ln \gamma_\nu
\end{aligned}
\tag{23}
$$

Here $G^L_A = G^L_{A_1}$, $G^L_B = G^L_{B_1}$ with ν = the stoichiometric coefficient of the respective species. Differentiating equation (23), remembering the condition of equation (18) which determines the equilibrium, yields the mass action law:

$$
\frac{(a_{A_1})^i \, (a_{B_1})^j}{(a_{A_iB_j})} = \frac{1}{K_{A_iB_j}} = \frac{1}{\exp\left(-\dfrac{\Delta G^O_{A_iB_j}}{RT}\right)}
\tag{24}
$$

where $\Delta G^O_{A_iB_j}$ is the molar Gibbs free energy of formation of the associates from the monatomic species. It now holds that

$$
\Delta G^O_{A_iB_j;} = \Delta H^O_{A_iB_j;} - T\Delta S^O_{A_iB_j;}
\tag{25}
$$

where $\Delta H^O_{A_iB_j}$ is the molar formation enthalpy of the associates and $\Delta S^O_{A_iB_j}$ is the molar formation entropy of the associates.

The mixing enthalpy of a liquid alloy with associate formation results, both from the interactions between the particular species, see equation (26a), and the interaction between atoms within the associates, i.e. the formation enthalpy of the associates, see equation (26b):

$$
\Delta H^{reg} = \frac{n_{A_1} \cdot n_{B_1}}{n} \cdot C^{reg}_{A_1,B_1} + \frac{n_{A_1} \cdot n_{A_iB_j}}{n} \cdot C^{reg}_{A_1,A_iB_j} +
$$
$$
+ \frac{n_{B_1} \cdot n_{A_iB_j}}{n} \cdot C^{reg}_{B_1,A_iB_j}
\tag{26a}
$$

and

$$
\Delta H^{as}_{A_iB_j} = n_{A_iB_j} \cdot \Delta H^O_{A_iB_j}
\tag{26b}
$$

In doing so it is presumed that the different species are statistically distributed in the solution. $C^{reg}_{A_1,B_1}$, $C^{reg}_{A_1,A_iB_j}$ and $C^{reg}_{B_1,A_iB_j}$ are the interaction parameters for the pair interactions between the respective species.

The integral enthalpy of mixing ΔH of the solution results from the sum of the single contributions

$$
\Delta H = \Delta H^{reg} + \Delta H^{as}_{A_iB_j}
\tag{27}
$$

Accordingly, for the mixing entropy ΔS of a liquid solution exhibiting associate formation the following applies:

$$\Delta S = - R \left[n_{A_1} \cdot \ln x_{A_1} + n_{B_1} \cdot \ln x_{B_1} + n_{A_iB_j} \cdot \ln x_{A_iB_j} \right] + $$

$$+ n_{A_iB_j} \cdot \Delta S^o_{A_iB_j} \qquad (28)$$

The interaction parameters $c^{reg}_{A_1,A_iB_j}$ and $c^{reg}_{B_1,A_iB_j}$ as numerous investigations, i.e. (10), (30),(31) have shown, are of lesser importance in many systems. In those cases they can be set approximatively equal to zero. The remaining three parameters $\Delta H^o_{A_iB_j}$, $\Delta S^o_{A_iB_j}$ and $c^{reg}_{A_1,B_1}$ can be obtained by simultaneously solving equations (24), (27), and (28). In doing so, the number of moles of the associates in the equilibrium, $n_{A_iB_j}$ is accessible because the numbers of moles are contained in the activities of the mass action law, equation (24), i.e.

$$a_{A_iB_j} = \frac{n_{A_iB_j}}{n} \cdot \gamma_{A_iB_j}, \qquad (29)$$

in which, $\gamma_{A_iB_j}$ is the activity coefficient of A_iB_j.

In order to obtain the Gibbs free energy of a supercooled melt its molar heat C_p (or the change $\overline{\Delta}C_p$ of its molar heat upon the transition from temperature T_1, at which C^L is known, to another temperature T_2) should be known. $\overline{\Delta}C_p$ is given by the temperature dependence of the enthalpy of mixing. As a rule, this value is equal to the mean change of the molar heat for a T_1 to T_2 transition with sufficient accuracy, i.e.

$$\frac{\Delta H_{T_1} - \Delta H_{T_2}}{T_1 - T_2} \simeq \overline{\Delta}C_p . \qquad (30)$$

For the determination of $\overline{\Delta}C_p$ one can proceed as follows. By fitting the parameters to the experimental $\Delta H(x)$ and $\Delta S(x)$ curves at T_1 = const., $\Delta H^o_{A_iB_j}$, $\Delta S^o_{A_iB_j}$, $c^{reg}_{A_1,B_1}$ and, thereby, also $n_{A_iB_j}$ are obtained as functions of the molar fraction x. Assuming that $\Delta H^o_{A_iB_j}$ $\Delta S^o_{A_iB_j}$, and $c^{reg}_{A_1,B_1}$ show practically no dependence upon the temperature (within the range between T_1 and T_2) the number of moles $n_{A_iB_j}$ for T_2 can be calculated as a function of x. Then, also $\Delta H(x)$ and $\Delta S(x)$ are derivable for T_2. Utilizing equation (30) $\overline{\Delta}C_p$ is obtained. The good consistency of $\overline{\Delta}C_p$ values thus derived becomes apparent when examining figure 7, in which experimentally determined C_p data of a liquid $Au_{72}Ge_{28}$ are compared to $\Delta C_p(T)$ values which were calculated using the association model (10).

5. ON THE GLASS FORMING ABILITY OF ALLOY MELTS

Numerous interesting approaches to the influence of thermodynamic and kinetic factors on the ability of an alloy melt to solidify as a glass upon rapid quenching have been proposed (see for example ref. 30). This paper will only touch upon some thermodynamic aspects. Good glass forming ability obviously requires poor crystallizability of the supercooled melt. In order to obtain a glass crystallization has to be subdued down to the glass temperature. In this respect $\Delta G = G^L - G^S$ at T_g, to be calculated with the aid of the association model, will give important indications.

The presence of associates stabilizes the melt with respect to a crystalline solid body. With increasing degree of association the Gibbs free energy G^L of the liquid alloy decreases, and for given G^S, ΔG becomes accordingly smaller. With falling temperature, the equilibrium shifts to the right side of equation (17), making the fraction of associates in the melt increase. The existence of an association equilibrium in liquid alloys is a prerequisite for their good glass forming ability. So far, no metallic system has been observed in which glass formation occurs on rapid cooling that does not simultaneously exhibit associate formation, i.e. in which $\Delta H > 0$. Within the scope of this rather simple consideration, however, it is not easily explained why in a given system glass formation in confined to a narrow range of concentration while at other concentrations of the same system crystal nucleation easily takes place. In order to investigate this problem it ought to be remembered that crystal nucleation is affected by a concentration dependence of G^S and G^L. Utilizing the association model, $G^L(x)$ is accessible for the glass temperature. These individual phenomena can thus be investigated.

In figure 11 Gibbs free energies of the liquid and solid phases in the Mg-Ca system are plotted as function of concentration for the glass temperature $T_g = 400$ K. The region of easy glass formation, i.e. at around $x_{ca} = 0.7$, is of particular interest. Upon crystallization a concentration change takes place. Depending on the concentration of the supercooled melt, either Mg_2Ca, a Ca-rich solid solution (β), or none of these phases are formed. The gain in Gibbs free energy, ΔG_n, on crystal nucleation is determined by the tangent touching the G^S - x curve at the concentration x_n. The distance ΔG_n between this tangent and the G^S - x curve of the respective solid produced by crystallization, at the concentration of the emerging nucleus, is the driving force of the crystallization process. Three cases are to be distinguished.

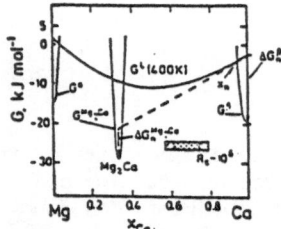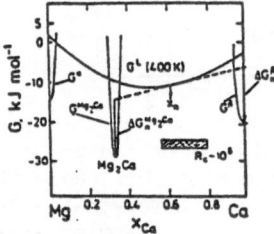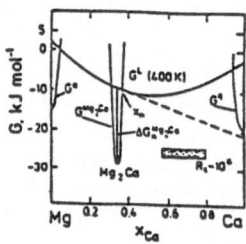

Figure 11: Dependance on alloy concentration of Gibbs free energy for the stable crystalline phases and the supercooled melt of Mg-Ca alloys at T_g = 400 K. The free energies of crystallization are shown for three different concentrations x_n of the supercooled melt (29). |XXXXX region of single phase glass, ///// = region of two phases (glassy and crystalline) for a cooling rate R_c of 10^6 K s^{-1}.

1. Ca-rich melts. G_n is large for the crystallization of β, yet small for Mg_2Ca. In addition, the formation of the β-solid solution can already be achieved by relatively few change-of-site processes, while the formation of Mg_2Ca requires more diffusion over longer distances. In this region, the nucleation of β is thus very probable.
2. Near the stoichiometric Mg_2Ca composition of the supercooled melt. ΔG_n is large for Mg_2Ca formation and small for β. The diffusion conditions are opposite to those described for case 1.
3. Concentrations approximately midway between Mg_2Ca and β. ΔG_n values are about equal and are both smaller than the respective largest ΔG_n-values for cases 1 and 2. A considerable amount of diffusion is also required in order to adjust the concentration locally, to such an extent that neither Mg_2Ca nor β-nuclei could be formed.

For case 3, therefore crystal nucleation is not very probable. Glasses can be more easily formed, in addition promoted by the fact that a eutectic occurs within this region of the phase diagram. Close to the eutectic the range of stability of the melt is extended to peculiarly low temperatures. Upon cooling therefore lower temperatures are also reached without driving forces for nucleation being present. Simultaneously atom mobility decreases and, thus, also does the rate of crystal nucleation controlled by diffusion.

6. ENERGETICS OF THE CRYSTALLIZATION OF METALLIC GLASSES

The equilibrium crystalline phases may form by crystallization of a glass, for example in the Mg-Cu binary system, figure 12. The composition range within which metallic glasses will be observed, have been marked specifically. It concentrates around the composition of the lowest melting eutectic. Upon crystallization of glasses of these compositions, a mixture of the equilibrium phases, $CuMg_2$ and Mg, is formed.

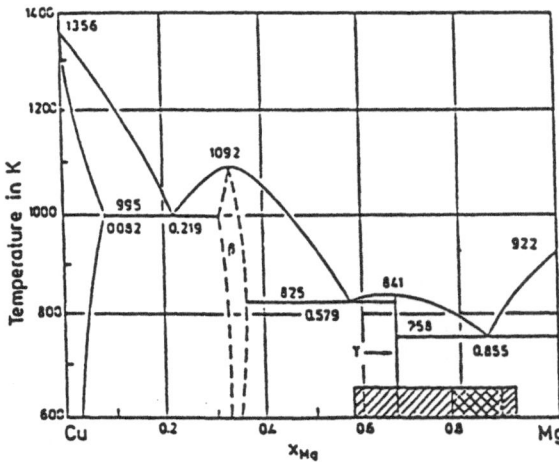

Figure 12: Cu-Mg phase diagram (31); indicating regions with one glassy phase XXXXX , with two phases, i.e. glassy plus crystalline phase ⧄.

In order to determine the formation enthalpy, ΔH^G, of the glass, two reaction processes characterized by the same initial and final states, need to be considered:

1. $Mg^S + Cu^S \xrightarrow[\Delta H^S_{CuMg_2}]{\substack{\text{formation of the} \\ \text{stable crystalline} \\ \text{phase at room} \\ \text{temp. } T_R}} CuMg_2^S + Mg^S \xrightarrow[]{\substack{\text{heating-} \\ \text{up to } T_e}} CuMg_2^S + Mg^S$

2. $Mg^S + Cu^S \xrightarrow[\Delta H^G]{\substack{\text{formation of the} \\ \text{glass at room} \\ \text{temp. } T_R}} \text{glass} \xrightarrow[\Delta H_c]{\substack{\text{heating-} \\ \text{up to } T_e}} CuMg_2^S + Mg^S$

In both cases, the sum of all enthalpy changes must be the same. Therefore it holds true that

$$3 \; x_{Ca} \; \Delta H^S_{CuMg_2} + \int_{T_R}^{T_e} c^S_p(x_{Cu}) dT = \Delta H^G(x_{Cu}) + \int_{T_R}^{T_e} c^G_p \; dT + \Delta H_c(x_{Cu}) \qquad (31)$$

Here, $\Delta H^S_{CuMg_2}$ is the formation enthalpy of the crystalline $CuMg_2$ as formed from the solid (= S) components at room temperature T_R; $\Delta H^G(x_{Cu})$ is the enthalpy of formation of the glassy phase with mole fraction x_{Cu} by crystallization from the glassy phase at x_{Cu}; $c^S_p(x_{Cu})$ is the molar heat of the glass at x_{Cu}; T_e is the temperature at which the crystallization process of the glass is entirely completed; Δc^{S-G}_p is the difference between the molar heat of the crystalline mixture of $CuMg_2$ and Mg and the glassy phase. Equation (31) can be rewritten in the following manner:

$$\Delta H^G(x_{Cu}) = 3 \; x_{Cu} \; \Delta H^S_{CuMg_2} + \int_{T_R}^{T_e} \Delta c^{S-G}_p \; dT - \Delta H_c(x_{Cu}) \qquad (32)$$

As the molar heats of crystalline solids and compositionally comparable glasses differ only slightly, Δc^{S-G}_p can be neglected. Equation (33) then can be simplified to

$$\Delta H^G(x_{Cu}) = 3 \; x_{Cu} \; \Delta H^S_{CuMg_2} - \Delta H_c(x_{Cu}) \qquad (33)$$

Relatively low values result for the formation enthalpies $\Delta H^G(x_{Cu})$. For a glass of composition $Cu_{13}Mg_{87}$, $\Delta H^G = -0.1$ kJ/g-atom and for a glass of concentration $Cu_{21.7}Mg_{78.3}$, $\Delta H^G = -0.8$ kJ/g-atom.

The formation enthalpy of a solid is usually described by the summation:

$$\Delta H = \Delta H_B + \Delta H_M + \Delta H_T \qquad (34)$$

Here, ΔH_B is the contribution from the change in bonding conditions upon synthesis from the components, H_M is the misfit enthalpy (lattice distortion) due to the difference in atomic radii, and ΔH_T is the enthalpy arising from the structural difference between the initial and the final states of the alloy formation reaction. This latter contribution is given for the Mg-Cu glasses considered here by:

$$\Delta H_T = x_{Mg} \cdot \Delta H^{C \to G}_{Mg} + x_{Cu} \cdot \Delta H^{C \to G}_{Cu} \qquad (35)$$

$\Delta H^{C \to G}_{Mg}$ and $\Delta H^{C \to G}_{Cu}$ are enthalpy terms required for transforming pure Mg or pure Cu into a (hypothetical) glass. Thus the formation enthalpy of such a glass may be hypothesized to be

$$\Delta H^G(x_{Cu}) = \Delta H(x_{Cu}) + x_{Mg} \; \Delta H^{C \to G}_{Mg} + x_{Cu} \; \Delta H^{C \to G}_{Cu} \qquad (36)$$

with $\Delta H(x_{Cu})$ as the mixing enthalpy of the melt made up of the bonding and the misfit terms (see above). Since $_\Delta H(x_{Cu})$ is known (39) for liquid alloys, the enthalpy required for the transformation of both crystalline Mg and Cu into their respective glasses can be calculated, provided that ΔH^G is also available for two or more concentrations, giving

$$\Delta H_{Mg}^{C \to G} = 1.8 \ kJ/g\text{-atom,}$$

$$\Delta H_{Cu}^{C \to G} = 13.3 \ kJ/g\text{-atom.}$$

A small discrepancy exists between these values and those given by Sommer (31) because newer $\Delta H(x_{Cu})$-values (39) were available for the calculations just presented. For comparison, the melting enthalpies at the equilibrium melting point are as follows:

$$\Delta H_m^{Mg} = 8.9 \ kJ/g\text{-atom.}$$

$$\Delta H_m^{Cu} = 13.0 \ kJ/g\text{-atom.}$$

In some cases the glass temperature T_g cannot be determined experimentally, because crystallization commence before reaching T_g. This is also true for Mg-Cu glasses. In these cases the glass temperature may be calculated from thermodynamic quantities of the system. For this purpose a circular process is considered with both initial and final state being the glass at the glass temperature T_g.

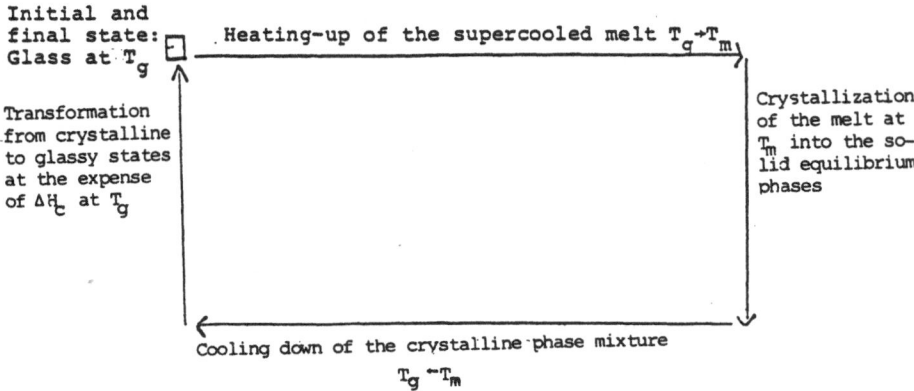

Initial and final state: Glass at T_g

Heating-up of the supercooled melt $T_g \to T_m$

Transformation from crystalline to glassy states at the expense of ΔH_c at T_g

Crystallization of the melt at T_m into the solid equilibrium phases

Cooling down of the crystalline phase mixture $T_g \leftarrow T_m$

The enthalpy balance for this circular process is:

$$\int_{T_m}^{T_g} c_p^L \ dT - \Delta H_m + \int_{T_m}^{T_g} c_p^S \ dT - \Delta H_c = 0 \qquad (37)$$

with ΔH_m equal to the melting enthalpy of the alloy.

For the $Cu_{14.5}Mg_{85.5}$ eutectic glass the value 380 K is obtained for T_g. The experimentally observed onset of crystallization has been found at $T_c = 374$ K, only slightly below the calculated value of T_g.

7. ENERGETICS OF METASTABLE CRYSTALLINE PHASE FORMATION

The occurence of metastable phases upon rapid cooling of metallic melts has different causes. Here, the effect of crystallographic structure of the pure solid components will be briefly considered.

Take the In-Sn system, figure 13 (32). Solid solutions with a face-centered tetragonal structure occur at the In-rich side, the axial ratio c/a deviating slightly from 1 (c/a > 1). The β-phase again is face centered tetragonal showing a slight deviation in the opposite sense: c/a < 1. Both phases can thus be regarded to be approximatively face centered cubic. The error resulting from this approximation has the very small value of 4 J/g-atom (33, 34) as calorimetric measurements have shown.

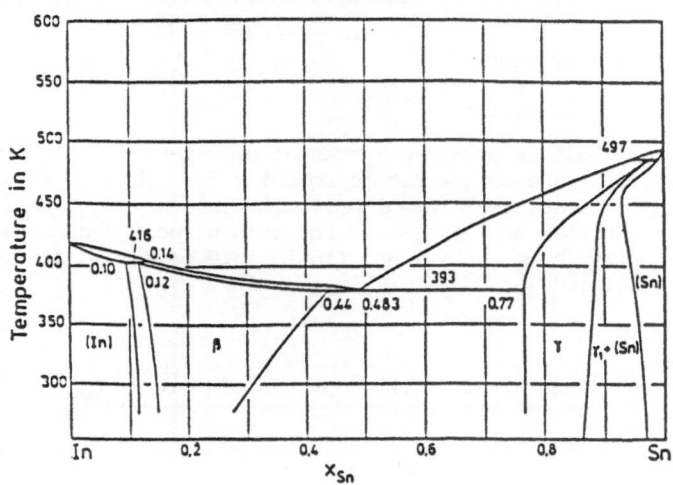

Figure 13: Phase diagram of the In-Sn system (32).

The lattice distortion enthalpy ΔH_M (see equation (34)) is small and can be neglected. The difference of the atomic radii is only 0.6 %.

The mixing enthalpy ΔH^L of the liquid In-Sn alloys thus ought to be due only to bonding differences. They are small, however, because ΔH^L exhibits a very low value. Correspondingly it may be expected that bonding energy differences will be similar on the formation of the solid phase, such that $\Delta H^L = \Delta H_\beta^S$. Furthermore, $\Delta H \simeq 0$. With this in mind

$$\Delta H^S = \Delta H^L + \Delta H_T^S \tag{38}$$

Because ΔH^L is small, ΔH^S must be predominantly determined by ΔH_T^S. Accordingly, see figure 14, $\Delta H^S(x)$ may be described by a straight line for the concentration range up to 50 at.-% Sn. If one substracts ΔH^L from $\Delta H^S(x)$, then $\Delta H_T^S(x)$ results:

$$\Delta H_T^S = x_{Sn} \cdot \Delta H_T^{Sn} \tag{39}$$

Thus, the transformation enthalpy ΔH_T^{Sn} of tetragonal Sn into its face centered cubic modification can be determined. It amounts to $\Delta H_T^{Sn} = 5.5$ kJ/g-atom (35).

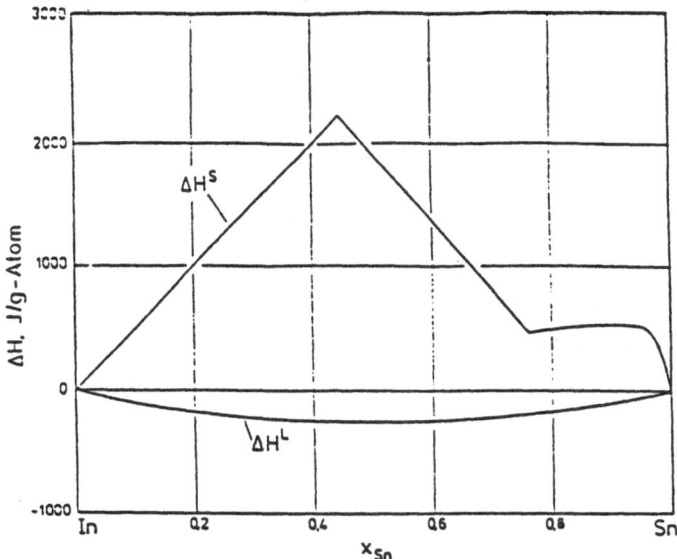

Figure 14: Dependence on concentration of enthalpy of
formation of solid (ΔH^S) and liquid (ΔH^L) phases in
In-Sn alloys (33).

In this way transformation enthalpies of hypothetical structural changes
for various metals can be determined. Such transformation enthalpies can
help to predict the occurrence of metastable phases. For instance, in the
Au-Si system no equilibrium Hume-Rothery phase is known to exist (36),
although such phases ought to be stable on the basis of electronic criteria. The liquid alloys hint distinctly at this tendency (see figure 15).
The maximal mixing enthalpies are found in the concentration regime in
which Hume-Rothery phases should be observed. In the solid state, however,
the formation of such phases with their high atomic packing densities is
impeded by the enormous transformation enthalpy of Si with $\Delta H_T^{Si} > 65$ kJ/g-
atom $(\Delta H_T^{Ge} = 65$ kJ/g-atom (38) and $\Delta H_T^{Si} > \Delta H_T^{Ge})$. This has to be provided if
Si is to transform from its diamond structure into a tightly packed metal-
lic structure, such as presupposed in the Hume-Rothery phases. In the melt,
this restriction is not effective. Here, a chemical short range order
prevails which complies with that characteristic for Hume-Rothery compound
forming systems.

Glasses are known to form within the Hume-Rothery prone concentration
range of the Au-Si binary system. Upon heating, such glasses first transform into a metastable γ-Hume-Rothery phase, rather than into the stable
solid solutions, Au and Si, see figure 15. This phase is also obtainable by
supercooling the melt at moderate coolings (39). Possibly the formation of
the γ-phase from the melt, which has an atomic structure resembling that
of the Au-Si glass, is preceded by a glassy phase with a short time of
existence.

In any case, the formation of γ-phase nuclei only requires relatively
few adjustments of atomic sites from the existing chemical short range
order to achieve the periodicity of the γ-phase lattice. For the nucleation
of Au and Si considerable concentration and structural changes would be
necessary. Therefore, this letter kind of nucleation is less probable than
that of the metastable γ-Hume-Rothery phase.

70

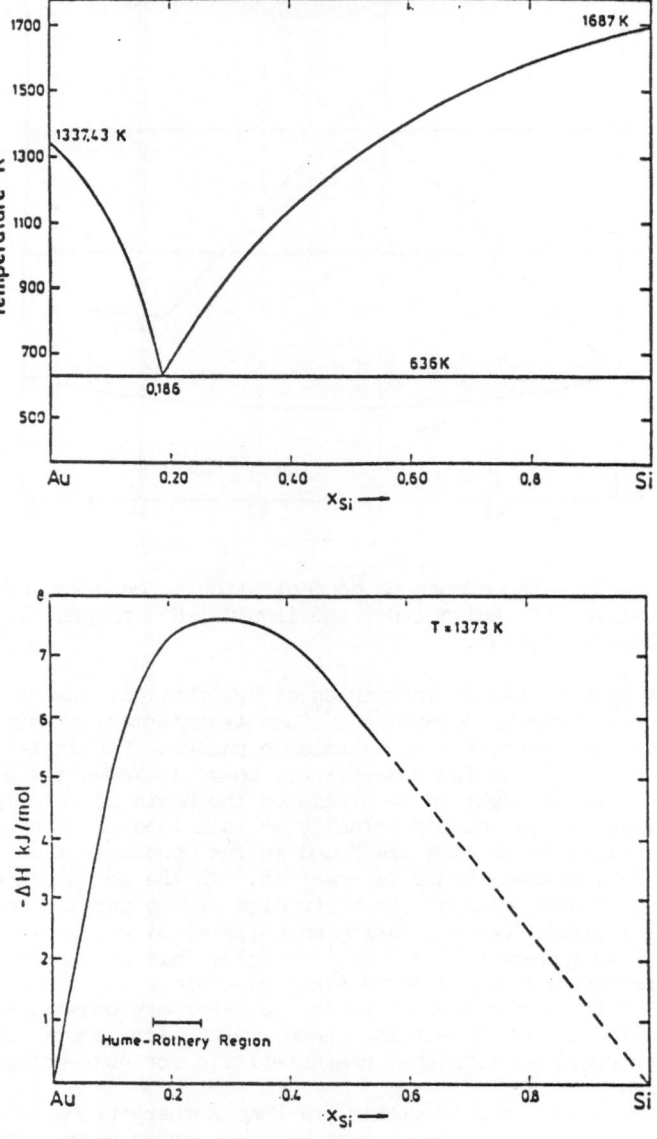

Figure 15: Phase diagram of Au-Si (36) and mixing enthalpies of liquid Au-Si alloys at 1373 K (37).

ACKNOWLEDGEMENT
    I am grateful to Dr. F. Sommer and Dr. I. Arpshofen for valuable
discussions.

REFERENCES

1. Chu M G, Shiohara Y and Flemings M C, Met. Trans. A, 1984, 15A, 1303.
2. Perepezko J H, Mat. Sci. Eng., 1984, 65, 125.
3. Chen H S and Turnbull D, Acta Met., 1968,16, 369.
4. Perepezko J H and Rasmussen D H, Met. Trans. A, 1978, 9A, 1490.
5. Kauzmann W, Chem. Rev., 1948, 43, 219.
6. Chen H S and Turnbull D, J. Appl. Phys., 1967, 38, 3646.
7. Dubey K S and Ramachandrarao P, Acta Met., 1984, 32, 91.
8. Allen J W, Wright A C and Connell G A N, J. Non-Crystalline Solids, 1980, 42, 509.
9. Schluckebier G and Predel B, Z. Metallkde., 1980, 71, 605.
10. Sommer, F, Z. Metallkde. 1982, 73, 77.
11. Hoffman J D, J. Chem. Phys., 1958, 29, 1192.
12. Thompson C V and Spaepen ,F Acta Met., 1979, 27, 1855.
13. Battezzati L and Garrone E, Z. Metallkde., 1984, 75, 305.
14. Jones D R H and Chadwick G A, Phil. Mag., 1971, 24, 995.
15. Singh H B and Holz A, Solid State Communications, 1983, 45, 985.
16. Perepezko J H and Paik J S, J. Non-Crystalline Solids, 1984, 61,62, 113.
17. Perepezko J H and Smith J S, J. Non-Crystalline Solids, 1981,44, 65.
18. Miura H, Isa S and Omuro K, J. Non-Crystalline Solids, 1984,61,62, 163.
19. Steeb S, Falch S and Lamparter P, Z. Metallkde., 1984,75, 599.
20. Fukunaga T and Suzuki K, Sci. Rep. RJTU, 1980, 28, 208.
21. Suzuki K, Fukunaga T, Misawa M and Masumoto T, Science Reports on the Research Institutes, Tohoku University, Amorphous Material Issue I, Sendai, Japan, 1976.
22. Kemeny T, Vincze J, Fogarassy B and Arajas S, "Structure and Crystallisation of Fe-B Metallic Glasses", Hungarian Academy of Science, Budapest, 1978.
23. Sommer F, Proc. 5th. Internat. Conference on Rapid Quenched Metals, 1984, in Press.
24. Bhatia A B and Singh R N, Phys. Chem. Liqu., 1984, 13, 177.
25. Hultgren R, Desai P D, Hawkins D T, Gleiser M and Kelley K K, Selected Values of Thermodynamic Properties of Binary Alloys, Amer.Soc. for Metals, Metals Park, Ohio, 1973.
26. Sommer F, Eschenweck D, Predel B and Schmutzler R W, Conf. Proceedings "Chemical Metallurgy - A Tribute to Carl Wagner", Gokcen N A Ed., Metallurgical Society AIME, Chicago, 1981,p.19.
27. Predel B and Oehme G, Z. Metallkde., 1979, 70, 68.
28. Sommer F, Mat. Res. Soc. Symp. Proc, 1983, 19, 163.
29. Sommer F, Ber. Bunsenges. Phys. Chem., 1983, 87, 749.
30. Sommer F, Z. Metallkde., 1981, 72, 219.
31. Sommer F, Bucher G and Predel B, J. de Physique, 1980, C-8, 563.
32. Predel B and Gödecke T, Z. Metallkde., 1975, 66, 654.
33. Alpaut O and Heumann Th, Acta Met., 1965, 13, 543.
34. Predel B, Z. Metallkde, 1964, 55, 117.
35. Heumann T and Wöstmann H, vgl. Diplomarbeit H. Wöstmann, Minster (Westf.), 1970.
36. Okamoto H and Massalski T B, Bull. Alloy Phase Diagrams, 1983, 4, 190.
37. Castanet R, Chastel R and Bergman C, Mat. Sci. Eng., 1978, 32, 93.
38. Predel B and Stein D W, Z. Naturforschung, 1971, 26a, 722.
39. Ellner M and Predel B, Z. Metallkde., 1980, 71, 364.
40. Sommer F, Lee J J and Predel B, Ber, Bunsenges. Phys. Che., 1983, 87, 792.

ABSTRACTED DISCUSSION OF THE PAPER BY B. PREDEL

Participants: R.W. Cahn, B. Cantor and J.H. Perepezko

Discussion commenced with a general statement about the nature of the ideal glassy state, and the temperature dependence of kinetics of relaxation processes which are ultimately responsible for crystallization on cooling from the liquid state. It was proposed by B. Predel that complex clustering effects persist in the liquid state for relaxation times about a hundred times longer than the mean atomic vibrational period, and that this clustering effect requires certain atomic sizes, ratios and stoichiometry for stabilization of the liquid (glassy) state. As a consequence of this liquid-state association certain systems become immune to fast crystallization at modest undercoolings. If the association is similar to that in an intermetallic compound found in the alloy system, then the glassy state is more difficult to attain on rapid cooling. Systems such as Mg-Ca, In-Sn, Au-Si and Pd-Si were discussed in this context. It was proposed that the presence of associated clusters in the liquid state would produce measurable changes in such physical properties as heat-capacity and viscosity, and B. Predel felt that there was adequate experimental data to support this effect when one compared liquid alloy systems with the pure component counterpart systems. Strong clustering and association effects were expected to produce increased liquid undercooling capability, and theoretical calculations of this clustering effect should then lead to a predictive capability for glass formation.

FORMATION AND STABILITY OF METALLIC GLASSES

R.W.CAHN
Clare Hall
Cambridge CB ODT, U.K.

1. INTRODUCTION

The purpose of this short paper is quite modest: it is to outline a few
of the principal extant models, criteria, calculations and fullblown theories that claim to interpret (never, as yet, to predict) the alloy combinations and compositions capable of forming metallic glasses by rapid solidification from the melt, and to point to some aspects which seem ripe for
further attention. Most attention is devoted to recent work.

2. PURE METALS

In a recent review on metallic glasses (1) the reader is reminded that,
over 30 years ago, Frank (2) presented arguments that the energetically
preferred topology of the SRO in monatomic liquids should be icosahedral
(incorporating fivefold symmetry), and the high difficulty of reconstruc-
ting such icosahedral clusters into a periodic crystal should, according to
Frank's ideas, lead to a high kinetic resistance to crystal nucleation in
such melts when supercooled. This high resistance has not been observed
with pure metallic melts, which nobody has yet succeeded in quenching to
form a glass. (However, it is striking that the recently announced creation
of non-periodic quasicrystals with fivefold symmetry, by RS of Al-14at.%Mn
and related alloys (3,4), implies the existence in the melt of icosahedral
clusters as suggested by Frank).

Motorin (5) has recently calculated the expected homogeneous nucleation
rate for the pure metals Ag, Cu, Ni and Pb, using known material parameters, and thence estimated the required minimum cooling rate from the melt
and the maximum feasible amorphous layer thickness. No account was taken
(it is difficult to see how it could be) of Frank's consideration. Allowing
for the great sensitivity of the conclusions to small errors in some of the
input parameters, for these four metals minimum cooling rates of 10^{12} -
10^{13} K/s were estimated, with limiting amorphous layer thicknesses of a few
tens of nanometres. On the basis of this analysis, Motorin agrees with the
criticism advanced by Wood and Akhurst (6) of Davies and Hull's (7) claim
to have vitrified pure nickel by cooling at circa 10^{10} K/s; Wood and
Akhurst advanced evidence that the glassy nickel must have been stabilised
by impurities.

Lin and Spaepen (8, 9) have reported experiments on attempts to vitrify,
by picosecond laser pulses, surface layers of specimens containing finely
spaced multilayers of two metals. In Fe-B and Ni-Pb specimens of varying
effective compositions, the glass-forming range of compositions was considerably extended in comparison with normal melt-spinning (cooling rate by
laser-treatment circa 10^{12} K/s, by melt-spinning, 10^5 - 10^6 K/s). However,
the pure metals could not be vitrified; in Fe-B, the minimum boron content
needed was 5 at.%, and in Ni-Nb, at least 18 at.% Nb was needed. It seems,
then, that Motorin's estimates of critical cooling rates for vitrification
of pure metals are, if anything, on the low side.

3. ROLE OF ATOMIC RADIUS

It has been clear from the earliest days of the study of bi-metallic glasses that for both glass formation from the melt and glass formation from the vapour, solvent and solute atoms have to differ sufficiently in size. In fact, this was first recognized in connection with amorphous surface layers made by co-evaporation of Cu-Ag and Co-Au on to amorphous substrates (10) (this early paper amply repays renewed study!). Mader was the first to simulate the process with a two-dimensional hard-ball model (11), and this was followed by model experiments with bubble rafts (12,13). These model experiments suggested that a radius mismatch approaching 15% is necessary to permit glass formation.

Model experiments have been followed by empirical correlations, of which there have been several. It has become customary to plot combinations of pairs of metallic elements on a graph showing radius mismatch between solvent and solute along one axis, and some measure of bond strength (typically the heat of mixing or the heat of evaporation) along the other (e.g., Giessen (14)). Careful examination of such graphs makes it clear that whereas the bond strength does play some part, this is a very "weak" variable, whereas the radius mismatch rigorously has to exceed 10% (rather than 15% as suggested by model experiments).

Very recently, Egami and Waseda (15) have broken quite new ground by establishing a correlation between the solvent/solute atom size mismatch and the critical solute concentration needed to allow vitrification by rapid melt-quenching. They found that, to remarkably close consistency, for 66 glass-forming systems, the relationship $c_B^{min} |(v_B - v_A)/v_A| \simeq 0.1$ was obeyed, where c_B^{min} is the minimum solute (B) concentration needed to allow a glass to be formed by quenching the melt, and the v's are atomic volumes. So it appears that the size mismatch not only has a strong effect on solubilities, but also on critical concentrations for glass formation. In fact, in a very detailed discussion of their findings, Egami and Waseda conclude that c_B^{min} represents a kind of ultimate solid solubility, beyond which the solid solution rapidly becomes unstable; the liquid prefers to precipitate another solid solution of different composition, and the associated need for solid-state diffusion slows down crystallization kinetics. This means that glass formation becomes relatively favoured over equilibrium crystallization.

The importance of this study lies in the fact that most attempts to establish a glass-formation criterion (including all earlier analyses of the role of atomic size) examined only what combinations of elements are able to form glasses, but not the compositions which are able to do so.

Some metals (e.g., nickel) have been amorphized by implantation of appropriate ions. Self-ion implantation is never effective, and it has in fact been established (16) that for successful amorphization, the radius of the implanted atom should be in the range $0.59r_A$ to $0.88r_A$, where r is the radius of the matrix atoms. Presumably an analysis similar to Egami and Waseda's is applicable here, since the disturbance exercised by the bombarding ions turns the metal into a supercooled, labile alloy liquid which has to choose between stabilising as a glass or crystallising.

4. FREE VOLUME AND VITRIFICATION

An early (i.e., 5-years old) attempt to interpret the composition ranges in which alloy melts are able to form glasses on rapid quenching was by Ramachandrarao, who has done so much to further the theoretical metallurgy of metallic glasses (17). In this important paper, the author uses an old analysis due to Varley which enables him to compute the mean atomic volume of liquid metallic solutions from the individual atomic radii and the compressibilities of the elements. There is generally a negative deviation, which can be quite large, from a "liquid Vegard's Law", or ideality. He finds that glasses can be formed by quenching only in a composition range encompassing the greatest deviations from ideality (i.e., the smallest mean atomic volume). He discusses the origins of this loss of atomic volume in terms of highly localised internal stresses, like those more recently analysed by Egami and his coworkers. In effect, low mean atomic volume implies a lack of free volume, hence high viscosity and low diffusivity. (It is not yet clear whether the very steep variation with composition of critical quenching rate for glass formation which has been measured in at least one system, Pd-Si-Cu, by Naka et al. (18) can be interpreted on the basis of Ramachandrarao's theory.)

A related but independent analysis is due to Yavari et al.(19). They established empirically that zero (or negative) change of specific volume on melting of a crystalline species favours glass formation on subsequent RS of the melt. In effect, the model is that if a crystal is denser than the melt from which it grows, then in growing it rejects free volume into the melt and thereby reduces its viscosity; thus crystal growth becomes self-catalytic. There is a problem here: we know that a grain boundary in a solid can act as a vacancy sink - so why should a solid/liquid interface not act as a sink for free volume, instead of the free volume being injected across the interface into the melt?

Again related to the same underlying ideas is an analysis due to Buschow (20). He calculated the energy required to form a "hole" in a liquid alloy, following Miedema's well-tested computational approach, and compared this with the crystallization temperature of the corresponding glass. The two variables cluster close to a line of mutual proportionality. The higher the hole formation energy, the smaller the free volume and the higher the viscosity (21,22), and so the higher the expected stability of the glass. So here we see that the glass-forming ability (limiting quenching rate required) and the stability of a glass once formed (crystallization temperature...generally very close to T_g) can be analyzed on the basis of essentially the same model.

5. METAL-METALLOID GLASSES

All of the foregoing theory has been applied only to metal-metal glass-forming alloys, and not to the important family of metal-metalloid alloys. This family has not attracted much serious theorising, perhaps because of the after-effect of the somewhat outdated, but very influential paper by Polk (23) which put forward the idea that the roughly 20 at.% of metalloid normally used to make such glasses "stuffs" and rigidifies a melt very effectively. Following recent analyses of the size and number of "Bernal voids" of different sizes in a DRP melt (24), it has become plain that even the smaller metalloid atoms cannot fit into more than a very small proportion of these voids, and certainly 20 at.% cannot be so accommodated. However, Turnbull's celebrated demonstration that metalloid atoms change their effective size readily as they dissolve in increasing amounts in a metal, both in glasses (25) and in crystalline solutions (26) suggests that Frost's analysis may not dispose as readily of Polk's simple model as is generally assumed.

One approach which has not been properly exploited as yet for analysing the size and concentration of the largest voids in a DRP glass stems from using hydrogen atoms as probes. Kirchheim et al. (27) first showed that the diffusity of hydrogen solutions in such a glass show relatively small hydrogen diffusity, because the hydrogen is firmly bound in large voids (this conception is well established by years of hydrogen embrittlement studies); Kirchheim et al. succeeded in analysing the steep concentration dependence of hydrogen diffusity accurately in terms of a postulated Fermi distribution of binding energies. (Berry and Pritchet (28) however point out the theoretical complications in understanding hydrogen diffusion in a metallic glass on any classical basis).

Very recently, Stolz, Nagorny and Kirchheim (29) measured volume changes resulting from the progressive dissolution of hydrogen in Pd-base glasses containing Si or P; they found that the partial molar volume of hydrogen increases sharply (from negative values for the lowest concentrations) with increasing hydrogen concentration, but only up to about 10^{-2} at.%, and then stabilises, which suggests that the larger voids of size comparable with metal atoms (which might be termed "quasivacancies") are present to the extend of about one in 10^4 atom sites. (This analysis has not yet found its way into statistical mechanical analyses of diffusivity in metallic glasses, of which the most comprehensive is that due to Lançon et al.(30)). It seems that conclusions about void sizes - at least for the larger voids - drawn from measurements on hydrogenated glasses might come to play a part in analysing the stability of metal-metalloid glasses.

In this connection,it should be noted that while a large radius mismatch is evidently crucial for a metalloid to facilitate formation of a metallic glass, if the mismatch is too great then the metalloid diffuses too readily and crystallization is thus not effectively inhibited. Hydrogen by itself certainly does not facilitate glass formation, and neither does carbon by itself; no direct measurements are extant for carbon diffusivity in a metallic glass (this is badly needed) but it is to be presumed that it is substantially higher than that of the larger boron atom, which is a good aid for glass formation.

A fault of most of the analyses made up to now of void size distributions in a DRP model glass is that they are made for the "ground state" and take no account of temperature. This is why the analysis of Cantor and Ramachandrarao (31) is so important. In this preliminary attempt at a statistical-mechanical analysis of the variation with temperature of the proportions of tetrahedra, octahedra, tetragonal dodecahedra, etc., in a hard-sphere DRP model (the glass equivalent of the familiar thermodynamic calculation of the variation with temperature of vacancy concentrations in crystals), the authors were able to show the expected increase with temperature of the number of larger assemblies (and associated voids), and as a bonus they were able to compute from their values a temperature variation of specific heat which compared quite closely with measured specific heats. In future attempts to analyse the stability of metal-metalloid glasses, this kind of variation of structure with temperature will need to be taken seriously.

6. GLASS FORMATION AND FAST DIFFUSION

An unexpected correlation which has recently been established is that between abnormally fast diffusion in certain metallic (crystalline) solid solutions and the glass-forming ability of melts of the corresponding compositions. Fast diffusion appears to be linked with the presence of a substantial proportion of interstitially located solute atoms. Turnbull (33) was the first to note that a weak dependence of interatomic (AB) potential to the separation near the minimum (a "soft" potential law) should both favour fast diffusion and glass formation. In a recent study, Tendler (34) analyses glass-forming ability, capacity for fast diffusion and solvent/solute radius ratio for numerous binary alloys based on zirconium, and finds an almost perfect correlation between fast diffusion and ready glass formation on rapid melt-quenching. The critical radius ratio proves to be close to 0.85.

Fast diffusion is also crucial to the feasibility of the remarkable "reaction-vitrification" process discovered by Johnson and his associates (35). Fast diffusion in at least one of the participating crystalline phases ensures kinetic preference for glass formation over the formation of a thermodynamically stabler crystalline phase.

7. THERMODYNAMICS AND CLUSTER THEORY

The use of computed free energy vs composition diagrams is increasingly being applied to interpret glass forming ability and may eventually lead to predictions (e.g., Perepezko (36)). Saunders and Miodownik (37) have applied this approach successfully to interpret the glass-forming composition ranges in several metallic systems. They showed how a glass-forming range can be bracketed between a composition at which (for a temperature close to T_g) free energies of crystal and liquid are equal, and another composition at which the tendency to compound formation becomes strong. This approach is as yet in its infancy. Another novel, purely thermodynamic treatment of the GFA problem is based on an attempt to calculate the "ideal" glass transition temperature (that at which the Kauzmann Paradox supervenes and glass and crystal have identical entropies) from measurable thermodynamic quantities (38).

Parallel to the mass of studies of metallic glass formation is another large population of papers that deals with covalently bonded, semiconducting glasses. Generally speaking, the two "worlds" have remained separate because the underlying theory is so different, but it may be that they are beginning to converge. Thus, Phillips (39) successfully analyses measurements of critical cooling rate for glass formation in the Ge-Se system (40) in the light of compound formation tendency and of the energy penalty for bond distortion in local Ge-Se clusters. The mechanical energy consideration arising from the presence of covalent bonds has no analogue in the study of metallic glasses, but the role of compositions tending to compound formation has.

There is no room here to discuss in detail the extensive literature which has built up in the last 5 years on the role of composition clusters in metallic melts on glass formation ability. Suffice it to say that the present, involved theoretical situation is very well presented in a recent paper by Ramachandrarao et al. (41).

It has also to be pointed out that most of the theorising with respect to glass-forming ability relates tacitly to the prevention of crystallization by homogenous nucleation. The importance in many systems of heterogeneous nucleation, and its prevention, is well attested by the recent spectacular experiments in which substantial ingots of $Pd_{40}Ni_{40}P_{20}$ were turned into glass by continous fluxing of the ingot surface (42). Methods are beginning to appear for establishing, ex-post-facto, whether in a particular system the predominant mechanism of nucleation is homogeneous or heterogeneous (43,44). However, this large issue will no doubt be treated in Perepezko's contribution to this conference.

8. CONCLUSION

This very partial discussion of the theoretical problem of interpreting which systems and which compositions within those systems are apt to form glasses on quenching - and the related but distinct problem of interpreting the variation of crystallization temperature with composition - has perhaps established one overriding factor - the atomic size. This factor, which rears its head over and over again in classical crystal chemistry, likewise refuses to go away in the theoretical examination of metallic

78

glasses, in spite of a repeated tendency to consider an insistence on this factor as being distinctly naive. Whenever atomic size has been looked at as a variable in association with other variables, it has always been atomic size which has emerged as the more determining factor. Consequently, with the closely related factor of free volume, atomic size considerations will continue to play a major part in glass-forming theory. Egami and Waseda's recent study (15) will perhaps prove the beginning of a new period of attention to this kind of analysis.

9. REFERENCES

1. Spaepen F and Turnbull D, Ann. Rev. Phys. Chem., 1984, 35, 241.
2. Frank F C, Proc. R. Soc. Ser. A., 1952, 215, 43.
3. Shechtman D, Blech I, Gratias D and Cahn J W, Phys. Rev. Lett., 1984, 53, 1951.
4. Levine D and Steinhardt P J, Phys. Rev. Lett., 1984, 53, 2477.
5. Motorin V I, phys. stat. sol.(a), 1983, 80, 447.
6. Wood J V and Akhurst K N, J. Mater. Sci., 1976, 11, 2142.
7. Davies H A and Hull J B, J. Mater. Sci., 1976, 11, 215.
8. Lin C J and Saepen F, in "Rapidly Solidified Metastable Materials", ed. Kear B H and Giessen B C, MRS Symposia Proc., Vol. 28, 1984, 75.
9. Lin C J, Spaepen F and Turnbull D, J. Non.-Cryst. Solids, 61 and 62, 767.
10. Mader S, Nowick A S and Widmer H, Acta Metall., 1967, 159, 203, 215.
11. Mader S, IBM J. Res. Dev., 1965, 9, 358.
12. Simpson A W and Hodkinson P A, Nature (London), 1972, 237, 320.
13. Argon A S, J. Phys. Chem. Sol., 1982, 43, 945.
14. Giessen B C, Proc. 4th. Int. Conf. on Rapidly Quenched Metals, Sendai, 1981, ed. Masumoto T and Suzuki K, Jap. Inst. Metals, Sendai, 213.
15. Egami T and Waseda Y, J. Non-Cryst. Solids, 1984, 64, 113.
16. Rauschenbach B and Hohmuth K, phys. stat. sol.(a), 1982, 72, 667; Hohmith K, Rauschenbach B, Kolitsch A and Richter E, Nucl. Inst. f. Meth., 1983, 209/210, 249.
17. Ramachandrarao P, Z. Metallkde., 1980, 71, 172.
18. Naka M, Nishi Y and Masumoto T, Proc. 3rd Int. Conf. on Rapidly Quenched Metals, 1978, ed. Cantor B, Metals Soc., London, 231.
19. Yavari A R, Hicter P and Desre P, J. Chim. Physique, 1983, 79, 572.
20. Buschow K H J, Solid State Comm., 1982, 43, 171.
21. Ramachandrarao P, Cantor B and Cahn R W, J. Non-Cryst. Solids, 1977, 24, 109.
22. Ramachandrarao P, Cantor B and Cahn R W, J. Mater. Sci., 1977, 12, 2488.
23. Polk D, Acta Metall., 1972, 20, 485.
24. Frost H J, Acta Metall., 1982, 30, 889.
25. Turnbull D, Scripta Metall., 1977, 11, 1131.
26. Turnbull D, Scripta Metall., 1982, 15, 1039.
27. Kirchheim F, Sommer F and Schluckebier G, Acta Metall., 1982, 30, 1059.
28. Berry B S and Pritchet W C, Nontraditional Methods in Diffusion, ed. Murch G E et al., Met. Soc. AIME, 1984, 83.
29. Stolz U, Nagorny U and Kirchheim R, Scripta Metall, 18, 347.
30. Lancon F, Billard L, Chambron W and Chamberod A, J. Phys. F, 1985, 15, in the press.
31. Cantor B and Ramachandrarao P, Proc. 4th. Int. Conf. on Rapidly Quenched Metals, ed. Masumoto T and Suzuki K, Japan Inst. Metals, Sendai, 1982, 291.
32. Vanfleet H B, Phys. Rev. B, 1980, 21, 4340.

33. Turnbull D, J. de Physique, 1974, 35, C4-1.
34. Tendler R H, J. Mater Sci., 1985, 20, in the press.
35. Johnson W L, Atzmon M, van Rossum M, Dolgin B P and Yeh X L, Proc. 5th Int. Conf. on Rapidly Quenched Metals, Würzburg, 1984, ed. Warlimont H, to be published by North-Holland, Amsterdam, 1985.
36. Perepezko J H, Mat. Res. Soc. Symp. Proc. Vol. 19, 1983, 223.
37. Saunders N and Miodownik P, Ber. Bunsenges. Phys. Chem., 1983, 87, 830.
38. Dubey K S and Ramachandrarao P, International J. Rapid Solidification, 1984, 1, 1.
39. Phillips J C, Physics Today, Feb. 1982.
40. Azoulay R, Thibierge and Brenac A, J. Non-Cryst. Solids, 1975, 18, 33.
41. Ramachandrarao P, Singh R N and Lele S, J. Non-Cryst. Solids, 1984, 64, 387.
42. Kui H W, Greer A L and Turnbull D, Appl. Phys. Lett., 1984, 45, 615.
43. Drehmann A J and Greer A L, Acta Metall., 1984, 32, 323.
44. Merry G A and Reiss H, Acta Metall., 1984, 32, 1447.

ABSTRACTED DISCUSSION OF THE PAPER BY R.W. CAHN

Participants: B. Cantor, H. Jones, B. Predel and F. Sommer

The discussion was concerned mainly with the possible unification of the
several   apparently   distinct criteria and approaches to glass formation.
Suggested possibilities included the unification of radius mismatch and
thermodynamic criteria for example by the Micdema model, i.e. relating the
atomistic to the macroscopic level. Even a kinetic model could not be the
last word in that the magnitudes of physical quantities used to make realistic predictions by kinetic modelling themselves require explanation in
order to complete understanding of why a particular material responds in
its own particular way. Only then could questions posed by apparent correlations between not obviously related phenomena, such as glass formability
and glass stability, be resolved.

MICROSTRUCTURE FORMATION IN RAPIDLY SOLIDIFIED ALLOYS

W.J. BOETTINGER and S.R. CORIELL
Metallurgy Division
National Bureau of Standards
Gaithersburg, MD 20899
USA

ABSTRACT
    In order to apply solidification theory to the interpretation of micro-
structures produced by rapid solidification, several modifications are
required. Different degrees of non-equilibrium occur during solidification
and constitute a hierarchy which is followed with increasing solidification
rate. Analytical expressions are given for a model of non-equilibrium
interface conditions which describe the temperatures and compositions at
the liquid solid interface as a function of solidification velocity. For
solidification at intermediate veloicities ($\approx$ 10 cm/s) the assumption of
local interfacial equilibrium remains valid but microstructures are often
produced under conditions where the solute Peclet number P_c = Vl/2D, is
greater than one. The parameters V, l and D are the solidification velocity, relevant microstructural length scale and liquid diffusion coefficient
respectively. This fact requires that several topics in solidification
theory be modified. Such a modification is presented for alloy dendritic
growth theory. For alloys solidifying dendritically into undercooled melts,
solute redistribution dominates the relatinship between growth rate and
initial undercooling when the initial undercooling is smaller than the
alloy freezing range (difference between liquidus and solidus temperatures). This fact has several important consequences for the eutectic
coupled zone boundaries and for arrayed dendritic growth.

1. INTRODUCTION
    The understanding, prediction and control of rapidly solidified micro-
structures has provided a major challenge to solidification theory over the
past two decades. While many of the basic principles are now in place for a
complete analysis, significant theoretical refinements and difficult experimental verifications remain to be performed. This paper will describe a
few of the basic principles required to understand microstructures produced
by rapid solidification and will point to directions where further work
is required.

1.1 Equilibrium vs. Non-Equilibrium
    Although rapid solidification is widely believed to be a "non-equili-
brium" process, it is clear that different degrees of non-equilibrium
constitute a hierarchy which is followed with increasing solidification
rate. This hierarchy is shown in Table 1.

Table I: Hierarchy of Equilibrium

Increasing Solidification
       Rate

      I. Full Diffusional Equilibrium
         A. No chemical potential gradients (composition of phases are uniform)
         B. No temperature gradients
         C. Lever rule

     II. Local Interfacial Equilibrium
         A. Phase diagram gives compositions and temperatures only at liquid-solid interface
         B. Corrections made for interface curvature (Gibbs-Thomson Effect)

    III. Metastable Local Interfacial Equilibrium
         A. Stable phase can not nucleate or grow sufficiently fast
         B. Metastable phase diagram (a true thermodynamic phase diagram missing the stable phase or phases) gives the interface conditions

    IV. Interfacial Non-Equilibrium
         A. Phase diagram fails at interface
         B. Chemical potentials are not equal at interface
         C. Free energy functions of phases still lead to criteria for the impossible

The conditions required for (I) for metals and alloys are usually obtained only in geological times. The conditions present in (II) constitute the vast majority of work on phase transformations excluding massive and martensitic transformations. The conditions required for (III) occur commonly in metallurgical practice. The change of cast iron from a gray form (austenite and graphite) to the white form (austenite and cementite) with increasing solidification rate is a familiar example. The eutectic temperature and composition for white cast iron is a measurable thermodynamic transformation temperature just like any stable eutectic temperature. The only extra requirement for the metastable phase diagram is the absence of graphite because its nucleation or growth is difficult. For the solidification of some phases these difficulties often become more pronounced at high solidification rate. Hence metastable phase diagrams become more important in describing interface conditions for many rapid solidification processes. The use of metastable diagrams to assist in the interpretation of rapidly solidified microstructures is described in detail in reference (1). Certain hierarchies also exist within (III) as described by Cahn (2).

Significant loss of interfacial equilibrium (IV), whether for a stable or a metastable phase, is thought to become important for simple metallic phases when the crystal growth rate exceeds the diffusive speed of solute atoms in the liquid phase. This diffusive speed is usually given as D/a_0 where D is the liquid diffusion coefficient and a_0 is the interatomic dimension. Experiments on doped Si (3) and on metallic alloys (4) have shown that significant interfacial non-equilibrium effects exist and solute is trapped into the solid for crystal growth rates of $\approx$ 5 m/s. As seen below, the free energy functions of the solid and liquid will restrict the range of compositions that can form at a given temperature and can, when other assumptions are made, predict the temperature of the liquid solid interface when deviations from the local equilibrium condition occur. One might consider (V) where the free energy functions for the phases must be abandoned. This will not be discussed in the present paper.

1.2 Liquid-Solid vs. Solid-Solid Transformations

The microstructure of rapidly solidified alloys, as with ordinary castings, combines the effects of solidification and solid state decomposition. At a practical level one may only be concerned with the microstructure as it exists in the specimen. However, to understand and control microstructure the distinction is quite important. Often the microstructure of rapidly solidified alloys are reported without any analysis of the origin of the structure, thus rendering the results only partially useful. Detailed analysis of the structure often reveals the origin. Two examples follow.

Figure 1 shows a TEM micrograph of an Al-1.5 wt% Fe-3.7 wt% Ni alloy which was melt spun (5). The structure consists of α-Al with a fine dispersion of particles of an intermetallic compound $Al_9(Fe,Ni)_2$. These particles were found to exist in all 24 variants of the orientation relation, $(100)Al_9(Fe,Ni)_2||(100)Al$ and $(031)Al_9(Fe,Ni)_2||(010)Al$. This fact combined, with the fine scale of the particles suggests that the particles formed by solid state precipitation and that the alloy existed as a supersaturated α-Al phase before solid state cooling. A direct product of solidification would not be likely to exhibit all 24 variants due to the directionality of solidification.

A second example shown in Figure 2 reveals the microstructure of the NiAl-Cr quasibinary eutectic composition when subjected to melt spinning (6). The structure shows a fine spinodal decomposition into compositions close to the β-NiAl and α-Cr phases. Clearly the alloy solidified as a supersaturated phase and subsequently decomposed. However, the initial alloy could have been α-Cr supersaturated with Ni and Al or β-NiAl supersaturated with Cr. Observation of antiphase domains with a size much greater than the scale of the spinodal separation suggests strongly that the phase which formed from the melt was β-NiAl supersaturated with Cr. This result is contrary to simple intuition which might suggest that rapid solidification should favour the more disordered phase.

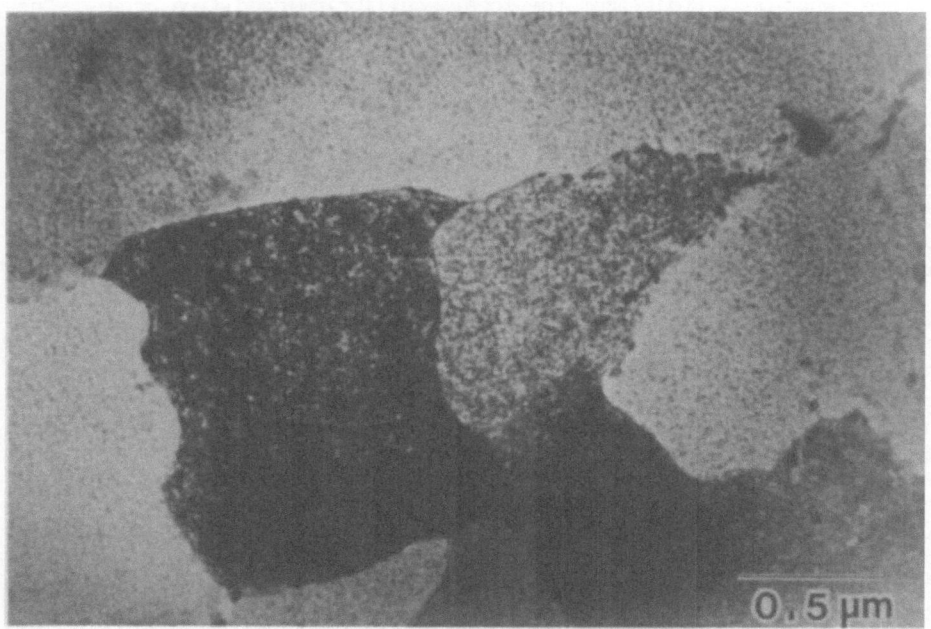

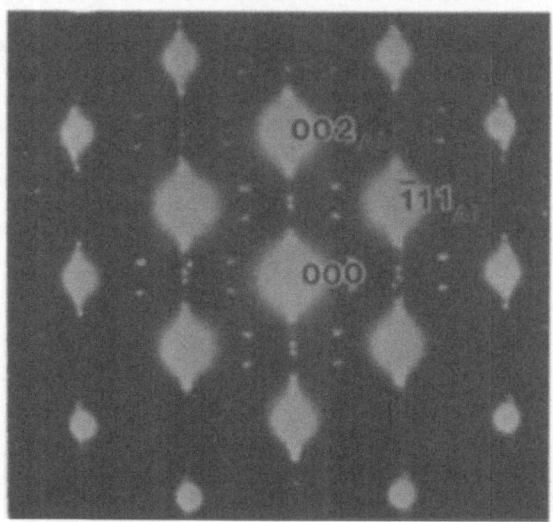

Figure 1: (a) Transverse section of columnar grains of α-Al containing uniformly distributed precipitates of $Al_9(Fe, Ni)_2$ in a melt spun Al-3.7wt% Ni - 1.5wt% Fe alloy. TEM. (b) SADP showing existence of many variant orientation relationships between α-Al and $Al_9(Fe, Ni)_2$.

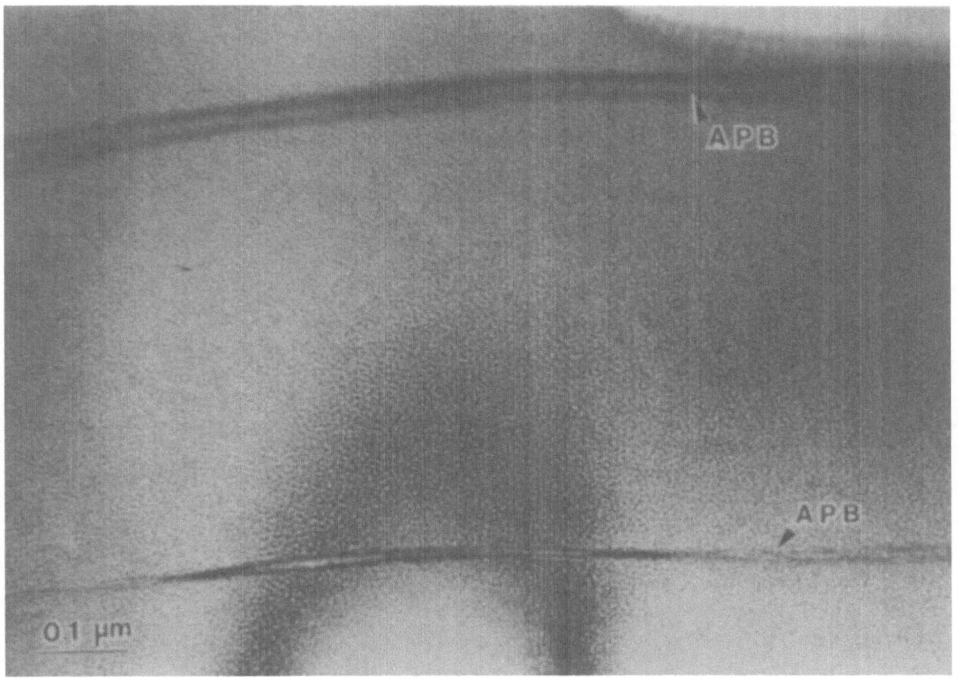

Figure 2: Transverse section of a columnar grain of melt-spun NiAl-Cr quasi-binary eutectic alloy showing a fine spinodal structure and antiphase domain wall defining domains larger than the scale of the spinodal.

2. NON-EQUILIBRIUM INTERFACE CONDITIONS

The analysis of rapid solidification microstructures requires the solution to a complex moving boundary problem for the solid-liquid interface. Diffusion equations for solute and heat must be solved subject to external conditions and conditions at the liquid-solid interface.

In 1971, Baker and Cahn (7) posed the interface conditions for solidification of a binary alloy using two response functions. One choice for the response functions describes the interface temperature T_I and the composition C_S^* of the solid at the interface (8). These response functions can be written as follows:

$$T_I = T(V, C_L^*) - T_M \Gamma K \qquad (1)$$

$$C_S^* = C_L^* k(V, C_L^*) \qquad (2)$$

where V is the local interface velocity, C_L^* is the composition of the liquid at the interface, $T_M \Gamma$ is a capillarity constant and K is the mean curvature of the solid-liquid interface. At zero velocity, the functions T and k are very simply related to the phase diagram: $T(0, C_L^*)$ is the equation for the phase diagram liquidus and $k(0, C_L^*)$ is the equation for the equilibrium partition coefficient k_E which can depend on composition. The dependence of k on curvature is neglected.

These response functions are constrained by thermodynamics. Figure 3 shows the molar free energy versus composition for a liquid and a solid phase at a fixed temperature. During solidification, the temperature and the compositions at the interface are constrained by the fact that there must be a net decrease in the free energy ΔG per mole needed to form an infinitesimal amount of solid of composition C_S^* from liquid of composition C_L^* (7). The free-energy change is shown in Figure 3 and is given by

$$\Delta G = (\mu_S^A - \mu_L^A)(1 - C_S^*) + (\mu_S^B - \mu_L^B)C_S^* \tag{3}$$

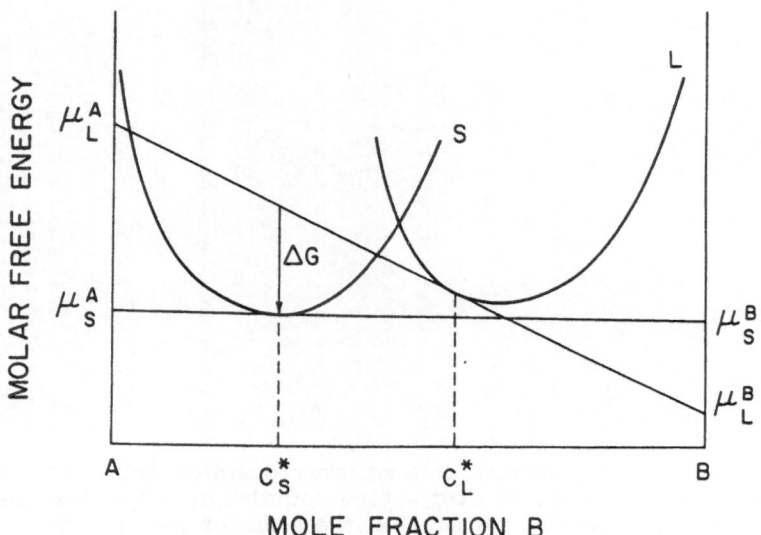

Figure 3: Construction to show the Gibbs free energy ΔG per mole for solidification of an infinitesimal amount of solid of composition C_S^* from a liquid of composition C_L^* under non-equilibrium conditions for which the chemical potentials $\mu_L^A \neq \mu_S^A$ and $\mu_L^B = \mu_S^B$.

Under the usual assumption of local equilibrium at the solid-liquid interface, believed to be valid at slow rates of solidification, the chemical potential μ_L^i of the liquid and the chemical potential μ_S^i of the solid for each species i (= A,B) must be equal. During rapid rates of solidification, this is not necessarily the case and non-equilibrium conditions at the interface can exist. The term "solute trapping" is loosely applied to conditions when the partition coefficient deviates from the equilibrium value but, as originally defined, is restricted to the case when the chemical potential of the solute increases during solidification. Baker and Cahn showed that for a planar interface the restriction that ΔG be negative could be represented on a figure where, for constant temperature, the ranges of possible solid and liquid concentrations at the interface are shown. Figure 4 shows an alternative representation of this restriction (8,9). The shaded region shows the range of possible solid compositions that can form from a liquid with interface composition C_L^* at various interface temperatures. When the temperature is equal to the liquidus temperature for the composition C_L^* , the only possible solid composition is the equilibrium solid composition, $k_E C_L^*$.

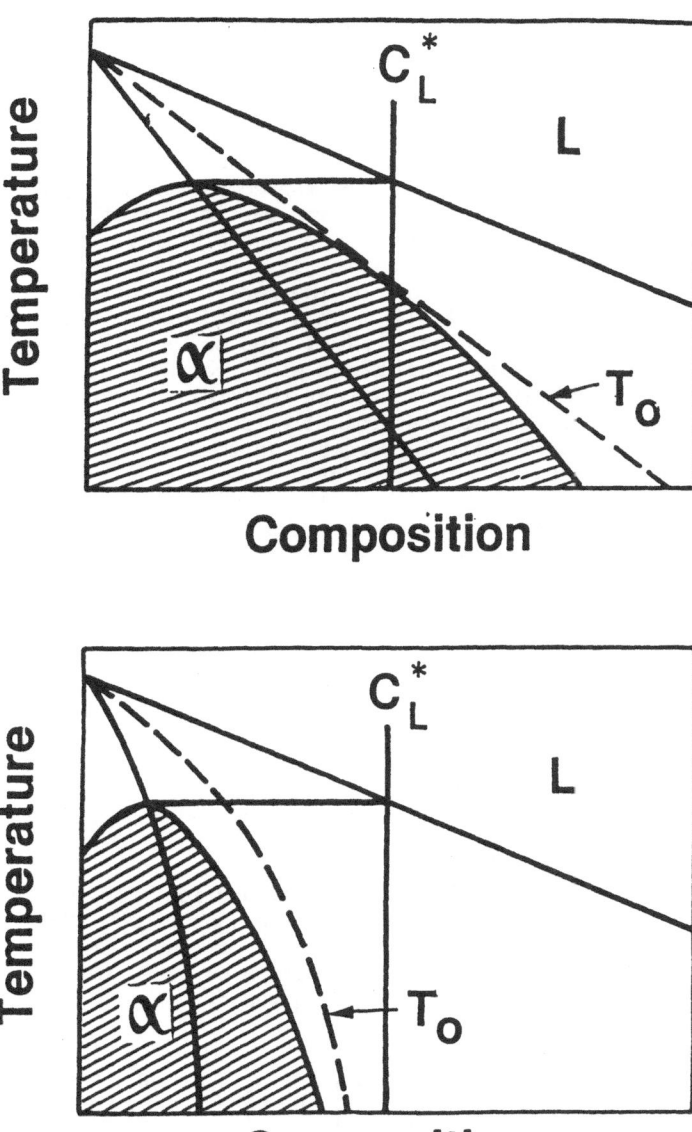

Figure 4: Regions (cross hatched) of thermodynamically allowed solid compositions that may be formed when solid solidifies from liquid of composition C_L^* at various temperatures. The value of T_0 is the highest temperature at which partitionless solidification of a liquid of given composition can occur. In (b) the T_0 curve plunges and partitionless solidification is impossible for liquid of compositions C_L^*. (From reference 9.)

For interface temperatures below the liquidus, the range of allowable solid compositions expands. Of particular interest is the T_0 curve which lies between the liquidus and the solidus. The T_0 curve is the locus of compositions and temperatures for which the molar free energies of the liquid and solid phases of the same composition are equal. When the interface temperature reaches the T_0 curve for the composition of the liquid at the interface, the solid can have the same composition as the liquid, i.e., partitionless solidification. Figure 4(b) shows a case where the T_0 curve plunges quite steeply. In this case, for the composition C_L^*, partitionless solidification is not possible, i.e., the partition coefficient k cannot approach unity at high velocities. The impossibility of partitionless solidification has been associated with glass formation (9).

Several models for the dependence of the partition coefficient on velocity have been formulated (10-14). This dependence on velocity is the second response function, equation (2). The model formulated by Baker (10,14) is quite general and includes a wide variety of possibilities. Figure 5 shows the results of more recent theories which predict that the partition coefficient changes monotonically from its equilibrium value to unity as the growth velocity increases. In these models the interface partition coefficient is a function of a dimensionless velocity $\beta = \beta_0 V$ where β_0 is the ratio of a length scale a_0 (normally the interatomic distance) to a diffusion coefficient D. The diffusion coefficient in the various models ranges between a liquid diffusion coefficient and a diffusion coefficient characteristic of the interface itself. The functional form of the model proposed by Aziz (11) and by Jackson et al. (12) for continuous growth is given as

$$k(V) = \frac{k_E + \beta_0 V}{1 + \beta_0 V}$$

(4)

where $\beta_0 = a_0/D$. At a velocity of $1/\beta_0$, the partition coefficient is the average of the equilibrium partition coefficient and unity. If we choose a liquid diffusion coefficient typical of metals (2.5×10^{-5} cm^2/s) and the length scale to be 0.5 nm, $1/\beta_0$ is 5 m/s.

To obtain an expression for the response function for interface temperature, equation (1), the quantity ΔG must be described. From Baker and Cahn (7), the change in free energy per mole of liquid of composition C_L^* which is transformed to solid of composition C_S^* at a temperature T_I, equation (3), for dilute solutions can be given by

$$\frac{\Delta G}{RT_I} = \left\{ \ln \left[\frac{(1 - C_s^*)}{(1 - C_L^*)} \cdot \frac{(1 - C_L^{eq})}{(1 - C_s^{eq})} \right] \right\} (1 - C_s^*) + \left\{ \ln \left[\frac{C_s^*}{C_L^*} \cdot \frac{C_L^{eq}}{C_s^{eq}} \right] \right\} C_s^*$$

(5)

where C_L^{eq} and C_S^{eq} are the equilibrium compositions of liquid and solid at the temperature T_I. For a phase diagram with liquidus and solidus curves which are straight lines with slopes m_L and m_S respectively

$$C_L^{eq} = (T_I - T_m)/m_L$$

(6)

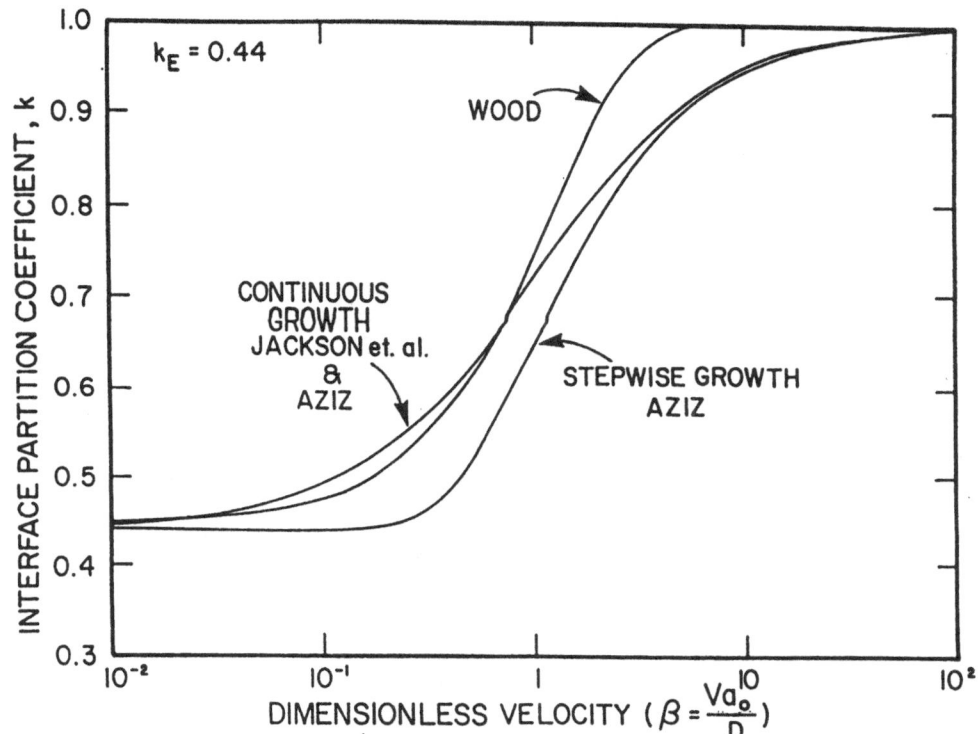

Figure 5: Curves showing the dependence of the interface partition coefficient k on velocity according to the models of various investigators (k_E =
0.44).

and

$$C_S^{eq} = (T_I - T_m)/m_S \qquad (7)$$

where $m_L/m_S = k_E$ is the equilibrium partition coefficient and T_m is the
pure component melting point. Equation (5) can be further reduced to

$$\frac{\Delta G}{RT_I} = \frac{1 - k_E}{m_L} (T_m + m_L C_L^* - T_I) + C_L^*(k_E - k(1 - \ln \frac{k}{k_E})) \qquad (8)$$

where $k = C_S^*/C_L^*$ which is not necessarily equal to k_E.

A familiar result used to approximate interface kinetics when $k = k_E$ is
obtained from equation (8); viz.,

$$\frac{\Delta G}{RT_I} = \frac{1 - k_E}{m_L} (T_m + m_L C_L^* - T_I) \qquad (9)$$

where $T_m + m_L C_L^* - T_I$ is the interface undercooling below the liquidus
temperature for the composition C_L^*. The expression $(1 - k_E)/m_L = -L/RT_m^2$
where L is the latent heat/mole of the pure component, R is the gas con-

stant and m_L has units of K/mole fraction.

If $\Delta G = 0$, and $k = 1$ ($C_S^* = C_L^* = C$), the equation for the T_0 curve is obtained from equation (8) as

$$T = T_m + \left(\frac{m_L \ln k_E}{k_E - 1} \right) C \tag{10}$$

where clearly $(m_L \ln k_E)/(k_E - 1)$ is the slope of the T_0 curve.

If $\Delta G = 0$, equation (8) will give an expression for $k = C_L^* / C_S^*$ as a function of T_I which is the boundary of the shaded region shown in figure 4. One is now in a position to write an analytical expression for the first response function, equation (1), mentioned above. Using the expression for ΔG given above and a linear kinetic law for the interface velocity V

$$V = -V_0 \frac{\Delta G}{RT_I} \tag{11}$$

one obtains the two response functions for a flat interface as follows:

$$T_I = T_m + m_L C_L^* + \frac{m_L C_L^*}{1 - k_E} \left(k_E - k(1 - \ln \frac{k}{k_E}) \right) + \frac{m_L}{1 - k_E} \frac{V}{V_0} \tag{12}$$

and

$$C_S^* = k C_L^* \tag{13}$$

where k is given by equation (4). Note that if $\beta_0 = 0$ and $V_0 = \infty$ then

$$T_I = T_m + m_L C_L^* \tag{14}$$

$$C_S^* = k_E C_L^* \tag{15}$$

which are the conditions for local interface equilibrium.

Figure 6 shows a composite plot of the two response functions obtained using equations (12) and (13) superimposed on a phase diagram including the liquidus, solidus and T_0 curves. The composition of the solid at the interface and the interface temperature are plotted along a curve parameterized by interface velocity for a given fixed concentration in the liquid at the interface. The figure is based on a phase diagram with $k_E = 0.44$ and $m_L = -5.6$ K/at%, a $1/\beta_0$ value of 5 m/s and two values of V_0 (infinity and $2 \cdot 10^3$ m/s). At zero velocity the composition of the solid lies on the solidus curve. At intermediate velocities (about 10 cm/s) the composition of the solid moves towards the composition of the liquid with a small increase in the undercooling. At high velocities the solid composition at the interface approaches the liquid composition at the interface near the T_0 curve. The broken curve shows the case where V_0 is infinite. This curve corresponds to the thermodynamic bound on the solid composition at the interface given in figure 4. The broken curve can also be obtained by letting $\Delta G = 0$ in equation (8). The full curve, for a finite value of V_0, shows the interface temperature plunging quite rapidly.

This analysis provides a pair of thermodynamically consistent response functions for the conditions at the liquid-solid interface. When combined with solute redistribution and heat flow analysis in the liquid and solid these functions permit the analysis of rapid solidification problems. For example, a more precise analysis of recalescence is possible. During recalescence the velocity of the interface decreases with time because of the evolution of latent heat. Also solute trapping can be added to the dendritic growth theory described below. Many assumptions are imbedded in this description which require experimental confirmation and measurement of β_0 and V_0.

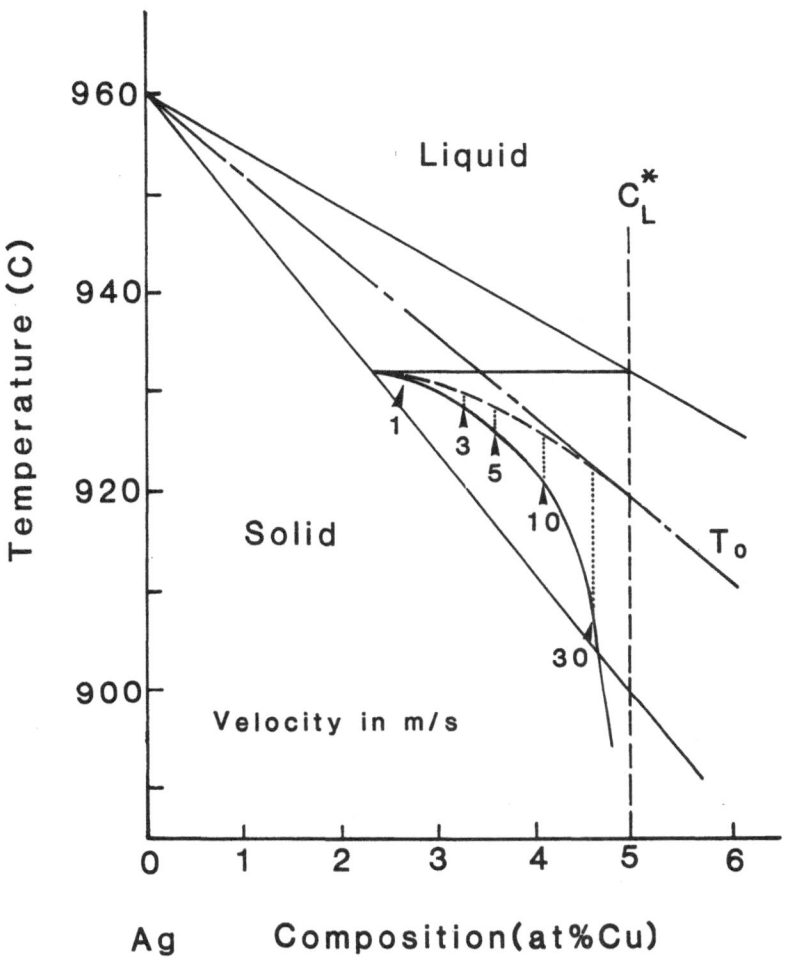

Figure 6: Interface temperature and solid compositions for solid that forms at the indicated velocities (in m/s) from a liquid of fixed composition C_L^* at the interface. Figure shown for Ag-Cu: $m_L = -5.6$ k/at%, $k = 0.44$, $C_L^* = 5$ at%, $1/\beta_0 = 5$ m/s. The dashed curve shows the case for $V_0 = \infty$ while the solid curve shows the case for $V_0 = 2 \times 10^5$ m/s.

3. SOLIDIFICATION CONTROLLED BY SOLUTE DIFFUSION

As seen in the previous section, local interfacial equilibrium holds reasonably well as long as the growth rate is significantly below 5 m/s. Here the solidification of alloys requires significant redistribution of solute in the liquid phase. Analysis of various solidification problems involves the solute Peclet number, $P_C = Vl/2D$ where V is the solidification velocity, l is a characteristic length scale typical of the microstructure being analyzed, and D is the liquid diffusion coefficient. Many solidification problems analyzed for slow solidification rates involve solutions to the diffusion equation for low Peclet numbers. However, many rapid solidification problems require the analysis of problems at high P_C. As summarized in Table II, various topics in solidification must be viewed differently at low and high values of P_C. It is instructive to add the topic of solute trapping to the Table, even though it is not treated by continuum diffusion theory.

Table II: Summary of the Effect of Solute Peclet Number on Various Solidification Topics.

Subject	l	$P_C \ll 1$	$P_C \simeq 1$
Interface Shape Stability	Interface Perturbation Wavelength	Constitutional Supercooling	Absolute Stability
Microsegregation	Primary Dendrite or Cell Spacing	Scheil-type analysis	Flat solute profiles(15)
Dendritic Growth	Dendritic Tip Radius, R	Trivedi (16) Lipton-Glicksman-Kurz (17)	?
Eutectic Growth	Eutectic Spacing, λ	Jackson-Hunt (18) analysis	No eutectic Ref. (19)
Solute Trapping	Interatomic Dimension, a_0	Local Equilibrium	Partition-less Solidification

3.1 Interface Stability

In a limited number of cases, absolute stability provides a mechanism whereby an alloy can solidify without microsegregation at high rate. Figure 7 summarizes results from Ag-1 wt% Cu and Ag-5 wt% Cu alloys solidified using electron beam surface melting and resolidification at the indicated velocities (19). A transition from cellular to cell-free structures is shown to agree fairly well with the theoretical curve. The alloys with no segregation have solidified with a planar interface. This kind of microstructure can be understood by including the effects of liquid-solid interfacial energy into constitutional supercooling analysis using the full morphological stability theory (20,21). From this theory a planar interface is stable for an alloy of composition C_0 if the growth rate V exceeds a critical value given by

$$V = \frac{m_L D(1 - k_E)C_0}{k_E^2 T_m \Gamma}$$

(16)

and if the net heat flow is into the solid. Hence this mechanism cannot produce a microsegregation-free solid when growth occurs into an under-cooled melt. The conditions for absolute stability are met in the electron beam experiments above. Typically, however, the velocity requirement of equation (16) is so high that the assumption of local equilibrium is invalid. The theory has been extended to include solute trapping (22).

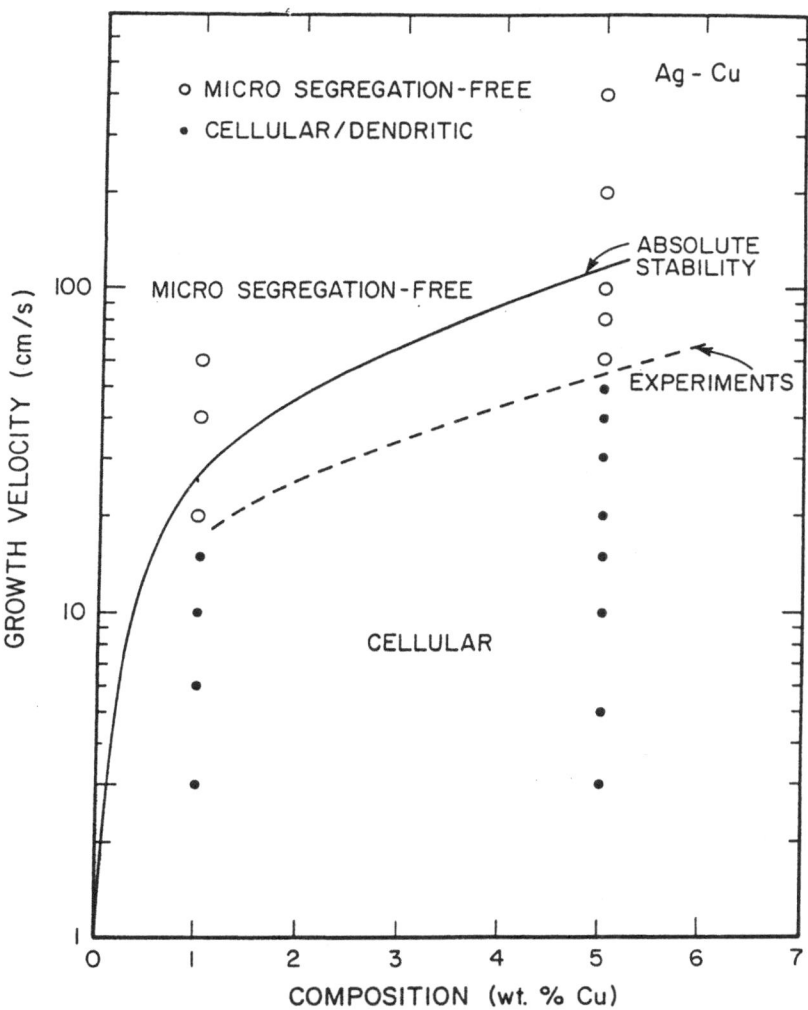

Figure 7: Summary of structures obtained by electron beam surface melting and resolidification of Ag-1 wt% Cu and Ag-5 wt% Cu alloys at various growth rates. The comparison of the absolute stability theory with experiments is shown.

3.2 Microsegregation

When microsegregation does occur in rapidly solidified alloys, it frequently occurs in a cellular morphology even though the alloy may solidify dendritically at lower velocity. These cellular structures obtained at high rate differ little geometrically from those observed at slow rates except for the scale. Figure 8 shows a transverse section obtained by TEM from a thin foil of a fine cellular structure of the Ag-rich phase in a Ag-15 wt% Cu alloy (19). However, the details of the microsegregation (composition) profile within cells, the volume fraction of intercellular material and/or the actual identity of phases found in intercellular regions may differ from those found in more slowly solidified alloys. In figure 8, most of the intercellular regions are filled with the Cu-rich phase not the eutectic of Ag and Cu.

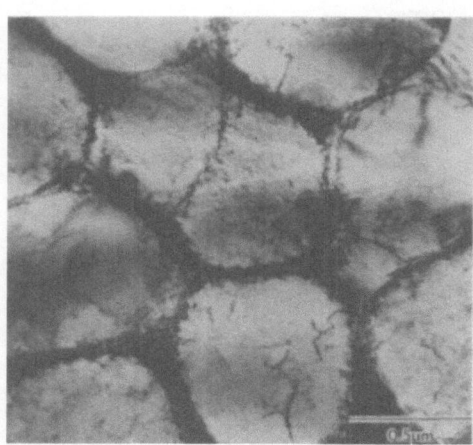

Figure 8: Cellular microsegregation pattern observed in Ag-15 wt% Cu alloy solidified at approximately 2.5 cm/s. TEM.

The amount of solute incorporated into the cell interiors has a strong influence on the volume fraction of intercellular phases and precipitation in the cell interiors during subsequent thermomechanical treatment. Several authors have measured cellular solute profiles in rapidly solidified alloys produced with unknown or calculated growth rates (23-26). Figure 9 shows solute profiles measured from samples where the growth rate is determined experimentally using the electron beam melting and resolidification technique (15, 19). One notes a general increase in the Cu content of the cells with increasing solidification rate, a trend expected from various theories for dendritic growth.

Except for the sample solidified at 0.1 cm/s, the solute profiles do not show the characteristic "U"-shaped solute profile expected from Scheil-type analysis where the assumption of no lateral concentration gradients in the liquid between solidifying cells is usually made. This assumption seems only valid when the cell spacing is much less than D/V, i.e., $P_c \ll 1$. Such is not the case in the present experiments where the solute Peclet number using the cell spacing and the measured growth rates ranges from 0.5 for the "U"-shaped profile to 25 for the sample solidified at ≈18 cm/s. Theoretical analysis of cell shapes and corresponding solute profiles is an active area of research (27).

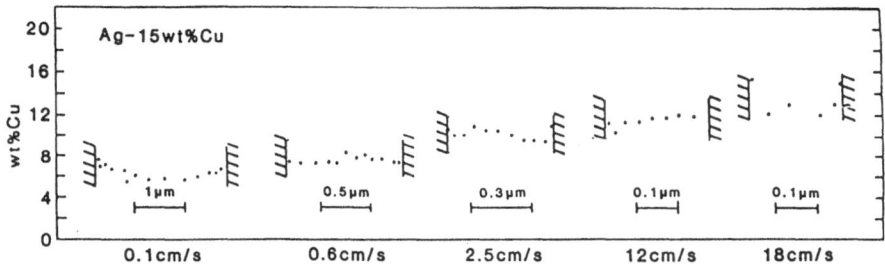

Figure 9: Microsegregation profiles measured by STEM across cells in Ag-15 wt% Cu alloys solidified at the indicated rates. The inter-cellular regions (cross-hatched) contain eutectic (V < 2.5 cm/s) or Cu-phase (V > 2.5 cm/s).

3.3 Eutectic Growth

For the case of eutectic growth the solute Peclet number based on the eutectic spacing is normally much less than one. However, using the electron beam rapid solidification technique mentioned above, eutectic spacings of about 20 nm have been observed for growth of the Ag-Cu and NiAl-Cr eutectic solidified at ≃ 2.5 cm/s (19,28). This spacing represents the minimum observed experimentally. At growth rates above 2.5 cm/s these alloys do not solidify with eutectic microstructures. They develop different growth morphologies. The Peclet number based on this minimum eutectic spacing and the measured growth rate is ≃0.1. Hence the transition of structure is again consistent with altered solutions to the diffusion problem at high P_C . In fact, the Jackson-Hunt analysis (18) of eutectic solidification assumes that the Peclet number is much less than one.

4. DENDRITIC GROWTH

Liquid alloys frequently undercool prior to solidification. This is particularly true in atomization where the potent catalytic sites for heterogeneous nucleation can be isolated into a few particles leaving the remainder of the particles free to undercool. Solidification into under-cooled melts is usually thought to be fully dendritic. In the following, several aspects of dendritic growth will be examined. Later in this paper microstructures involving cellular growth in undercooled powders will also be discussed.

A recent theory by Lipton, Glicksman and Kurz (LGK) (17) will be examined. This theory predicts the growth rate of an isolated dendrite tip as a function of undercooling, ΔT, and alloy composition, C_0 , assuming local equilibrium at the liquid-solid interface. The assumption of local interface equilibrium will be retained as long as the predicted growth rate is much less than 5×10^2 cm/s as described above. The theory includes the effects of solute redistribution, latent heat removal and interface curvature at the tip and uses the concept of marginal stability to determine the operating condition at the tip. Here the undercooling is the difference between the liquidus temperature for the bulk alloy composition, T_L , and the temperature of the liquid far from the dendrite tip, T_∞ (the bath or nucleation temperature). The (bath) undercooling is composed of three parts: the thermal undercooling, ΔT_t , the solute undercooling, ΔT_s , and the curvature undercooling, ΔT_C according to

$$\Delta T = \Delta T_t + \Delta T_s + \Delta T_c \tag{17}$$

where the undercoolings are given by

$$\Delta T_t = I_v(P_t)L/C \tag{18}$$

$$\Delta T_s = k_E \Delta T_0 \left\{ \frac{I_v(P_c)}{1 - (1 - k_E)I_v(P_c)} \right\} \tag{19}$$

and

$$\Delta T_c = 2T_m \Gamma/R \tag{20}$$

where L is the latent heat per unit volume, C is the heat capacity per unit volume and ΔT_0 is the alloy freezing range, $m_L C_0(k_E - 1)/k_E$. The function $I_v(P)$ is given by

$$I_v(P) = Pe^P E_1(P) \tag{21}$$

where the function $E_1(P)$ is the exponential integral. The parameters P_t and P_c are the thermal and solute Peclet numbers and are given by $VR/2\alpha$ and $VR/2D$, respectively where R is the tip radius, α is the liquid thermal diffusivity and D is the solute diffusion coefficient.

For a given value of ΔT, the dendrite growth rate V and tip radius, R, are not uniquely specified by equations (17–21). Lipton, Glicksman and Kurz (17) employ an approximation to the marginal stability condition of Langer and Mueller-Krumbhaar (29) by equating the operating tip radius with the minimum unstable wavelength result of linear stability theory of a planar interface at low velocity. The wavelength is given by

$$R = \left\{ \frac{T_m \Gamma/\sigma^*}{m_L G_c - \overline{G}} \right\}^{1/2} \tag{22}$$

where G_c is the concentration gradient and $\overline{G}$ is the conductivity-weighted temperature gradient at the dendrite tip. The parameter σ^* is $1/4\pi^2$.

However, equation (22) only holds for small k_E or small P_c . A rigorous expression for the minimum wavelength, including large P_c (but small P_t), can be obtained from Mullins and Sekerka (20) and is given by

$$R = \left\{ \frac{T_m \Gamma/\sigma^*}{m_L G_c(1 + g) - \overline{G}} \right\}^{1/2} \tag{23}$$

where

$$g = \frac{2k_E}{1 - 2k_E - (1 + \frac{1}{\sigma^* P_c^2})^{1/2}} \tag{24}$$

Note that for $k_E = 0$ or $P_c \ll 2\pi$, g approaches zero to reduce equation (23) to equation (22). However, as will be seen, dendritic solidification of alloys frequently involves bath undercoolings of several hundred degrees, a situation where $P_c \simeq 2\pi$ and this modification appears important. Using equation (23) and substituting, as did Lipton, Glicksman and Kurz (17,30), for the values of G_c and $\bar{G}$ at a dendrite tip one obtains

$$R = \left\{ \frac{T_m\Gamma/\sigma^*}{(L/C)P_t + \dfrac{2k_E P_c \Delta T_0(1 + g)}{1 - (1 - k_E)I_v(P_c)}} \right\} \tag{25}$$

Equations (17-21) and (25) can also be written in a dimensionless form using

$\overline{\Delta T} = \Delta T(C/L)$ = dimensionless bath undercooling
$\overline{\Delta T}_0 = \Delta T_0 (C/L)$ = dimensionless freezing range or dimensionless composition
$\bar{R} = (R/T_m\Gamma)(L/C)$ = dimensionless tip radius
$\bar{V} = (1/2)(VT_m\Gamma/\alpha)(C/L) = P_t/\bar{R}$ = dimensionless velocity

as

$$\overline{\Delta T} = I_v(P_t) + k_E\overline{\Delta T}_0 \left\{ \frac{I_v(P_c)}{1 - (1 - k_E)I_v(P_c)} \right\} + \frac{2}{\bar{R}} \tag{26}$$

and

$$\bar{R} = \frac{1/\sigma^*}{P_t + \dfrac{2k_E\overline{\Delta T}_0 P_c(1 + g)}{1 - (1 - k_E)I_v(P_c)}} \tag{27}$$

Figures 10 and 11 show the effect of the factor g on the dimensionless growth rate and tip radius for various dimensionless compositions for the case of $k_E = 0.1$, $\alpha/D = 10^4$ and $\Delta T = 0.04$. For aluminium alloys a dimensionless growth rate of 10^{-5} corresponds to 248 cm/s, a dimensionless bath undercooling of 0.1 corresponds to 36.4 K, a dimensionless radius of 10^2 corresponds to 2.7 x 10^{-6} cm and a dimensionless composition of 0.1 corresponds to a composition having a freezing range of 36.4 K. One can observe a significant difference in the velocity and radius for very dilute alloys (where P_c is highest) and almost no effect for concentrated alloys.

The reason for this effect can be seen by examining equation (27) for values of P_c near 2π where $(1 + g)$ is rapidly changing from one to zero, and $I_v(P_c) \simeq 1$. In this case

$$\bar{R} = \frac{1/\sigma^*}{P_t(1 + 2\overline{\Delta T}_0(\alpha/D)(1 + g))} \tag{28}$$

Because α/D is very large, a small addition of solute (ΔT_0 increasing from zero) has a major effect in reducing $\bar{R}$ and increasing the growth rate as seen in figures 10 and 11 without the correction (i.e., g = 0). However, inclusion of the factor g greatly mitigates the tip sharpening effect of a minor solute addition. Because $\alpha/D \simeq 10^2$ for the succinonitrile alloys studied by Lipton, Glicksman and Kurz (17), the tip sharpening effect of solute is much less significant than for metals where $\alpha/D = 10^4$. The large tip sharpening effect of solute for metallic alloys if g is excluded leads to extremely large curvature undercoolings $\Delta T > (1/2)\Delta T$, which may negate the use of the modified Ivantsov model. Figure 12 shows the distribution of the bath undercooling, ΔT, into its three parts ΔT_t , ΔT_S and ΔT_c all normalized by L/C versus dimensionless composition for the same case shown in figures 10 and 11. When g = 0, the large value of ΔT_c can be seen for intermediate compositions in figure 12a.

Other modifications of dendritic growth theory for high P_c are most likely necessary. Many high P_c situations also seem to require the inclusion of non-equilibrium interface conditions. Furthermore, the application of the minimum unstable wavelength from planar interface stability theory to approximate the marginal stability concept at high P_c has not been verified.

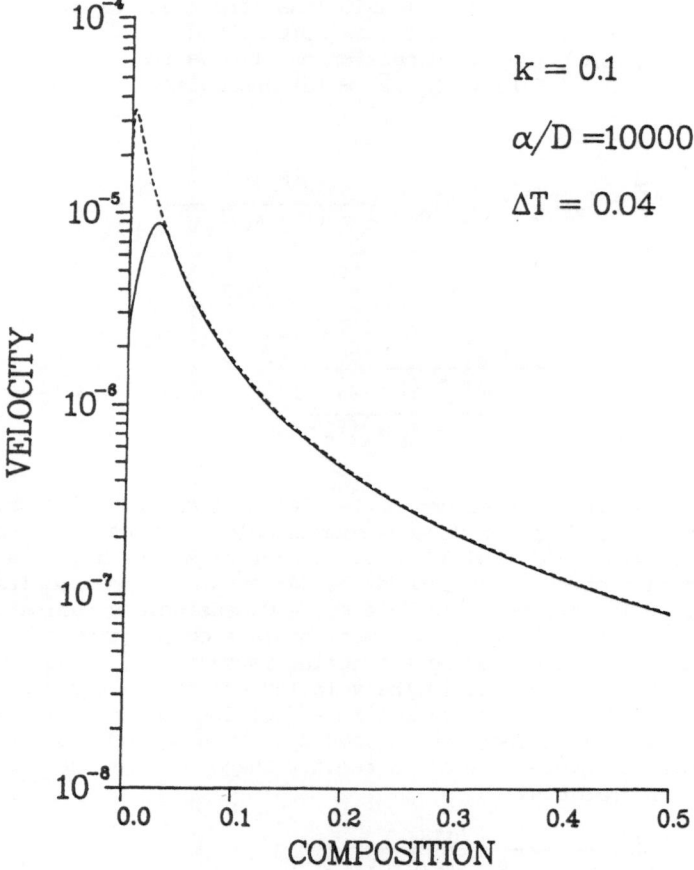

Figure 10: Calculated dimensionless dendritic growth velocity versus dimensionless composition for $k_E = 0.1$ and a dimensionless bath undercooling of 0.04 (0.04 L/C) using LGK theory (dashed) and the modification for high Peclet number (solid).

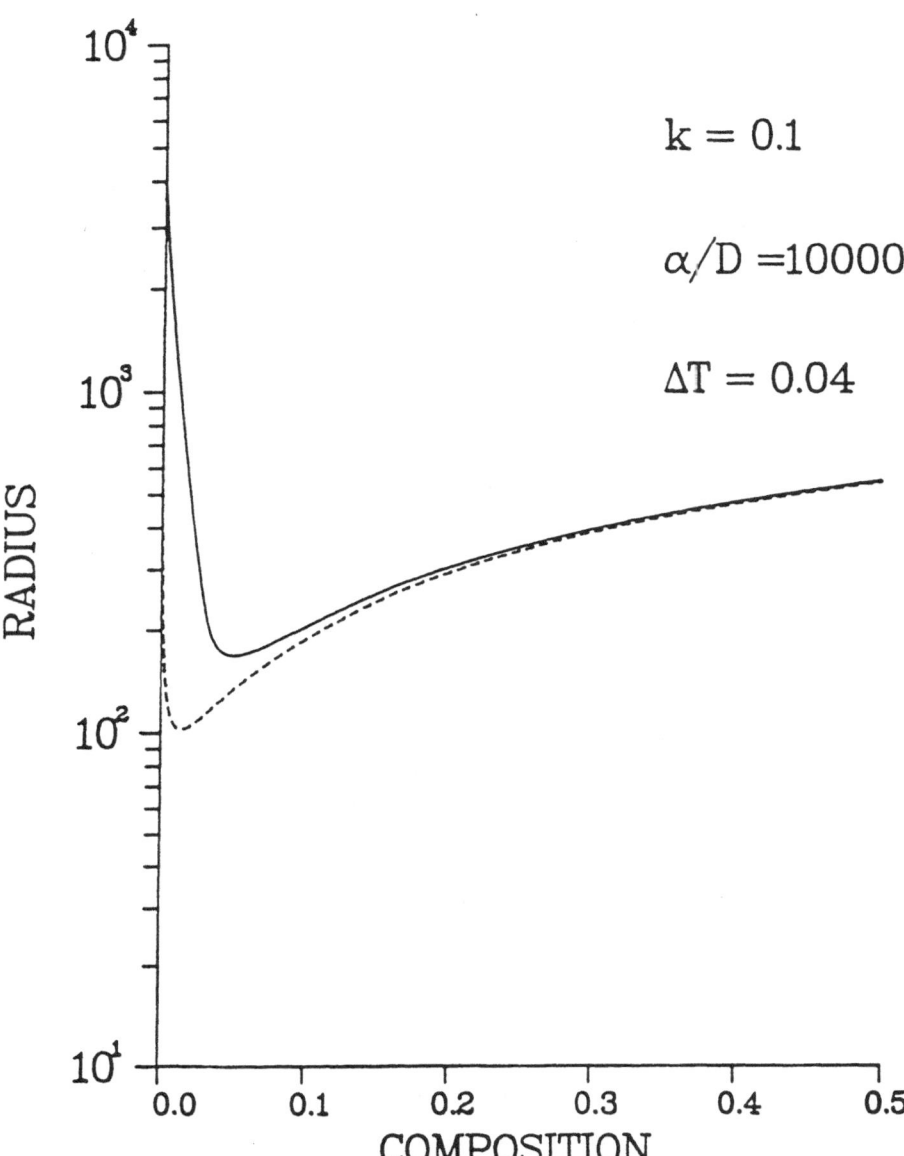

Figure 11: Calculated dimensionless dendrite tip radius versus dimensionless composition for $k_\ell = 0.1$, and a dimensionless bath undercooling of 0.04 (0.04 L/C) using LGK theory (dashed) and the modification for high Peclet number (solid).

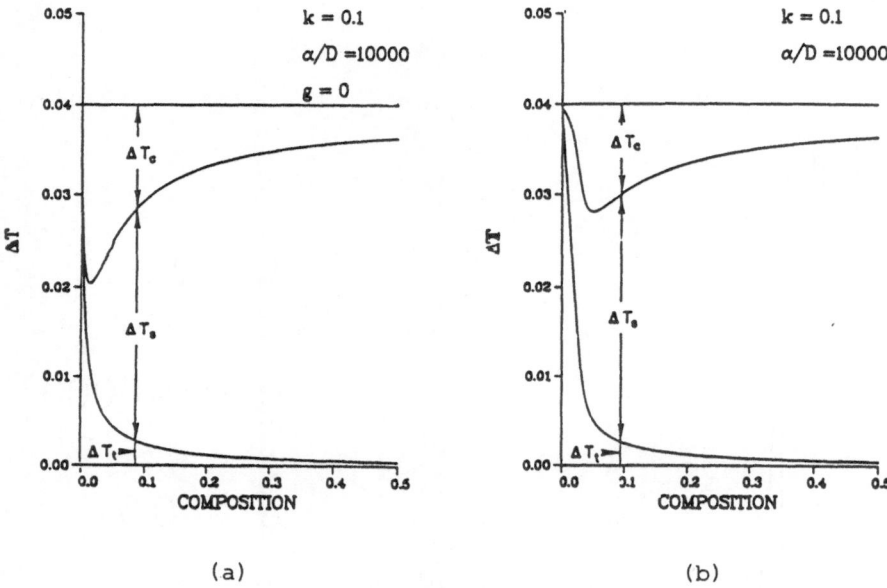

(a)   (b)

Figure 12: Relative size of the curvature undercooling ΔT_c, solute
undercooling, ΔT_s and thermal undercooling, ΔT_t for a dimensionless bath undercooling of 0.04 and $k_f = 0.1$ (a) using LGK theory;
(b) using the modification for high Peclet number.

4.1 Dendritic growth in concentrated alloys

Regardless of the modification of the theory described in the previous
section, the dendritic growth rate of concentrated alloys is very low
compared to the growth rate of the pure component as seen in figure 10.
This slow growth is clearly caused by the difficulty of solute redistribution.

In figure 13 the effect of composition on the three parts of the bath
undercooling when $\Delta T = 0.1$ is superimposed on a phase diagram for $k_f = 0.1$.
Note that when the curve for the bath temperature crosses the solidus curve
the distributions of undercooling change significantly. On the left where
$\Delta T > \Delta T_0$, ΔT_t is large. On the right where $\Delta T < \Delta T_0$, ΔT_t is small. Similar
plots occur for different values of k_f.

When ΔT_t is small the temperature of the dendrite tip, T^*, is very close
to the temperature of the undercooled melt far from the tip, T_∞. The predictions of the LGK theory can usually be fitted by an equation of the form

$$V = \mu_n \Delta T^n \tag{29}$$

where $\Delta T = T_L - T_\infty$. When $\Delta T < \Delta T_0$, T_∞ and T^* can be used interchangeably to
calculate ΔT for use in equation (29). In other words, when $\Delta T < \Delta T_0$,
solute effects dominate the dendritic growth problem, and the bath undercooling is almost identical to the tip undercooling. Equivalently, the term
for $\overline{G}$ in equation (22) or (23) plays almost no role in the selection of the
dendritic tip radius. This has interesting implications in comparing the
results of solidification experiments in undercooled and non-undercooled
melts.

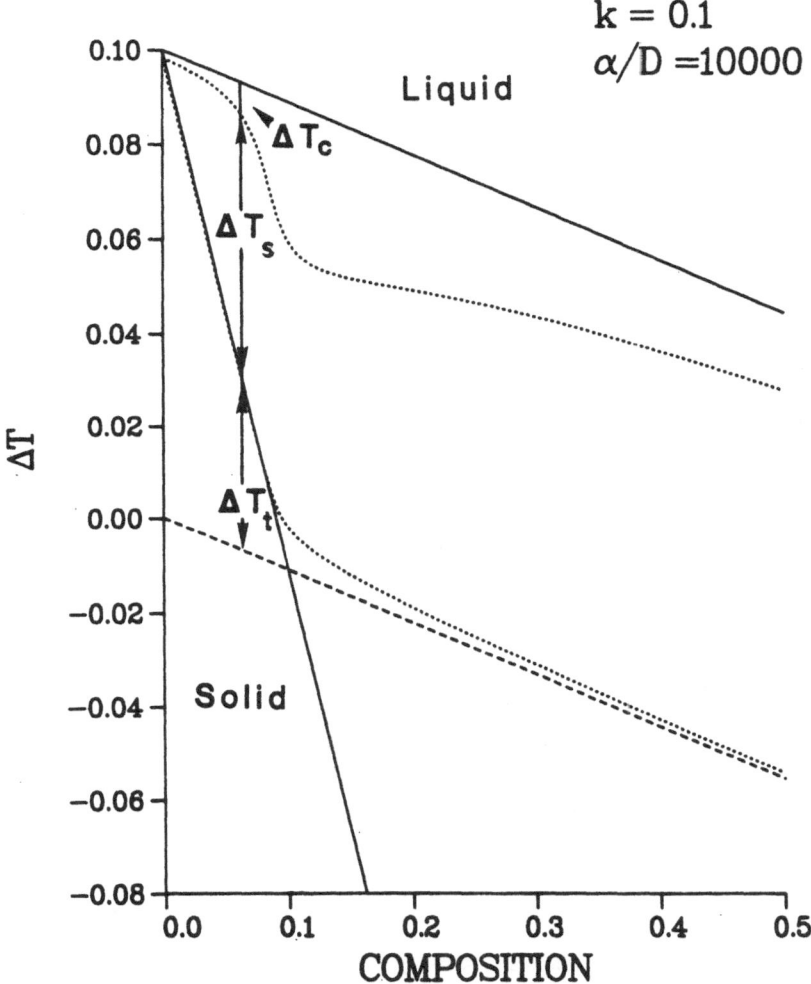

Figure 13: Relative size of the curvature undercooling ΔT_c, solute undercooling, ΔT_s, and thermal undercooling, ΔT_t, for a bath undercooling of 0.1 superimposed on a dimensionless phase diagram. For compositions where the bath undercooling exceeds the alloy freezing range ΔT_t is not small. For compositions where the bath undercooling is less than the alloy freezing range, ΔT_t is small.

4.2 Coupled Zone

The range of alloy compositions which undergoes eutectic solidification without primary phases at a given growth rate may or may not include the thermodynamic eutectic composition. Frequently the range of compositions, referred to as the coupled zone, expands and shifts in composition as the growth rate increases. Originally coupled zones were measured as a function of bath undercooling. Later they were measured as a function of growth velocity (and interface temperature) by directional solidification. The most successful method for analyzing the location of coupled zone boundaries (31) employs a competitive growth analysis. By comparing the rate of growth of dendrites and eutectic at fixed interface temperature for any alloy composition, the fastest growing structure is determined and microstructure predicted for that composition.

The above discussion of dendritic growth has suggested that the role (even the sign) of the temperature gradient has little effect on the dendritic growth rate of concentrated alloys. Also the eutectic growth rate for fixed interface temperature is not significantly effected by the temperature gradient. Hence the coupled zone boundaries should be the same regardless of whether the growth is into a positive or negative gradient as long as one compares zones measured at the same growth rate or interface temperature. This last statement is difficult to guarantee experimentally due to the effect of recalescence described below. However, this principle is extremely useful once a thermal analysis of the recalescence is performed.

In a recent study (32) of the microstructure of Al-8wt% Fe powder atomized by the Homogeneous Metals Co., transitions of microstructure as a function of powder diameter were found to be extremely consistent with the previous work of Hughes and Jones (33) using directional solidification. They showed that the microstructure of Al-6 wt% Fe changes from a primary intermetallic structure of Al_3Fe to fully eutectic structure of α-Al + Al_6Fe and finally to a cellular structure of α-Al in the range of velocity between 0.1 and 1 cm/s. Slightly higher transition velocities would be expected for Al-8 wt% Fe powders. An identical change in microstructure is observed for Al-8wt%Fe powders with diameters between 50 and 10 μm where the estimated growth rate changes as a function of powder diameter from 0.2 to 1 cm/s as analyzed for the Homogeneous Metals process (32). Additionally Hughes and Jones (33) measured the $\lambda^2 V$ constant which relates the eutectic spacing, λ, to the growth rate for the α-Al + Al_6Fe eutectic as a function of iron content. Extrapolating their results to Al-8 wt% Fe gives a $\lambda^2 V$ constant of 3.8×10^{-11} cm^3/s. The range of eutectic spacings seen in powders with diameters between 20 and 40 μm is 90 to 180 nm .The spacings indicate growth rates between 0.5 and 0.1 cm/s which are consistent with estimated growth rates in powder between 20 and 40 μm in diameter where eutectic structures are most common. An example of a fully eutectic structure in an Al-8 wt% Fe powder is shown in figure 14. Hence the occurence of the cellular, eutectic, and primary intermetallic structures in the various size particles is consistent with directional solidification results even though solidification in the powders involves very different temperature gradients than the directional solidification.

5. THE MICROSTRUCTURE OF INITIALLY UNDERCOOLED POWDERS

When a liquid alloy is broken up into a large number of fine droplets, as occurs in atomization, the most potent nucleation sites can become dispersed such that large liquid undercoolings below the equilibrium liquidus may be achieved prior to nucleation. When nucleation does occur in an undercooled droplet, the initial solidification rate is extremely rapid. This initial rate, however, depends strongly on the solid-liquid interface morphology. When the initial undercooling below the relevant liquidus temperature, ΔT, is less than the ratio L/C, solidification can occur in three different morphologies as shown in figure 15 for the case of nucleation on the powder surface. In figure 15a, a dendritic interface moves at a nearly constant velocity across the entire particle to form a mixture of liquid and solid at the end of recalescence. The interdendritic regions solidify later by removal of latent heat to the powder exterior usually at a much lower rate. In figure 15b a smooth interface moves partially across the powder to produce a particle which is composed of two zones, one solid and one liquid at the end of recalescence. Solidification is again finished by removal of latent heat to the powder exterior usually at a much lower rate to produce the coarse segregated strucure shown on the right. In each case the volume fraction solid at the end of recalescence is ΔT (C/L). In the case shown in figure 15a the microstructure is relatively uniform across the powder dimensions. In the case shown in figure 15b the internal heat flow and changes in the liquid-solid interface temperature, velocity and shape combine to produce a strongly time-dependent solidification process and a two-zone microstructure in the droplet. This latter situation has been observed and analyzed in detail by Levi and Mehrabian (34,35). However, the solidification of the first zone does not always occur with a smooth interface. The first zone can solidify with a cellular interface as shown in figure 15c. At the end of recalescence the volume fraction of the powder which contains the cellular structure is ΔT (C/L) where L is a fraction of the ordinary latent heat given by the volume fraction solid within the cellular structure at the end of recalescence. The coarser structure to the right in figure 15c again forms after recalescence is complete.

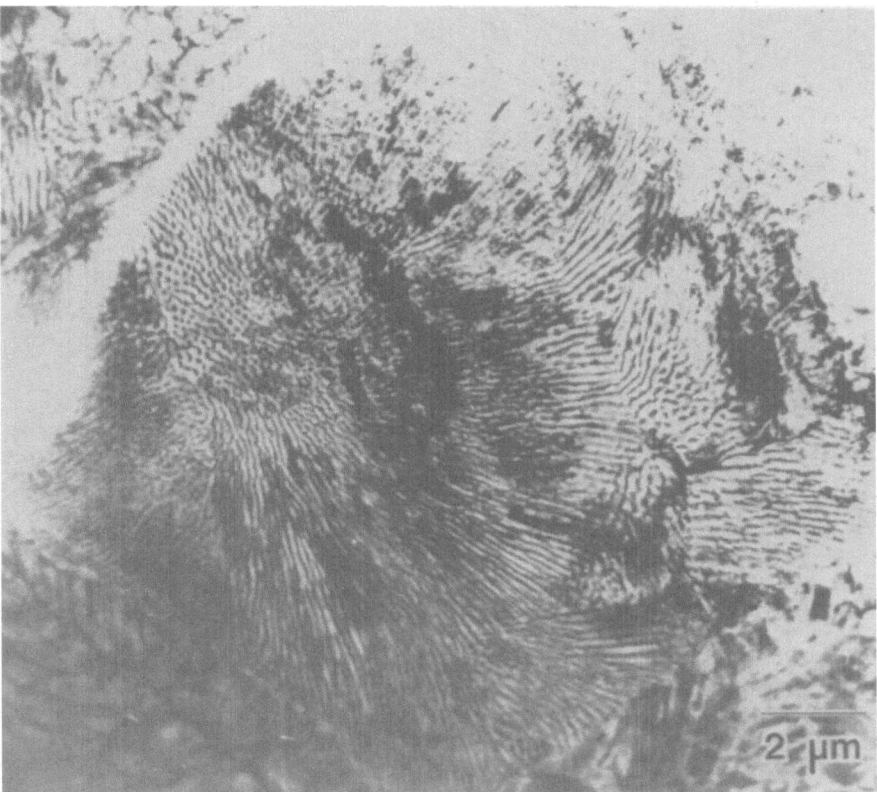

Figure 14: Fully eutectic structure seen in a 20 μm diameter powder particle of Al-8 wt% Fe. The metastable phase Al_6Fe occurs as rods in a matrix of α-Al to form the eutectic structure.

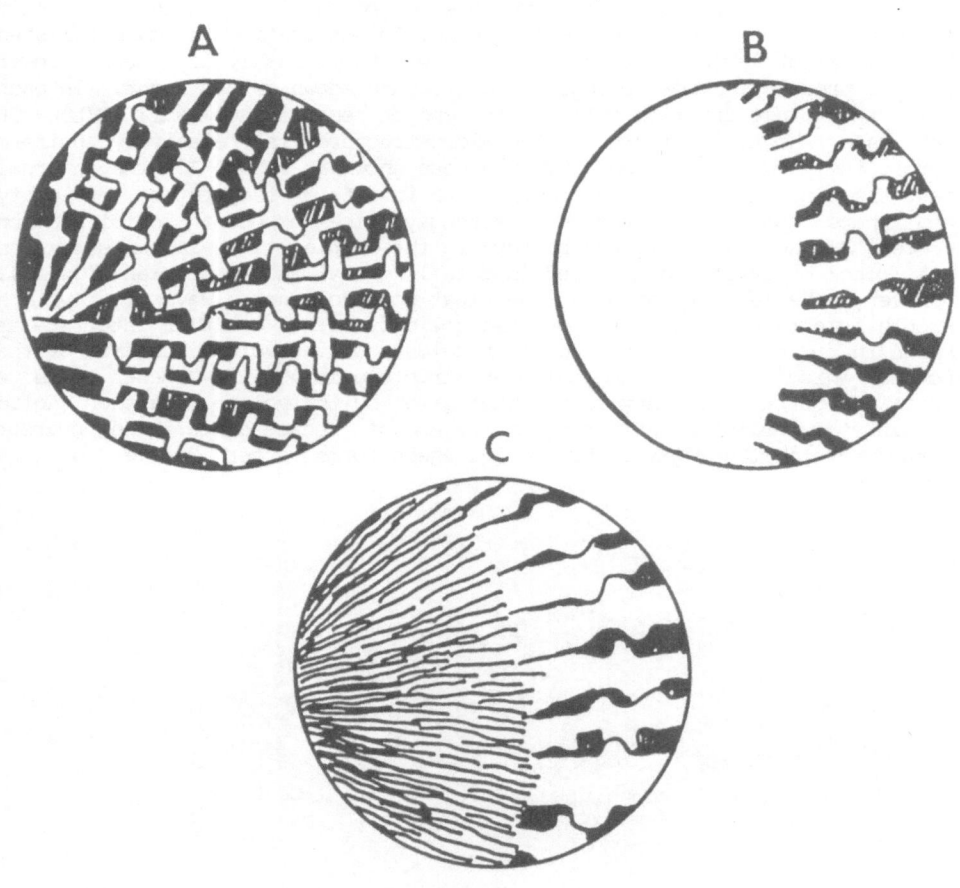

Figure 15: Schematic representations of the solidification of initially undercooled powder particles for nucleation on the surface when the initial interface is (a) dendritic, (b) smooth, (c) cellular.

The structure, shown in figure 15a, is quite commonly observed in Ni-base superalloys, which frequently have small freezing ranges. The structure, shown in figure 15b, is typical of the sub-micron powders examined by Levi and Mehrabian (34). The case, shown in figure 15c, is observed in Al-Fe alloys. Figure 16 is a TEM micrograph of the entire cross-section of an Al-8 wt% Fe powder particle. The particle consists primarily of FCC α-Al with a single crystallographic orientation. Nucleation appears to have occurred at a single site on the surface of the particle on the left and the solidification front has passed from left to right. The microstructure near the point of nucleation is extremely fine and is termed microcellular (32). The microstructure on the right is a more usual coarse cellular structure of α-Al. Examined from a point of view of dendritic growth theory one wonders about the reasons for the three structures shown in figure 15. Figure 15a presents the ordinary view of dendritic growth. One might presume that the structures shown in figures 15a, b and c represent situations of low, high and intermediate undercooling. However, at the present time solidification theory cannot predict such transitions for growth into an undercooled melt. In the case of Figure 15b, Levi and Mehrabian, assuming a smooth interface and kinetics controlled by interface attachment, have calculated and observed that the growth rate is high enough for partitionless solidification. However, the interface should have been thermally unstable and formed dendrites (albeit with no microsegration).

The role of high solute concentration on dendritic growth in undercooled powders has been examined to understand the microstructure seen in figure 15c or 16 and is described in detail in what follows. For the case of an isolated dendrite of α-Al with an alloy composition of Al-8 wt% Fe, the LGK theory predicts a growth rate totally dominated by solute redistribution. For initial undercooling, between 1 and 300 K, $\Delta T_t < 0.02 \Delta T$. This is due to the fact that $\Delta T_0 = 740K$ for α-Al at the composition Al-8wt% Fe. ($k_E = 0.038$)(36). Hence, $\Delta T < \Delta T_0$.

Under conditions where the growth rate of an isolated dendrite is dominated by solute redistribution, it seems reasonable to assume that the primary spacing on an array of dendrites or cells should also be determined by solute redistribution. If the primary spacing is set at a spacing where the solute fields of adjacent dendrite tips overlap slightly, the thermal fields from adjacent tips will overlap significantly. This is due to the fact that the thermal diffusivity is four orders of magnitude larger than the solute diffusivity D. When the growth rate of a dendrite is dominated by thermal effects (i.e. for small freezing range alloys), neither the thermal nor the solute fields will overlap significantly and the structure seen in figure 15a results.

When the thermal fields from adjacent dendrite tips overlap significantly, the motion of an advancing array of dendrite or cellular tips can be modelled by a thermal analysis of the motion of an instantaneously isothermal smooth surface which emits an effective latent heat L'. The temperature of this surface as a function of velocity is given by the kinetic law for an isolated dendrite from equation (29). The divorce of the thermal and solute problems is only valid when the growth rate of an isolated dendrite is dominated by solute redistribution. In this case the tip undercooling and the bath undercooling are identical as described above. This approach has recently been used by Dustin and Kurz (37) to model equiaxed dendritic growth. A similar method can be used for eutectic growth in an initially undercooled alloy where obviously redistribution of solute dominates the relationship between interface undercooling and grwoth rate.

The latent heat L' emitted by this effective interface will be a fraction of the normal latent heat governed by the volume fraction solid in the cellular or dendritic array. It is not clear at the present time what factors determine the volume fraction in the cellular or dendrite array. For a pure material, the final volume fraction is the dimensionless undercooling $\Delta T(C/L)$. For solution growth, the volume fraction is the dimensionless supersaturation (see reference (17)). In the present case the volume fraction can be estimated as 0.7 as seen from Figure 16 and hence $L' = 0.7L$.

The separation of the thermal and solute problems permits an understanding of how a two-zone microstructure can occur in an undercooled particle when the initial growth is cellular or dendritic. Two-zone structures develop when the latent heat from a freezing interface raises the temperature of the powder in such a way as to diminish the growth rate. For thermally controlled dendritic growth (i.e., in small freezing range alloys), the thermal field in front of a dendritic array is confined to the proximity of each dendrite tip and the growth rate is constant across a powder. For solute controlled growth the thermal field in front of a dendritic array has a characteristic dimension of the entire array. This causes the latent heat to spread throughout the powder and slow growth causing a two-zone microstructure.

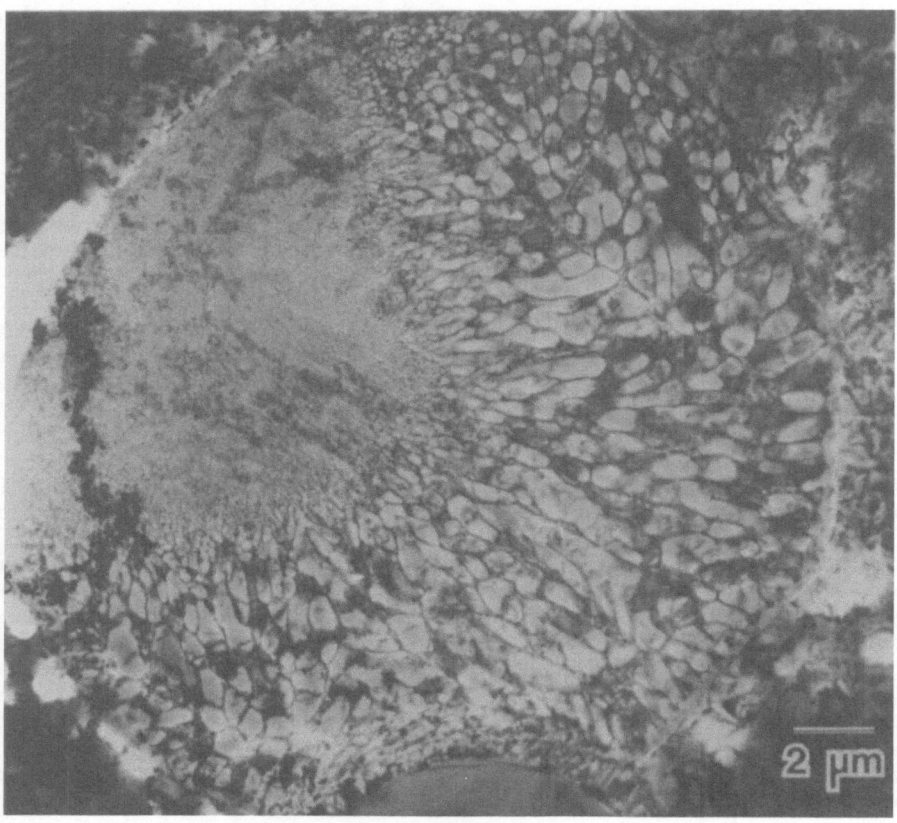

Figure 16: TEM micrograph of a cross-section through a rapidly solidified Al-8wt% Fe powder particle. A single nucleation site of the undercooled particle (left) initiates solidification at a high interface rate to produce a fine cellular structure. Recalescence slows the interface rate to produce the coarser cellular structure at the right. This structure is very common in powders less than 10 μm in diameter.

6. CONCLUSION

1. Different degrees of non-equilibrium occur in solidification and constitute a hierachy which is followed with increasing solidification rate.

2. Analytical expressions are given for one model of non-equilibrium interface conditions which describe the temperature and compositions at the interface as a function of solidification velocity.

3. Even when interface conditions do not deviate significantly from local equilibrium, many rapidly-solidified microstructures are produced

under conditions where the solute Peclet number, P_c, is large. This fact requires that many solidification theories developed for slow speeds be modified.

4. The Lipton, Glicksman, Kurz theory of alloy dendritic growth into undercoooled melts is modified for high P_c using the results of morphological stability theory for a planar interface at high P_c to describe the tip operating conditions.

5. Solute redistribution, rather than latent heat redistribution, controls the dendritic growth rate of concentrated alloys into undercooled melts. This fact permits a direct comparison of results obtained from growth into undercooled melts and from directional solidification. It also permits a divorce of the analysis of solute redistribution from latent heat redistribution during the growth of a dendritic array in an undercooled powder particle.

8. ACKNOWLEDGMENT

The authors would like to thank L. Bendersky, M. E. Glicksman, J. H. Perepezko, R. F. Sekerka, R. J. Schaefer, and D. Shechtman for many helpful discussions.

9. REFERENCES

1. J.H. Perepezko and W.J. Boettinger, Mat.Res.Soc. Symp. Proc. 19, 1983, 223.
2. J.W. Cahn, in Rapid Solidification - Principles and Technologies II, ed. by R. Mehrabian, B.H. Kear and M. Cohen, Claitor's Pub. Co., Baton Rouge, 1980, 24.
3. C.W. White, D.M. Zehner, S.U. Campisano and A.G. Cullis, in Surface Modification and Alloying, ed. by J.M. Poate, G. Foti and D.C. Jacobson, Plenum Press, NY, 1983, 287.
4. S.T. Picraux and D.M. Follstaedt, in Surface Modification and Alloying, ed. by J.M. Poate, G. Foti and D.C. Jacobson, Plenum Press,NY, 1983, 287.
5. L. Bendersky, Met. Trans. 16A, 1985, 683.
6. D. Shechtman, W.J. Boettinger, T.Z. Kattamis, F.S. Biancaniello, Acta Met. 32, 1984, 749.
7. J.C. Baker and J.W. Cahn, in Solidification, American Society for Metals, Metals Park, OH, 1971, 23.
8. W.J. Boettinger, S.R. Coriell and R.F. Sekerka, Mats. Sci. & Eng., 65, 1984, 27.
9. W.J. Boettinger, in Rapidly Solidified Amorphous and Crystalline Alloys, Proc.Symp. of Materials Research Society, B. H. Kear, B.C. Giessen and M. Cohen, eds., North-Holland, Amsterdam, 1982, 15.
10. J.C. Baker, Interfacial Partitioning During Solidification, Ph.D. Thesis, Massachusetts Institute of Technology, 1970, ch. 5.
11. M.J. Aziz, J. Appl. Phys., 53, 1982, 1158.
12. K.A. Jackson, G.H. Gilmer and H.J. Leamy, in Laser and Electron Beam Processing of Materials, Proc.Symp. of Materials Research Society, C. W. White and P.S. Peercy, eds., Academic Press, NY, 1980, 104
13. R.F. Wood, Phys. Rev. B. 25, 1982, 2786.
14. J.W. Cahn, S. R. Coriell and W.J. Boettinger,in Laser and Electron Beam Processing of Materials, Proc. Symp. of Materials Research Society, C.W. White and P.S. Peercy, eds., Academic Press, NY, 1980, 89.
15. L.A. Bendersky, W.J. Boettinger, in Rapidly Quenched Metals, S. Steeb, H. Warlimont, eds, North Holland, Amsterdam, 1985, 887.

16. R. Trivedi, J. Cryst. Growth 49, 1980, 219.
17. J. Lipton, M.E. Glicksman and W. Kurz, Mats.Sci. & Eng., 65, 1984, 57.
18. K.A. Jackson and J.D. Hunt, Trans. TMS-AIME 236, 1966, 1129.
19. W.J. Boettinger, D. Shechtman, R.J. Schaefer and F.S. Biancaniello, Metall. Trans. A. 15, 1984, 55.
20. W.W. Mullins and R.F. Sekerka, J. Appl. Phys., 35, 1964, 444.
21. S. R. Coriell and R.F. Sekerka, in Rapid Solidification Processing: Principles and Technologies II, ed. by R. Mehrabian, B.H. Kear and M. Cohen (Claitor's, Baton Rouge, LA, 1980), 35-49.
22. S. R. Coriell and R.F. Sekerka, J. Cryst. Growth 61, 1983, 499.
23. L.J. Mazur and M.C. Flemings, Proc. 4th Int. Conf. on Rapidly Quenched Metals, Vol. 2, T. Masumoto, K. Suzuki, eds, Japan Inst. of Metals, Sendai, 1982, 1557.
24. T.Z. Kattamis and R. Mehrabian, J. Mat. Sci. 9, 1974, 1446.
25. T.F. Kelly, G.B. Olson and J.B. Van der Sande, in Rapidly Solidi-fied Amorphous and Crystalline Alloys, B.H. Kear, B.C. Giessen and M.Cohen, eds., North Holland, 1982, 343.
26. H. Palacio, M. Solari and H. Biloni in Physical Metallurgy, Part 1, ed. by R.W. Cahn and P. Haasen, North Holland, Amsterdam, 1983, 526.
27. G.B. McFadden and S. R. Coriell, Physica D, 12D, 1984, 253.
28. W. J. Boettinger, D. Shechtman, T.Z.Kattamis and R. J. Schaefer, Rapidly Quenched Metals, S. Steeb and H. Warlimont, eds, North Holland, Amsterdam, 1985, 871.
29. J.S. Langer and H. Mueller-Krumbhaar, Acta Met. 26, 1978, 1681.
30. M.E. Glicksman, private communication, 1985.
31. See for example, D.J. Fisher and W. Kurz, Acta Met. 28, 1980, 777.
32. W.J. Boettinger, L.A. Bendersky and J.G. Early, Met. Trans A, submitted for publication.
33. I.R. Hughes and H. Jones, J. Mat.Sci., 11, 1976, 1781.
34. C. Levi and R. Mehrabian, Met. Trans, 13A, 1982, 221.
35. C.Levi and R. Mehrabian, Met. Trans, 13A, 1982, 13.
36. J. Murray, unpublished research, NBS, 1985.
37. I. Dustin and W. Kurz, Met. Trans A., 1985, submitted for publica-tion.

ABSTRACTED DISCUSSION OF THE PAPER BY W.J. BOETTINGER

Participants: C.M. Adam, J. Ågren, R.W. Cahn, B. Cantor, H. Fraser,
M.E. Glicksman, H. Jones, L. Katgerman, R. Trivedi and
J.V. Wood

Discussion commenced with the questions about the accuracy of extrapolating dendrite arm spacings--cooling rate relationships for assessment of rapid solidification cooling rates. It was generally agreed that such extrapolations are invalid, particularly when well developed secondary arms are absent, and that more useful microstructural, such as retained solute level, should be parameters related to interfacial growth velocity. In discussing absolute stability concepts at high solidification velocities, M.E. Glicksman felt that the recently published analysis by R. Trivedi and W. Kurz (Acta Met., submitted for publication) indicated that a substantial modification to the absolute stability concept had now been quantified. Under high supercooling conditions the thermal Peclet numbers approaches unity, and Laplace s equations for both solute and thermal diffusion must be solved. At velocities slightly lower than that required for complete solute trapping, some lateral diffusion was presumed to handle the excess solute, although W.J. Boettinger felt that surface tension forces effectively prevented interfacial breakdown to a dendrite structure, and a cellular interface resulted. The distinction between cellular interface structures found at high solidification velocities in Al-8Fe alloys, for example, which have been described as Zone A (H. Jones, 1970) and later as microeutectic structures (C.M. Adam, 1980) was discussed. W.J. Boettinger felt that a fully coupled thermal and solute diffusion field, characteristic of the Jackson-Hunt eutectic solidification, could not exist for solidification velocities greater than a few centimeters per second. R. Trivedi pointed out that it is possible that the Jackson-Hunt analysis underestimates coupled solidification velocities at high interfacial undercoolings, and that the distinction between microcellular and microeutectic solidification is largely semantic. Discussion regarding thermal stability of the icosohedral quasi-crystals revealed that most workers have found the crystal structure to be remarkably stable; i.e. transformation had not occurred to Al_6Fe or Al_6Mn within 1 hour at 450°C.

MAGNETIC AND SUPERCONDUCTING MATERIALS

J. DURAND
Universite de Nancy 1
Laboratoire de Physique du Solide
F-54506 Vandoevre-les-Nancy, France

SUMMARY

1. INTRODUCTION

As compared with conventional materials obtained in their equilibrium state, typical materials produced by rapid solidification differ both in terms of chemical composition and atomic arrangement. Among the novel magnetic or superconducting properties of these new materials, some are clearly related to composition, others are directly connected to structure; while in some cases, the predominant effect (structure or alloying) is not clearly identified . In addition, RS materials can be obtained in a wide variety of structural conditions ranging through purely amorphous alloys or amorphous alloys with various amount of crystalline precipitates to meta-stable crystalline compounds with various microstructures, and, in bet-ween, some materials the structure of which is not properly crystalline nor amorphous (ultrafine particles, "nanocrystals"..). These different aspects of composition and structure were discussed along with a schematic review of the magnetic and superconducting properties of RS materials.

2. MAGNETIC PROPERTIES
2.1. Basic magnetic properties (saturation moment, curie temperature T_c, critical behaviour, magnetic excitation, magnetostiction).
- Predominant influence of composition.
- Role of atomic structure.
- Discussion of the "size effect" in ultrafine particles.
- Magnetic properties of compositionally modulated alloys or metallic superlattius.

2.2. Soft magnetic materials
- Purely amorphous alloys.
- Role of crystalline precipitates.
- The crystalline RS FeSi alloys.

2.3. Permanent magnetic materials: new crystalline materials (such as NdFeB) obtained by different routes, including that of rapid solidifi-cation to produce an amorphous precursor material.

3. SUPERCONDUCTING PROPERTIES
3.1. Basic superconducting properties of pure (bulk) amorphous materials.
- Influence of composition.
- Influence of disorder.
- Size effects.

3.2. Application oriented materials.
- Role of crystalline precipitates.
- Perspectives for composite materials.

ABSTRACTED DISCUSSION OF THE PAPER BY J. DURAND

Participants: H.W. Bergmann, R.W. Cahn, B. Cantor, W. McCallum and
                R.E. Maringer

The discussion touched on
- soft magnetic materials
  and centered on
- permanent magnetic and superconducting RST materials.

Little knowledge is available on the effect of small alloy additions on
soft magnetic matrices in particular what affects the conventional hystere-
sis losses. Desired textures of Fe-6.5Si, for example, can however, be
rather well controlled if suitable annealing is utilized - with (101)
becoming parallel to the spinning direction. Presently, technological
shortcomings such as joinability of ribbons to each other or magnetic
isolation in ribbon winding operations represent the bottle-neck with
respect to applications.

The appropriateness of RST for permanent magnetic materials of the Fe-Nd-B-
type is less obvious unless powders are included. Nevertheless, there
appear to be optimal solidification speeds for achieving maximal coercive
force fields, mainly depending on grain sizes produced. Results obtained
from an American source show grain sizes of 3 μm, those from a Japanese
source 500 to 800 Å.

RAPID SOLIDIFICATION BY LASER BEAM TECHNIQUES

G. SEPOLD and R. BECKER
Bremer Institut für angewandte Strahltechnik (BIAS)
D-2820 Bremen, F.R. Germany

1. INTRODUCTION

CO_2 -lasers are high power heat sources for surface modification of structural components.

Scanning laser beams over the surface of bulk materials leads to melting and solidifying of thin layers within very short periods. Dependent on high cooling rates which result from heat conduction into the underlying material, new phases and structures are produced in suitable alloys which offer new physical and chemical properties. In this paper we report on melting and solidification in an intermediate cooling regime of 10^3 - 10^5 K/s yielding microcrystalline structures as well as on phenomena in the ultra-high cooling regime (above 10^6 K/s) producing amorphous layers.

2. LASER MELTING OF SURFACES

The principle of laser melting is shown in figure 1. The laser beam impinges on the specimen surface which is moved with a high velocity. Dependent on power density and processing speed the surface is melted and solidified within a wide range of cooling and solidification rates.

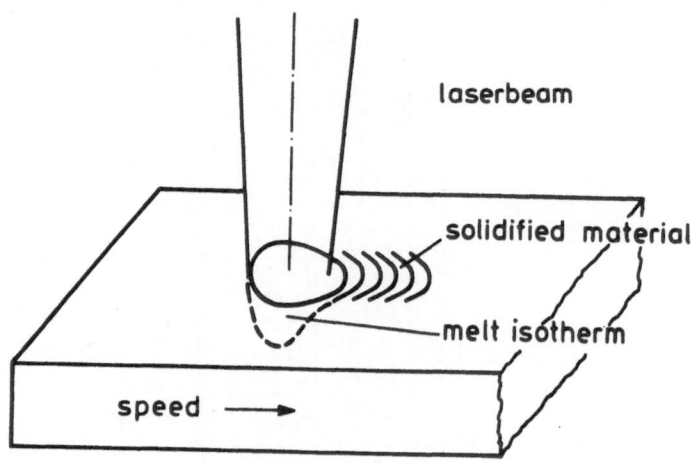

Figure 1: Principle of laser melting.

Besides power density and processing speed other parameters have to be considered as, for example, the position of the focus, the focussing angle, the intensity distribution, and the spatial and temporal stability of the beam. An exact adjustment and controlling of the laser beam parameters is very important for a successful laser heating experiment (1).

Fundamentally, heating with laser beams is possible according to two different principles:

In the first one energy is transferred through heat conduction into the inside (heat conduction mode), heating a small surface area to temperatures above the melting point. The necessary power densities are in the range of 10^3 to 10^5 W/cm^2. An upper limit is given by the boiling point. The essential disadvantage of laser heating by conduction is the low process efficiency which depends to a high degree on the optical properties of the surface. In extreme cases more than 90% of the laser power is lost due to the reflecting metal surfaces. Blackening the surfaces by using graphite layers will increase process efficiency.

The second principle includes boiling of the material and anomalous absorption. The resulting deep penetration of the laser beam into the material leads to high absorption. In this case a well focussed laser beam is entrapped in a small vapor capillary and at high power densities of more than 10^5 W/cm^2 - depending on the speed - it penetrates into the material. It vaporizes, condenses at the walls of the capillary, and finally solidifies creating a zone similar to a welded seam. It is a special characteristic of such a melting zone to be considerably deeper than wide. In our experiments we used both principles, heat conduction and deep penetration.

The application of a focussed beam with Gaussian intensity distribution is not the only procedure of laser heating. For example, it is also possible to integrate the laserbeam over a defined area with a homogeneous intensity distribution by means of an optical converter or by a high frequency deflection ("sweeping") of the laser beam over a large trace.

Depending on the laserbeam parameters and thermophysical properties of the material, high cooling rates are reached in the resolidifying melted layer (2). Theoretical calculations show that cooling rates up to 10^7 K/s in thin surface layers of some 10 µm thickness may be attained by heat conduction into the underlying solid material. This dependence was calculated and is shown in figure 2 for different beam diameters (absorbed beam power: 0.2 - 1.0 kW).

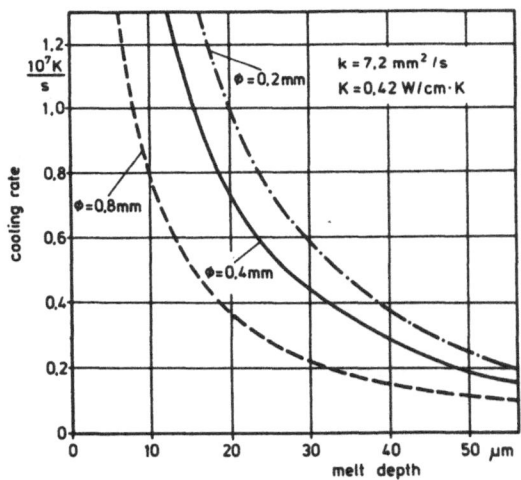

Figure 2: Cooling rate versus depth.

Those cooling rates may be sufficient to produce microcrystalline structures with metastable or even amorphous phases. In the following, we investigated the production of metastable, micro-crystalline structures by melting cast irons and producing a variety of phases reaching into the amorphous state by applying various alloy compositions.

3. PRODUCTION OF MICROCRYSTALLINE, METASTABLE PHASES BY LASER MELTING OF CAST IRON

By melting of cast iron, hard, wear-resistant ledeburitic layers are produced. Therefore melting by conventional heat sources like the TIG-welding process is a commercially accepted practice and is applied to components such as crankshafts, rocker arms and camshafts (3). The application of lasers seems to be favorable because much higher cooling rates could be achieved (300 K/s for TIG versus 10^7 K/s for laser processing) resulting in the formation of new structures and new phases depending on cooling conditions (4).

In the cooling range up to some 1000 K/s a refinement of the ledeburitic structure was found with an increase of hardness due to the formation of martensite. This is shown in figure 3 for different cooling rates in the lower cooling regime up to 2400 K/s. An increase of hardness is accompanied

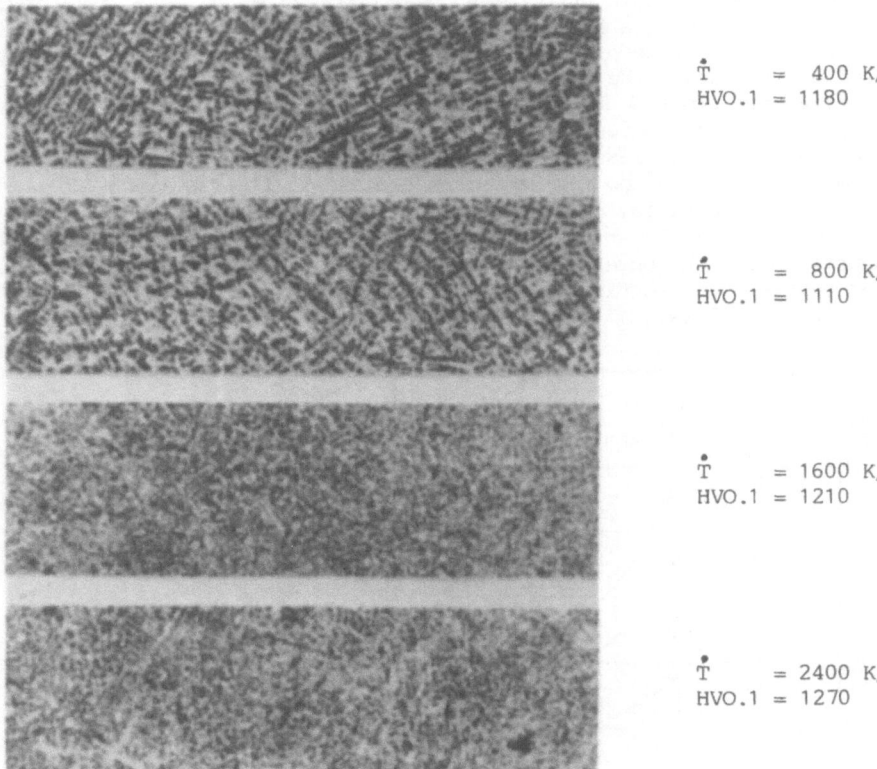

$$\overset{\bullet}{T} \quad = \quad 400 \ K/s$$
$$HVO.1 = 1180$$

$$\overset{\bullet}{T} \quad = \quad 800 \ K/s$$
$$HVO.1 = 1110$$

$$\overset{\bullet}{T} \quad = \quad 1600 \ K/s$$
$$HVO.1 = 1210$$

$$\overset{\bullet}{T} \quad = \quad 2400 \ K/s$$
$$HVO.1 = 1270$$

Figure 3: Structure of laser-melted globular cast iron with ferritic matrix for different cooling rates, calculated for cooling conditions in the center of the melting zone. $p = 3\text{-}16 \ kW/cm^2$, $t_i = 1\text{-}5$ s, rectangular beam 25 mm.

by an improvement of wear resistance, as outlined in figure 4. However, increased hardness is only observed in the range of a few 1000 K/s. Above this, residual austenite is formed in the ledeburitic structure. This is the reason for decreasing hardness at higher cooling rates. Best abrasive wear resistance was found for those structures. It is expected that residual austenite is transformed to martensite under the abrasive wear loading, leading to higher wear resistance (5). As compared to conventional methods laser quenching of cast irons offers the chance of a remarkable improvement of the wear resistance.

Technological application is more and more in progress, mainly for the automotive industry. Examples are camshafts, crankshafts and rocker arms, see figure 5. Besides metallurgical advantages, the industry is mainly

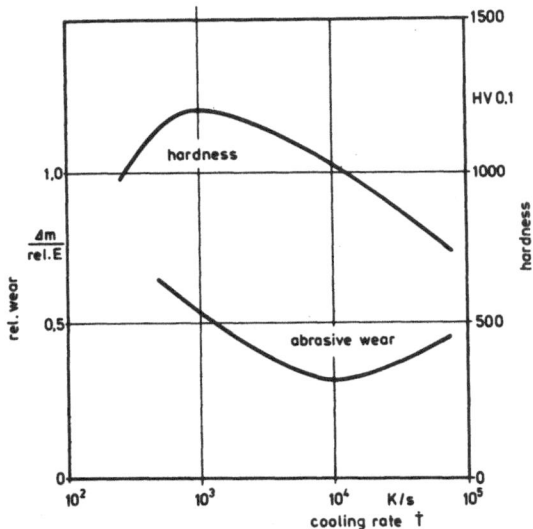

Figure 4: Relative abrasive wear and hardness of laser melted cast irons versus cooling rate for globular cast iron.

Figure 5: Laser melting of a cast iron rocker arm; left: top view; right: cross-section with ledeburitic

interested in the low and locally well defined heat loading of the component. Furthermore, the smooth and flat melting zones reduce the necessity
of post-treatment processing of the work piece.

4. PRODUCTION OF AMORPHOUS STRUCTURES
4.1 Laser vitrification
Under the conditions of higher power densities of more than 10^6 W/cm^2
and short interaction times of less than 1 ms, cooling rates between 10^5 –
10^7 K/s are achieved in thin layers of some 10 µm thickness. For suitable
alloys not only microcrystalline structures, but also amorphous layers are
then obtained (6). There are, however, two possible effects which might
prevent the attainment of a homogeneous, amorphous surface layer:

Firstly, the underlying crystalline material being in contact with the
melt serves as a nucleation site for crystalline phases which may grow so
rapidly that they disrupt the amorphous layer. Secondly, when vitrifying a
large surface area with overlapping laser traces, recrystallization of the
amorphous material starts in the heat-affected zones.

The problematic nature of overlapping laser traces is shown in figure 6,
which is a cross section of an overlapping vitrification experiment in
$Fe_{80}B_{20}$ material. Crystallization and growth from the underlying material is
seen in zone B, which is limited by a small seam of unknown microstructure.
The region above (zone A) was identified by electronmicroscopic examination to be fully amorphous. In the heat affected zones (C) partial crystallization was observed.

Nucleation and growth from the underlying crystalline material and
crystallization in the heat affected zones have to be considered as critical factors of the laser vitrification process. However, the given example

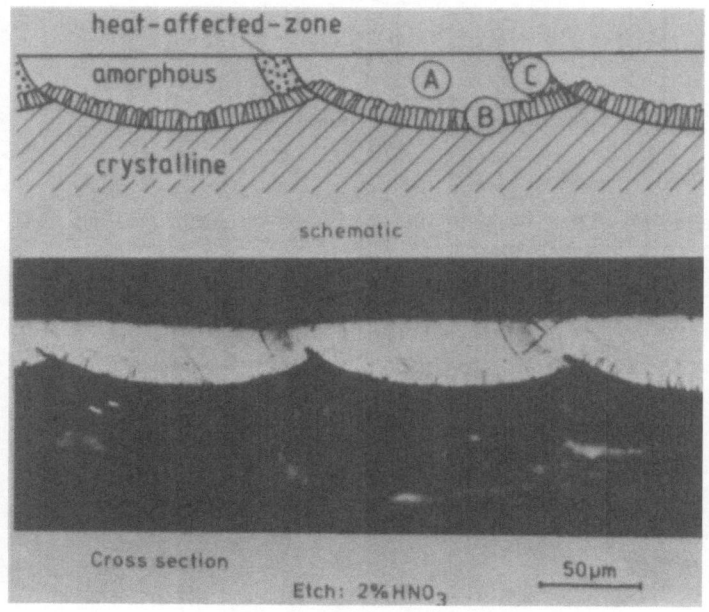

Figure 6: Overlapping laser vitrification in cristalline $Fe_{80}B_{20}$.

of melted $Fe_{80} B_{20}$ proves the applicability of the laser vitrification process even for alloys which need high critical cooling rates of up to 10^5 K/s for amorphous solidification. Favorable conditions are given if heterogeneous nucleation is retarded and growth rates are slowed down, which for example, is well known for $Ni_{60}Nb_{40}$.

Overlapping melt traces were produced in crystalline $Ni_{60}Nb_{40}$ and examined electronmicroscopically, figures 7a and 7b.

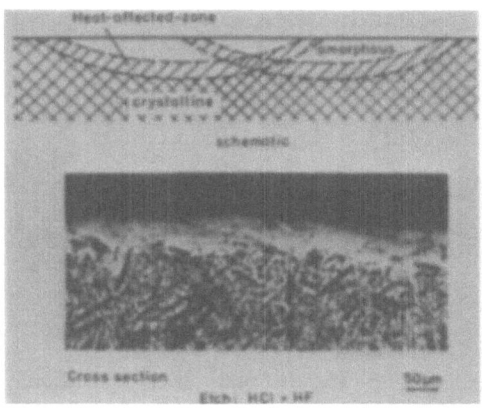 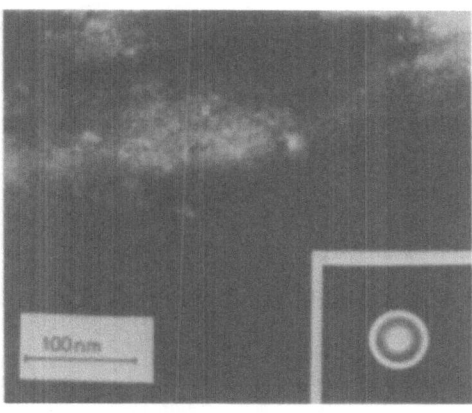

(a)   (b)

Figure 7: Overlapping laser vitrification in $Ni_{60}Nb_{40}$, (a) cross section, (b) transmission electron micrograph.

The micrograph is shown together with a selected area electron diffraction pattern. The structure is basically amorphous, as shown by the diffuse diffraction ring, and contains only a few crystalline particles of about 10 nm in diameter. Some areas with isolated crystallites of somewhat larger diameter (up to 100 nm) were found in the critical heat affected zones embedded in the amorphous matrix.

From these experimental results we conclude that only alloys with lower nucleation rates and/or low growth rates are candidates for fully amorphous, laser vitrified coatings.

4.2 Laser layer technique for joining amorphous materials

The production of amorphous layers by laser vitrification is limited to a thickness of approximately 50 μm. Thicker amorphous coatings may be obtained by a special laser joining technique which has been developed at BIAS. Commercially available melt-quenched ribbons are used and arranged in layers by a combined high speed roller welding technique. The principle is shown in figure 8.

Two amorphous ribbons are pressed together between feed rolls and laser welded in the contact zone. Under the conditions of high power densities our new joining technique may be applied in several ways: Cladding of amorphous ribbons to crystalline components and laser joining of amorphous foils for the production of tubes. A layer-by-layer technique offers the opportunity of producing amorphous structures of larger thickness which is desired for several technological applications. The processing speed is high enough (more than 3 m/s) to retain the amorphous structure in the welding zone. So the formation of undesired brittle crystalline phases is

x

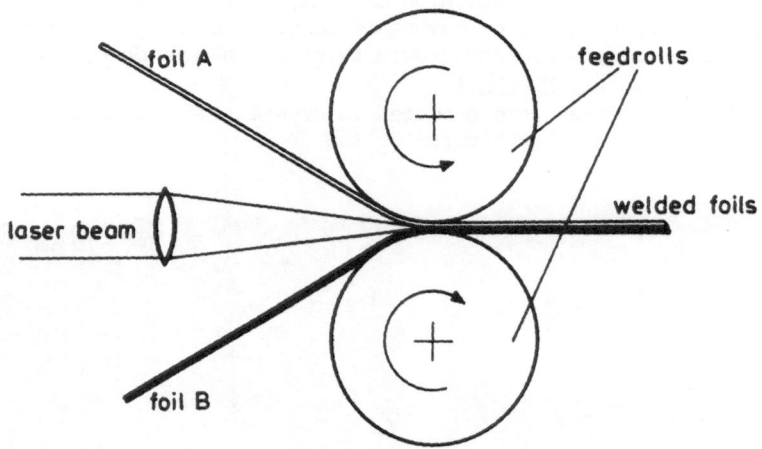

Figure 8: Technique for laser welding of amorphous foils.

prevented. For the time welded zones of 5 mm in width and smaller than 1 mm in thickness have been produced. In a next step we intend to join several amorphous foils to get thicker materials.

Our first results are promising. For example, figure 9 shows a cross section of laser welded Co-based amorphous ribbons (VITROVAC 6025). With the exception of some crystallites, no crystalline phases could be made out by metallographic etching in HF/HNO. Ductile bending of the specimen was possible at a small bending radius (r = 30 μm) without fracture or the formation of cracks, figure 10.

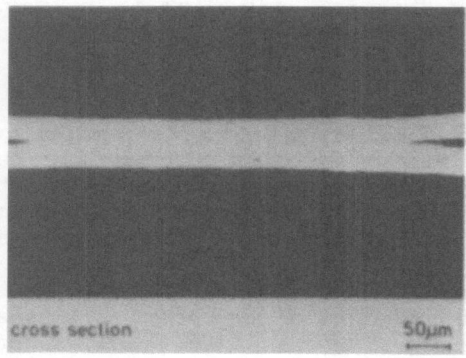

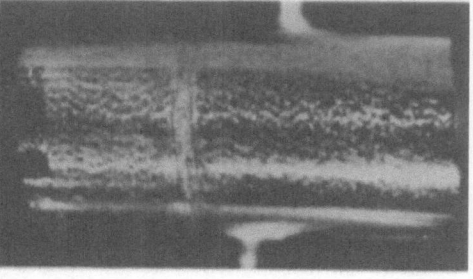

Figure 9: Laser-joined amorphous ribbons.

Figure 10: Bent, laser-welded amorphous ribbons.

5. SUMMARY

- Laser beam processing techniques lead well defined melting zones with low heat input into the components.
- The extent of these zones can be calculated, and rough estimations on the cooling rates are posssible.
- Depending on cooling rates in the range of $10^2 - 10^5$ K/s, different structures were produced in laser melted globular cast irons.
- Results of vitrification of crystalline $Ni_{60}Nb_{40}$ and $Fe_{80}B_{20}$ are shown for cooling rates larger than 10^7 K/s.
- A new technique for the production of thick amorphous layers is presented using a new layer joining method.

6. REFERENCES

1. Sepold G, Schweissen und Oberflächenveredeln mit dem Co -Laser, VDI-Bildungswerk, BW 5640, 1-21.
2. Jüptner W, Kreis Th, Mathematisches Modell zur Beschreibung der Wechselwirkung zwischen Strahlen hoher Intensität und technischen Werkstoffen, Forschungsbericht, VW-Stiftung, 1980.
3. Reinke F H, TIG Hardening of Cast Iron Camshafts, Heat Treatment of Metals, 1, 1981, 17-23.
4. Bergmann H W, et al., Laser Surface Melting on Cast Iron Containing Undercooled Graphite, Zt. Werkstofftechn., 14, 1983, 237-244.
5. Gahr K H zum, Werkstoffgefüge und abrasives Verschleißverhalten metallischer Werkstoffe, HTM 35, 4, 1980, 182-191.
6. Breinan E M, Kear B H, Rapid Solidification Laser Processing of Materials for Control of Microstructure and Properties, in Rapid Solidification Processing, Mehrabian R, Kear B H and Cohen M, eds. Claitors Publishing Division, Baton Rouge, La, 1978, 87-103.

120

ABSTRACTED DISCUSSION OF THE PAPER BY G. SEPOLD

Participants: H.W. Bergman, R.W. Cahn, S. Hock, R.E. Maringer, P.R. Sahm, T. Sheppard and J.V. Wood

Discussions centered on three main topics:
- microstructures of laser remelted surfaces,
- in situ-control of the laser beam,
- methods of welding both amorphous metals and crystalline materials and combinations.

Microstructure formation by surface laser melting: the depth of penetration of the laser beam was one aspect of the discussion. The finer laser beams are focussed more highly to permit increased penetration into the laser melted substrate. The question of how much could be done in terms of combining dissimilar surface melted and substrate materials could not be conclusively answered because too little experience had so far been gained in this field. Similarly, control of surface smoothness as a function of types of laser beams utilized was not answerable either, again due to lack of concise knowledge and experiments in this field.

In situ-control of the laser beam: this type of endeavor has not been widely practised. It appears particularly difficult to suggest clearcut models because too many parameters are unknown. For example, the question of the interaction between the laser beam and the absorbtivity of the formed gas phase, i.e. the plasma generated by the evaporated part of the metal, is hardly known at all.

Joining of amorphous metals: the point of joining amorphous metals by utilizing laser welding was discussed at length. No definitely purely amorphous phase weld has so far been obtained. However, encouraging results were reported. Specific questions came up which touched on experiences with spot-welding or roller-edge-welding, and it was, in effect, asked whether a similar technique could possibly be utilized employing a laser for melting together the two edges of amorphous foils. Naturally, the discussion deviated somewhat into directions of introducing a fusible layer between the two sheets to be joined. Electrochemical plating was offered as well as a certain type of wire-brush-welding or friction welding. The discussion swayed back to questions already touched upon above, namely the synthesis of surface alloying by some type of technology mix, for example, combining plasma-spray-methods with laser remelting or else, powder feeding and laser melting a substrate so that dissimilar materials are combined.

All in all, the discussion indicated that the field of surface melting appears to be in its infancy and will definitely offer in the future not only
- new or modified processes of surface treatment but also, directly connected with it,
- the synthesis of new materials which possibly cannot be obtained in any other way.

CONTINUOUS PRODUCTS IN RAPID SOLIDIFICATION

L. KATGERMAN
Alcan International Ltd
Banbury Laboratories
Banbury OXI6 7SP, U.K.

ABSTRACT
The characteristics of some selected melt quenching processes to produce
continuous ribbons or tapes was discussed.
To understand the process physics and to quantify the effects of process
variables on ribbon geometry and as-quenched microstructure mathematical
models have been formulated. Because ribbon formation is a complex process
it is difficult to give an accurate (mathematical) description. Therefore
empirical correlations have been established between process variables and
ribbon geometry or the occurrence of geometrical defects.
In the mathematical models both fluid and heat transfer aspects are
considered and the ribbon formation is characterized by the propagation of
the thermal and momentum boundary layers.
The existing models only successfully describe the experimental observations to a limited extent. A rigorous mathematical model of these processes, including the development of the as-quenched microstructure, requires
quantitative descriptions of solute redistribution and nucleation and
growth of alloys at high quench rates taking into account the various
thermodynamic and kinetic factors.
Difficulties and limitations in incorporating these effects in the
mathematical models will be discussed.

1. INTRODUCTION
In the last decades a variety of techniques that involve direct
quenching of liquid metals and alloys have been applied in order to obtain
improved microstructure and properties. Most of these rapid solidification
methods and general characteristics of their products have been reviewed
recently by several authors (1-4). Rapid solidification processes in
general can be divided into three categories depending on the quenching
medium (1):
    I quenching by gases (atomization methods)
    II quenching by a solid substrate (chill methods)
    III self-quenching (surface methods).
Whilst there is a wide variety of atomization techniques to produce powders
or particulate, the fabrication techniques to produce continuous products
from the melt are rather limited. During fabrication of powders or particulates a range of particle sizes is produced which is a consequence of the
mechanisms of melt fragmentation (5). This particle size distribution can
result in a non-uniform microstructure because of the difference in thermal
history between powder particles. The fabrication of rapidly quenched
materials by continuously making the product from the melt results in an
improved sample geometry and consequently a more uniform quench rate. In
the present paper current research on characterization and modelling of
processes to produce continuously tapes, ribbons and strips will be summarized. To understand the process physics and to quantify the effects of
process variables on ribbon geometry and as-quenched microstructure, mathe-

matical models have been formulated. Because ribbon formation is a complex process (e.g. melt puddle length and melt-substrate interfacial contact are not fully understood), it is difficult to give an accurate mathematical description. Therefore empirical correlations have been established between process variables and ribbon geometry or the occurrence of geometrical defects.

In the mathematical models both fluid flow and heat transfer aspects are considered and the ribbon formation is characterised by the propagation of the thermal and momentum boundary layers.

2. PROCESS METHODS

Because of the low viscosity and high surface tension of molten metals a liquid metal stream will become unstable after some distance and droplets are formed (6,7). Since a continuous ribbon type product is required solidification must occur before break-up of the molten stream takes place. This phenomenon limits drastically the variety of alternative methods to produce continuous ribbons, strips or fibers. The most successful processes to produce continuously rapidly quenched ribbons or strips are chill block melt spinning (CBMS) and planar flow casting (PFC) or melt drag (MD) (1-4).

Chill block melt spinning (CBMS) involves the formation of a molten jet impinging on a rotation disk, figure 1. The melt puddle which results from

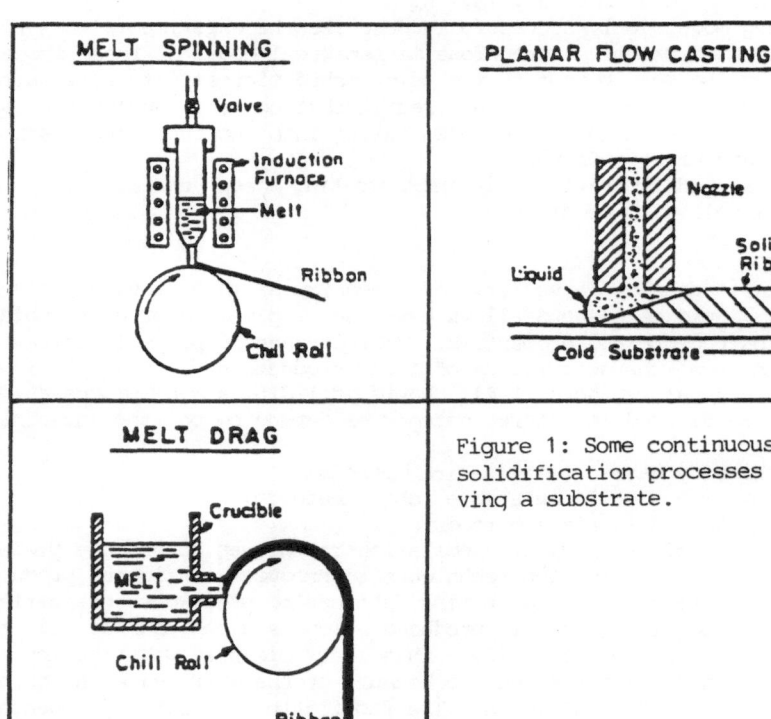

Figure 1: Some continuous rapid solidification processes involving a substrate.

continued delivery of molten metal serves as a reservoir from which ribbon is continuously formed and quenched. The quench rate and ribbon geometry depend upon the characteristics of the melt puddle. A stable melt puddle during CBMS results in good quality, uniform ribbon (8,9). Although different moving substrate configurations can be applied (1-4), impingement outside a rotating wheel is commonly used. Because of its relative simplicity CBMS is widely used on a laboratory scale to produce research quantities of uniform continuous ribbon. The maximum ribbon width during CBMS is about 3 mm and ribbon thicknesses vary between 20-200 microns depending on substrate velocity.

Planar flow casting (PFC) is a development of the CBMS process to produce tapes wider than is possible with CBMS. The process is different from CBMS in that the nozzle which feeds the molten alloy is situated very close to the moving substrate (10,11). The melt puddle is confined between substrate and nozzle wall which results in an improved ribbon quality, figure 1. By applying a slot type nozzle wider tapes can be formed and strips with widths up to 15 cm have been processed this way. Ribbon thicknesses of 20-200 microns are basically the same as in CBMS and consequently the quench rates are similar (10^5 - 10^6 K/s).

The melt drag process is very similar to PFC but in this case the slot nozzle is situated horizontally rather than vertically (12), figure 1. The metal is removed from the slot nozzle by contact of the meniscus with the rotating wheel. Because MD is a gravity fed process (CBMS and PFC are pressure fed processes) a wider range of thicknesses (as a function of wheel velocity) can be obtained (25-1000 microns), although under certain process conditions surface defects originating from meniscus break-up occur. The obtained ribbon widths are similar to PFC (about 15 cm).

All processes mentioned above (CBMS,PFC,MD) have been applied to produce amorphous alloys as well as microcrystalline alloys. Apparently good ribbon quality and adequate process control is easier to obtain for glass-forming alloys than for crystallizing alloys. Although the drastic difference in liquid-solid transformation is quite obvious, its effect on ribbon quality and ribbon geometry is not completely clear.

One of the disadvantages of melt spinning methods is the relatively small production rate as compared with most atomization methods. Substantial scale-up of the fabrication process is required to overcome this problem, which implies technical solutions to difficulties associated with e.g. efficient collection of the spun product, nozzle and wheel erosion, surface quality and smooth edges. Because the applications of as-spun rapidly quenched ribbons are somewhat limited, especially in the case of microcrystalline material, current emphasis is focussed on production of wide strip of widths up to 1000 mm.

Another recent development is composite melt spinning by additions of powder or particulate in the melt crucible or into the melt puddle to produce metal matrix composite ribbons (13-15).

3. PROCESS CHARACTERISATION

To understand the process physics and to quantify the effects of process variables on ribbon geometry and as-quenched microstructure mathematical models have been formulated. Because ribbon formation is a complex process initially empirical correlations between the main process variables have been established. These experimental relations give only limited insight of the mechanisms of ribbon formation. Therefore more rigorous mathematical models have been applied based on fluid flow and heat flow phenomena to analyse the process.

In the next paragraphs the different approaches will be discussed in

more detail. Most analyses have been carried out for CBMS and PFC but not
for MD explicitly. Because the mechanisms of ribbon formation in all three
processes are very similar most of the modelling approaches are not exclu-
sively applicable to one specific process.

3.1 Effect of process variables on the formation of ribbon geometry

As mentioned in section 2, the stability of the melt puddle is of
great importance for control of the final ribbon dimensions (16). PFC
and MD achieve this stability partly by confining the melt puddle with a
gas stream (8,9,17). By narrowing the distance between orifice and substrate perturbations caused by capillarity waves and vibrations (6,16) can
be minimized and molten jet break-up can be avoided (7,18).

The main process variables during ribbon formation are substrate
velocity U_0 , ejection pressure P, nozzle cross-sectional area A_j , and
temperature of the melt. Correlation between the process parameters can be
obtained by application of macroscopic balances of energy and matter. By
combining Bernoulli's equation

$$v_j^2 = 2 P/\rho \qquad (1)$$

with the equation for the conservation of matter

$$Q = A_j v_j = A_R U_0 \qquad (2)$$

where v_j is the velocity of the jet, A_R the cross-sectional area of the
ribbon, ρ the liquid metal density and Q the volumetric flow rate, a relation for the ribbon cross-sectional area can be obtained (19):

$$A_R = C_1 A_j /U_S (2P/\rho)^{1/2} \qquad (3)$$

A_j is usually taken as $\pi D_0^2/4$ in which D_0 is the orifice diameter. Allowing
for friction and energy losses a coefficient C_1 has been included in equation (3). It was found empirically that the ribbon width w is a function of
the volumetric flow rate Q for amorphous alloys (20)

$$w = w_0 + Q v_p \qquad (4)$$

The constant v_p was interpreted as an average dynamic viscosity of the
solidifying liquid by Liebermann while the constant w_0 is dependent on the
orifice diameter. By combining equations (1)-(4) correlations between ribbon thickness d and width w, and process variables volumetric flow rate,
substrate velocity and/or ejection pressure can be obtained.

In a more detailed analysis based on thermal transport and fluid flow
Kavesh (18) obtained independent expressions for ribbon thickness and
width.

$$d = C_2 \, Q^{1-n} \, /U_S^n \tag{5}$$

$$w = 1/C_2 Q^n \, /U_S^{1-n} \tag{6}$$

To obtain these expressions for the ribbon dimensions Kavesh assumed that the puddle contact area on the moving substrate remains constant. This assumption has been criticized by Vincent et al. based on experimental observations (21-24). Assuming that the puddle shape remains constant Vincent et al. derive alternative expressions for equations (5) and (6) being,

$$d = C_3 \, Q^m /U_S^{2m} \tag{7}$$

$$w = 1/C_3 \, Q^{1-m} \, /U_S^{2m} \tag{8}$$

Experimental data (18-29) show reasonable agreement with the derived empirical equations, figures 2-4.

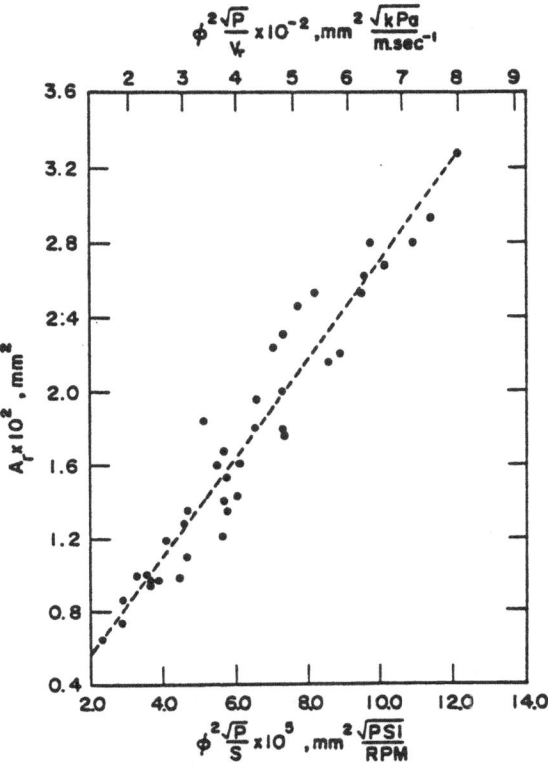

Figure 2: Experimental variation of cross-sectional area of Fe40Ni40B20 ribbon with melt delivery parameters (after Liebermann and Graham (19)).

126

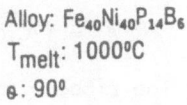

Alloy: $Fe_{40}Ni_{40}P_{14}B_6$
T_{melt}: 1000°C
θ: 90°

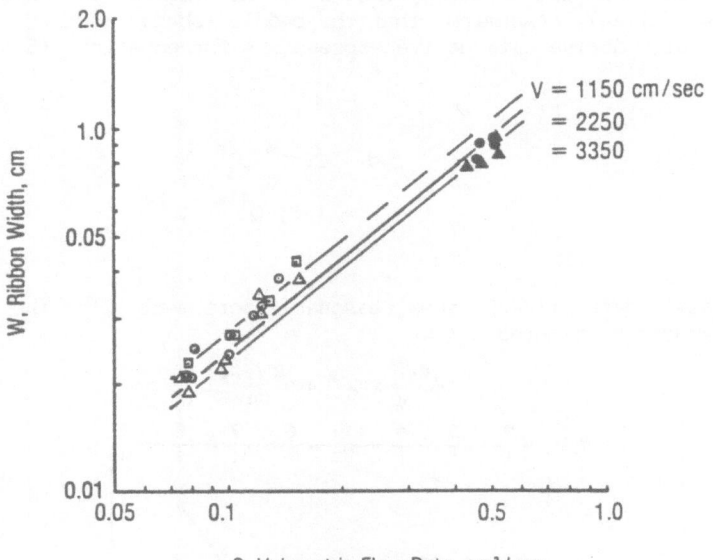

Figure 3: Melt spun ribbon width as a function of material flow rate at various substrate velocities (after
Kavesh, (18)).

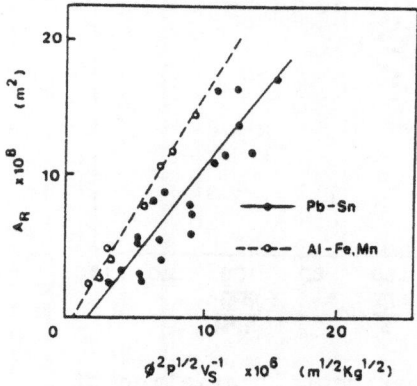

Figure 4: Experimental ribbon
cross-sectional areas of eutectic Pb-Sn and Al-alloys
as a function of melt delivery variables (after Charter
et al., (28).

Although the last derived expressions might indicate an improved fit with experimental data it is impossible to make a final discrimination between the expressions. Experimental results by Charter et al. (28) clearly indicate that the expressions originally derived for amorphous alloys can be extended to crystalline materials, although the constants in the equations differ substantially.

As indicated before, the derived equations can only be used quantitatively and in themselves do not reveal any information about the mechanisms of ribbon formation. The exponents in equations (5)-(9) obtained by regression analysis of the experimental data can be explained either by dominating thermal or momentum transport. Models based on principles of momentum and heat transfer are required to predict more quantitatively the ribbon formation process.

3.2 Fluid flow

The momentum transport mechanism during melt spinning can be regarded as a Blasius problem (30), which describes the development of the momentum boundary layer along a flat plate. In the melt spinning process the boundary conditions are different from the original Blasius problem but the governing equations are the same.

continuity $\qquad \dfrac{\partial u}{\partial x} + \dfrac{\partial v}{\partial y} = 0$ (9)

motion $\qquad u\dfrac{\partial u}{\partial x} + v\dfrac{\partial u}{\partial y} = \nu\dfrac{\partial^2 u}{\partial y^2}$ (10)

with boundary conditions

$\qquad u = U_o \quad v = 0 \qquad$ at $y = 0$ (11)

and

$\qquad u = 0 \qquad\qquad$ at $y = \infty$ (12)

By application of the similarity transform $\eta = y(\nu x/U_S)^{1/2}$ and substitution of the stream function $\emptyset = (\nu U_0 x)^{1/2} f(\eta)$, equations (9) and (10) can be written in terms of the dimensions stream function $f(\eta)$ as

$\qquad 2\,f''' + f''f = 0$ (13)

with boundary conditions

$\qquad f = 0 \quad f' = 1$ at $\eta = 0$ (14)

and

$\qquad f' = 0 \qquad\qquad$ at $\eta = \infty$ (15)

This problem can be solved numerically by Runge-Kutta methods similar as for the original Blasius problem (32,32,34).

Approximate analytical solutions for the velocity profile can be obtained by use of the integral momentum equation for the boundary layer (32,33a). The advantage of this solution is that the velocity profile expressed in polynomial form can readily be substituted in the heat flow and diffusion equations. In a simpler way the velocity profile can be approximated by Rayleigh's method (32) giving

$$u = U_O \text{ erfc } (\eta/2) \tag{16}$$

To characterize the velocity profile the concept of the boundary layer thickness is used (30). Usually the boundary layer thickness is defined as that thickness for which u is 1% of the substrate velocity U_O. As pointed out by Vincent et al. (21) it is more appropriate to use the displacement thickness in order to relate boundary layer thickness to ribbon thickness. The displacement thickness d_1 is defined as (30):

$$U_O \, d_1 = \int_0^\infty u(x,y) \, dy \tag{17}$$

To relate the calculated profile to the ribbon thickness it has been sugge-. sted alternatively (33b) that the liquid metal has to be dragged out of the melt puddle and consequently the dynamic pressure of the liquid metal has to exceed the surface tension. In this case the characteristic distance d_2 is defined, following (33b), as

$$1/2 \, \rho \, u^2 (x,d_2) = 4\sigma_{Lv}/L \tag{18}$$

in which σ_{Lv} is the surface tension of the metal and the puddle curvature is characterized by the puddle length L. The characteristic distances obtained by the different methods are summarised in Table 1. The boundary layer thicknesses based on the surface tension effect are not included because of its dependence of the actual values of the surface tension involved. It will be clear from the Table that the approximate analytical solution obtained by the author (33a) gives the best agreement with the exact solution which is easily obtained by numerical methods (32,34). At this point it should be noted that Vincent et al. (21) used the characteristic distances from the original Blasius problem which leads to the wrong values of the numerical factors 5.0 (1% boundary layer) and 1.72 (displacement thickness) respectively.

TABLE 1 : Boundary layer and displacement thickness for different solution methods.

approximation	ref.	u/U_0 profile	$d(1\%)/(\nu x/U_0)^{1/2}$	$d_1/(\nu x/U_0)^{1/2}$
exact	32	numerical	6.37	1.62
integral	32	$1-2\eta + 2\eta^3 - \eta^4$	4.68	1.40
integral	33a	$(1-\eta)^3$	6.48	1.61
rayleigh	32,33c	$\text{erfc}(\eta/2)$	3.64	1.12

3.3 Heat flow and solidification front velocity

The heat flow against a chill surface during rapid solidification has been modelled extensively and various calculations were carried out (1, 35-38). Depending on the nature of the interfacial contact between chill surface and solidifying metal two limited cases can be considered:
(1)-Newtonian cooling. Heat flow resistance at the chill surface (characterised by the heat transfer coefficient h) is controlling the heat flux and the constant solidification front velocity can be written as (1):

$$R = h \, (T_M - T_W)/\rho H \tag{19}$$

where T_M is the melting temperature, T_W the substrate temperature and H the latent heat of solidification.

(2)-Ideal cooling. No heat flux constraints at the chill surface and the heat flux is controlled by the thermal conductivity of the solidifying metal resulting in a front velocity of the form (39):

$$R = C_4 \, (a/t)^{1/2} \tag{20}$$

in which a is the thermal diffusivity, t the solidification time and the constant C_4 is calculated from the heat balance at the solid-liquid interface (39).

Approximate analytical solutions have been derived for intermediate conditions (40) and are of the form:

$$R = C_5 \, h \, (T_M - T_W)/(1 + h \, S \, /k) \tag{21}$$

where S is the solidified thickness, k the thermal conductivity and the constant C_5 depends on the choice of the approximate thermal profile (40). More accurate solutions can be obtained by numerical methods including the effect of melt superheat (35). In general it is believed that heat flow during rapid quenching against a chill surface is Newtonian and typical values of h = 10^4 - 10^6 W/m^2/K are estimated from microstructural observations (1). In the derived expressions so far it was implicitly assumed that the solid-liquid interface is plane and at equilibrium, which means that no effects of nucleation or undercooling prior to or during rapid quenching are considered.

Recently Clyne (41) has taken nucleation and undercooling into account and thermal histories for rapid quenching with non-equilibrium solidification growth kinetics have been calculated. In this analysis the solidification growth rate takes the form (41)

$$R = C_6 \, (\Delta T)^m \tag{22}$$

in which ΔT is the undercooling and the exponent m depends on the growth mechanism (42). The start of the solidification was calculated from the classical nucleation theory for solidification (42). For pure metals some quantitative data exist for the growth rate and the thermal profiles can be calculated, figure 5 (41).

3.4 Interaction between moving solidification front and fluid flow

For a more rigorous description of the melt spinning process the coupled effects of fluid flow and heat transfer have to be taken into account. In addition to equations (9)-(12) for fluid flow, the convective equations of heat flow have to be solved.

$$(\text{liquid}) \quad u \frac{\partial T}{\partial x} + v \frac{\partial T}{\partial y} = a \frac{\partial^2 T}{\partial y^2} \tag{23}$$

$$(\text{solid}) \quad u_0 \frac{\partial T}{\partial x} = a \frac{\partial^2 T}{\partial y^2} \tag{24}$$

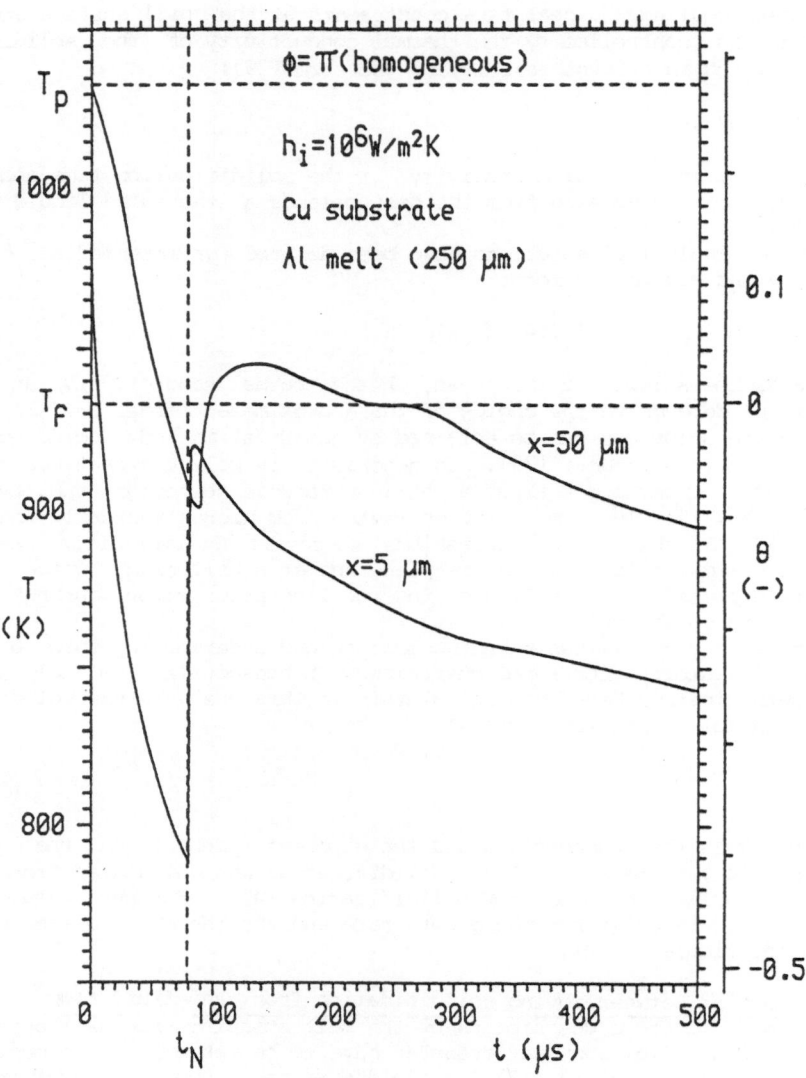

Figure 5: Thermal profiles at two points within an aluminium melt derived from non-equilibrium growth kinetics (after Clyne, (41)).

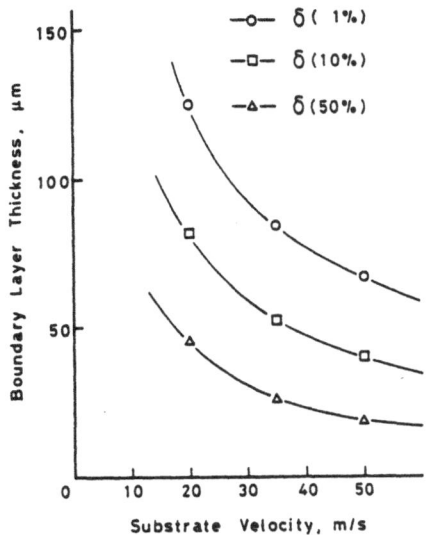

Figure 6: Effect of substrate velocity on momentum boundary thickness of Fe40Ni40P14B6; $h = 4.2 \ 10^5 \ W/m^2/ \ K$, melt temperature 1268 K (after Takeshita and Shingu, (45)).

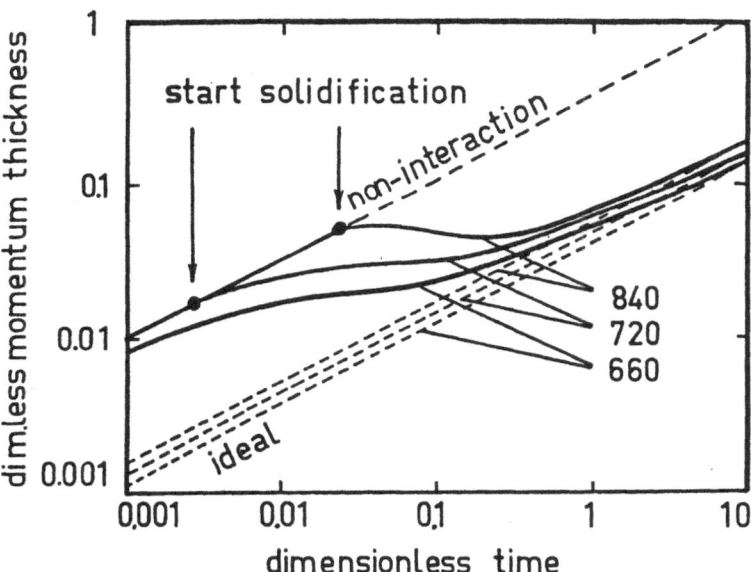

Figure 7: Momentum boundary layer thickness of aluminium as a function of residence time; $h = 10^5 \ W/m^2/K$ for different melt temperatures, (33d).

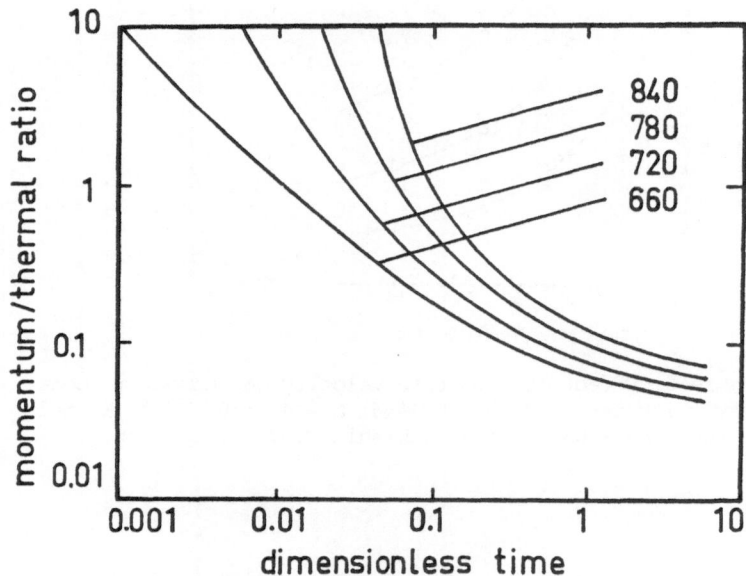

Figure 8: Calculated momentum to thermal thickness ratio as
a function of residence time for different melt temperatures;
$h = 10^5$ W/m^2/K, (33d).

Assuming equilibrium the thermal boundary conditions are

$$y= 0 \quad T= T_W \quad \text{(ideal)} \quad k\ dT/dy= h(T - T_w) \tag{25}$$
$$y= S \quad T= T_M \tag{26}$$
$$y= \infty \quad T= T_P \tag{27}$$

and a additional boundary condition for the conservation of heat at the
solid-liquid interface.

For ideal cooling conditions numerical solutions can be obtained by
application of the similarity transform analogue as described in section
3.2. Solutions have been obtained for glass-forming alloys (solidification
without latent heat evolution) (43) and for crystalline alloys (44). More
realistic calculations with regard to melt spinning process conditions were
carried out by Takeshita and Shingu (45) for a constraint heat flux condition at the chill surface, (equations (25)) in the case of glass-forming
alloys, figure 6.

For crystalline alloys the coupled effects were calculated in the Rayleigh approximation (33c,d) and recently in the Blasius formulation (33a). The effect of the moving solidification front on the momentum boundary layer thickness is clearly demonstrated in figure 7. Because of the short residence time in the melt puddle (about 10^{-4} s) this effect is somewhat limited. Application of the calculations to ribbon thickness formation indicates that although momentum transport plays an important role, heat transfer makes a substantial contribution to the ribbon thickness, figure 8. To relate calculated boundary layer thicknesses to ribbon thicknesses quantitative data for the melt puddle length are required. Up to now no theoretical derived expressions between puddle length L and substrate velocity U are available. In all calculations empirical correlations between L and U are used. Incorporation of non-equilibrium growth kinetics during melt spinning have not been carried out so far.

3.5 Further considerations for sophisticated melt spinning models

For the development of a rigorous physically consistent model describing melt spinning processes further study is necessary in specific areas to obtain a better understanding of the following process aspects:

(1)-Melt puddle formation.

To interpret the calculated profiles in terms of ribbon thicknesses a quantitative description of the mechanisms of puddle formation is indispensable. This is a complex hydrodynamic problem which has not been completely solved yet and requires an accurate description of the mechanisms of a hydraulic jump on a fast moving surface. The actual description should include surface tension, viscosity and shear flow aspects in order to predict puddle length as a function of substrate velocity.

(2)-Non-equilibrium solidification.

By assuming equilibrium solidification it is impossible to describe rapid solidification processes adequately, especially if predictions regarding microsegregation and as-quenched structural morphology are being made. Quantitative data on growth rates of highly undercooled melts are difficult to obtain experimentally (46-48) so theoretical expressions for high cooling rates have been derived (49, 50). Only recently Clyne (41) has included non-equilibrium growths kinetics during rapid solidification against a chill surface. To incorporate non-equilibrium kinetics and nucleation effects in melt spinning models, accurate theoretical expressions are required and the effect of convection as imposed by the moving substrate on growth kinetics should be studied in detail.

(3)-Microsegregation and as-quenched microstructure.

For a quantitative description to predict microsegregation and as quenched microstructure, the governing equations of fluid flow and heat flow have to be extended with the equations for solute transport. Mechanisms of microsegregation-free solidification (51-53) together with the effects of convection and growth rate on interface stability (54-56) should be included. Although some attempts have been made in a semi-equilibrium way to predict microsegregation and morphology (33a, 57, 58) realistic solidification models under non-equilibrium conditions have to be incorporated to enable quantitative predictions of the variety of microstructures that occur in melt spun ribbons to be made.

4. SUMMARY AND CONCLUSIONS

Considerable progress has been made in characterisation and modelling of ribbon formation during melt spinning processes as a function of the main process parameters. The various approaches to describe the experimental observations are only successful to a limited extent and further study is required to develop a physically consistent model of the mechanisms of ribbon formation. For a rigorous mathematical model of the melt spinning process incorporating the development of the as-quenched microstructure, quantitative descriptions of solute redistribution, nucleation and growth of alloys at high quench rates are indispensable.

5. REFERENCES

1. Jones H, Rapid Solidification of Metals and Alloys, The Institution of Metallurgist, Monograph No. 8, London, UK, 1982.
2. Savage S J and Froes F H, J. Metals, 1984, 36(4), 20-33.
3. "Amorphous and Metastable Microcrystalline Rapidly Solidified Alloys: Status and Potential, Report NMAB-358, National Materials Advisery Board, Washington DC, 1980.
4. Liebermann H H in "Amorphous Metallic Alloys", ed. Luborsky F E, Butterworths, London, 1983.
5. Cliff R, Grace J R and Weber M E, "Bubbles, Drops and Particles", Academic Press, New York, 1978.
6. Anthony T R and Cline H E, J. Appl. Phys., 49, 1978, 829.
7. Liebermann H H, J. Appl. Phys., 50, 1979, 6773.
8. Liebermann H H, J. Mater. Sci., 15, 1981, 2771-2776.
9. Pavuna D, J. Non.Cryst.Solids., 37, 1980, 133-137.
10. Narasimhan M, (a) US Patent No. 4212343, 1980; (b) US Patent No. 4221257, 1980.
11. Narasimhan M, US Patent No. 414257, 1979.
12. Huber J, Mollard F and Lux B, Z. Metallk., 64, 1973, 835.
13. Narasimhan M, US Patent 4330027, 1982.
14. Zielinsky P G and Ast D G, J. Mater.Sci.Letters, 2, 1983,, 495-498.
15. Kimura H, Cunningham B and Ast D G in "Rapidly Quenched Metals IV", eds. Masumoto T and Suzuki H, The Japan Inst. of Metals, Sendai, 1982, 1385.
16. Anthony T R and Cline H E, J.Appl.Phys. 50, 1979, 245.
17. Liebermann H H in "Rapidly Quenched Metals III", ed. Cantor B, The Metals Society, London 1978, 34.
18. Kavesh S in "Metallic Glasses", eds. Gilman J J and Leamy H J, ASM, Ohio 1976, 36.
19. Liebermann H H and Graham C D, IEEE, Trans. Magn.,12, 1976, 921.
20. Liebermann H H, Mater.Sci.Eng., 43, 1980, 203.
21. Vincent J H, Davies H A and Herbertson J G in "Continuous Casting of Small Cross Sections", eds. Murty Y V and Mollard F R, TMS-AIME, Warrendale Pa. 1981, 103.
22. Vincent J H, Davies H A and Herbertson J G, as ref. 15, 77.
23. Vincent J H, Davies H A in "Solidification Technology in the Foundry and Casthouse", The Metals Society, London 1983, 153.
24. Vincent J H, Davies H A and Herbertson J G, J. Mater. Sci. Letters 2, 1983, 88.
25. Hillmann H H and Hiltzinger, as ref. 17, 22.
26. Pavuna D, J. Mater.Sci., 16, 1981, 2419.
27. Huang S C, General Electric Report No. 81 CRD 152, 1981.
28. Charter S J B, Mooney D R, Cheese R and Cantor B, J. Mater. Sci. 15,1980, 2658.

29. Fiedler H, Muhlbach H and Stephani G, J. Mater.Sci., _19_, 1984, 3229.
30. Schlichting H, "Boundary Layer Theory", Mc Graw-Hill, London 1979, ch. 2.
31. Carnahan, Luther H A and Wilkes J O, "Applied Numerical Methods", Wiley New York 1969, 407.
32. Sakiadis B C, A.I.Ch.E. Journ., _7_, 1961, 221.
33. (a) Katgerman L in "Modelling of Casting and Welding Porcesses II", eds. Dantzig J A and Berry J T, TMS-AIME, Warrendale Pa. 1984, 135; (b) Katgerman L and van den Brink P J, as ref. 15, 61; (c) Katgerman L , Scripta Met. _14_, 1980, 861; (d) Katgerman L and Zalm WE in "Numerical Methods in Laminar and Turbulent Flow III", eds. Taylor C A et al., Pineridge Press Swansea 1983, 157.
34. Shingu P H, Kobayashi K, Suzuki R and Takeshita K, as ref. 15, 57.
35. Ruhl R C, Mater.Sci. Eng., _1_, 1967, 313.
36. Shingu P H and Ozaki R, Met.Trans., _6A_, 1974, 33-37
37. Clyne T W and Garcia A, J. Mater.Sci. _16_, 1981, 1643.
38. Burden M H and Jones H, J.I.M. _98_, 1970, 249.
39. Carslaw H S and Jaeger J C, "Conduction of Heat in Solids", Oxford University Press 1959, ch. 11.
40. Jones H, J.I.M. _97_, 1969, 38.
41. Clyne T W, Met. Trans. _15B_, 1984, 369.
42. Flemings M C, "Solidification Processing", Mc Graw-Hill New York 1974, ch. 9
43. den Decker P and Drevers A in "Metallic Glasses: Science and Technology, Budapest, Hungary 1980, 181.
44. Kuiken H K, Int. J. Heat Mass Transfer _20_, 1977, 309.
45. Takeshita K and Shingu P H, Trans.Jap.Inst.Met. _24_, 1983, 529.
46. Colligan G A and Bayles B J, Acta Met. _10_, 1962, 895.
47. Perepezko J H and Rasmussen D H, Met. Trans _9A_, 1978, 1490.
48. Chu M G, Shoihara Y and Flemings M C in "Chemistry and Physics of Rapidly Solidified Materials", eds. Berkowitz B J and Scattergood R O,
49. Turnbull D and Bagley B G in "Treatise on Solid State Chemistry", eds. Hannay N B, Plenum New York, 1975, Vol. 5, 513.
50. Coriell S R and Turnbull D, Acta Met. _30_, 1982, 2135.
51. Hillert M and Sundman B, Acta Met. _25_, 1977, 11.
52. Aziz M J, J.Appl. Phys. _53_, 1982, 1158.
53. Boettinger W J and Coriell S R, Mat.Sci. Eng. _65_, 1984, 27.
54. Baker J C and Cahn J W, Acta Met. _17_, 1969, 575.
55. Coriell S R and Sekerka R F in "Rapid Solidification Processing: Principles and Technology II", Claytors Baton Rouge USA, 1980, 35.
56. ibid., J.Cryst.Growth _61_, 1983, 499.
57. Midson S P and Jones H, as ref. 15, 1539.
58. Katgerman L, Scripta Met. _17_, 1983, 537.

ABSTRACTED DISCUSSION OF THE PAPER BY L.KATGERMAN

Participants: C.M. Adam, W.J. Boettinger, B. Cantor, M.E. Glicksman,
                R.E. Maringer, P.R. Sahm and J.V. Wood

The discussion was concerned with two points:

- modelling of solidification microstructure by including a description of
  fluid flow   conditions close to the solid-liquid interface,
- nucleation behavior in a strongly undercooled or rapidly solidified melt.

Solidification morphology modelling: this topic was discussed in depth. The
modelling presented had mainly presupposed normal stationary process conditions and parameters considered were thus: growth rate v, temperature
gradient G, boundary layer thicknesses δ etc. It was contended that convection phenomena close to the solidifying interface must not be left out for
realistic modelling. It was agreed that turbulent conditions certainly
prevail in melt spinning but probably not in planar flow casting. It was
also stated that realistic modelling of the actual conditions would not be
easy. No simple analytical models appear to be possible, but rather numerical solutions would have to be obtained. In fact, some sort of multiparameter treatment would be required. However, the point was also made that
by modifying well-known physical constants by effective factors (i. e.
effective diffusion coefficients) could help more easily to tackle the
problems for a start.

The fact that Reynolds numbers determined for the microscopic flow conditions described in the paper were above 2000, suggesting that turbulent
conditions had been prevailing, was not a helpful statement. Due to experiments performed on model systems it seemed clear that critical Reynolds
numbers Re_c ($Re_c \geq$ 2000 for the onset of turbulence in macroscopic pipe-
flow) would certainly be quite different for the conditions at a growing
solid-liquid interface. Possibly also the treatment of highly undercooled
liquids should follow different types of laws because one may not simply
assume Newtonian behavior, and at least one has to consider substantially increased viscosities.

It was agreed both by discussion participants and the speaker that the
treatment presented was more applicable to melt drag or planar flow casting
than to melt spinning.

Heterogeneous nucleation behavior: the introduction of heterogeneous nuclei
into a rapidly solidifying melt was briefly touched upon. It had been
intended to show that the undercooling could be decreased significantly by
the raised nucleation level. The point was made that, with such studies,
the existence of a heterogeneous nucleation spectrum valid for the
solidification of any type of metallic alloy, could be forwarded. Naturally, also the shift of heterogeneous nucleation spectra to higher temperatures must be taken into account.

THE COMPACTION OF RAPIDLY SOLIDIFIED MATERIALS

H. FISCHMEISTER and E. ARZT
MPI für Metallforschung
Institut für Werkstoffwissenschaften
D-7000 Stuttgart, F.R. Germany

1. INTRODUCTION

Powder compaction has two equally important aspects: the elimination of pores and the creation of bonds between the particles. Important bonding mechanisms are cold welding and mechanical interlocking of particle surfaces. The first requires plastic flow in order to break the surface oxide, the second depends on irregular surface geometry. Hard, smooth particles do not usually permit the action of either of these mechanisms; after compaction, they simply fall apart again. Such powders must be contained in capsules and, after compaction, must be bonded by sintering or hot pressing. This has in fact become the normal fabrication route for superalloy and tool steel powders.

Powders of rapidly solidified metals are either obtained as spherical particles, by solidification of melt droplets, or in the form of flakes or splat, or by the mechanical comminution of ribbons (usually after embrittlement). All of these belong to the difficult-to-bond category, and they tolerate only limited heating without loss of the special properties for whose sake the rapid solidification process was used. Consequently, hopes and efforts have been directed at developing thermomechanical treatments which provide enough force for complete densification with limited heat exposure - either in the direction of "warm" (as opposed to hot) working, or in the direction of very fast ("dynamic") compaction with high energy input.

Regarding their heat tolerance, rapidly solidified materials can be divided into two categories: those whose properties depend on the amorphous or microcrystalline state, and those which merely rely on a high degree of solid supersaturation for subsequent precipitation hardening. The latter are less sensitive, especially if the heating suffered during consolidation can be utilized for the precipitation treatment. The former category, however, appears more exciting because its property potential in the compacted state is as yet virtually unexplored. Here, dynamic compaction is of particular interest because it has been found that, in favourable circumstances, the particles are bonded by a thin layer of melt which freezes rapidly enough to become amorphous or microcrystalline, without serious property degradation in the interior of the particles. It is difficult, however, to achieve this favourable combination of circumstances throughout the volume of the compact and to obtain complete densification at the same time. "Warm" working, while less powerful, is easier to control.

In the first part of this paper, the theory of the compaction of particle systems will be reviewed; in the second, techniques of dynamic compaction. The third part discusses particle bonding, and the final sections deal with the problem of processing windows for dynamic and "warm-slow" compaction.

2. A MODEL FOR THE COMPACTION OF SPHERICAL POWDERS

The compaction behaviour of powders is usually characterized by plotting the pressure required to achieve desired density levels. These curves have an unexciting shape and are easily fitted by combinations of simple functions with adjustable parameters. Such fit formulae are reviewed in references (1) to (4); we will pass them by because they do not contribute to physical understanding.

A useful descriptor for the progress of compaction is the relative density, D, normalized to the density (m/v) of the pore-free material:

$$D = \frac{(m/v) \text{ porous compact}}{(m/v) \text{ pore-free-material}} = \frac{v_m}{v_m + v_p} \qquad (1)$$

(v = volume; m = mass; indices: m = 'material', p = 'pores')

The relative density is directly related to the volume fraction of pores, or void fraction, V:

$$V = 1 - D \qquad (2)$$

The first rational model of the compaction process was based on the idea that compactability is proportional to the void fraction (5) and (6):

$$\frac{dV}{dP} = \beta V \qquad \text{with P = pressure} \qquad (3)$$

Integration leads to

$$P = \frac{1}{\beta} \ln \frac{V_0}{V} \qquad \text{with } V_0 = \text{void fraction in uncompressed} \qquad (4)$$
$$\text{powder.}$$

It is interesting to note that the same expression is obtained when the compaction process is modelled by the plastic collapse of a hollow sphere, with a volume ratio of void-to-shell equal to that of pores-to-solid in the powder compact (7).

Equation 4 describes the compaction behaviour of many powders in a limited range of pressures, relevant for the molding of sintered parts (which have about 15% porosity), but the agreement turns out to be fortuitous. The constant β is inversely proportional to the flow stress of the particle material, but the factor of proportionality derived from the hollow-sphere model is wrong by a factor of 4.5.

Later work (4),(8)-(12) has shown that the compaction of a powder mass has three stages:

(i) TRANSITIONAL RESTACKING - reduction of the total pore volume by sliding of particles past each other; this is brought to an end by the frictional locking of particle contacts as the pressure increases. It has been shown that restacking plays only a minor rôle in the compaction of spherical, deformable powders (12).

(ii) PLASTIC DEFORMATION AT PARTICLE CONTACTS (figure 1). This mechanism predominates until the porosity has been brought down to slightly below 10% (13). From then on, it is increasingly replaced by

(iii) PLASTIC COLLAPSE OF RESIDUAL PORES entirely surrounded by solid.

An exact model has been constructed for the second stage, figure 1a and b, and combined, with an approximate model for the third stage (13). An important feature is the increase of the number of particle contacts as all particles are brought closer together, cf. figure 1c. This can be described (14) by recourse to the radial distribution function (RDF) of the particle packing. The RDF of a random close packing of spheres of equal size is well known (15); within the narrow interval of interparticle distances of interest, it can be replaced by a linear term (figure 1d).

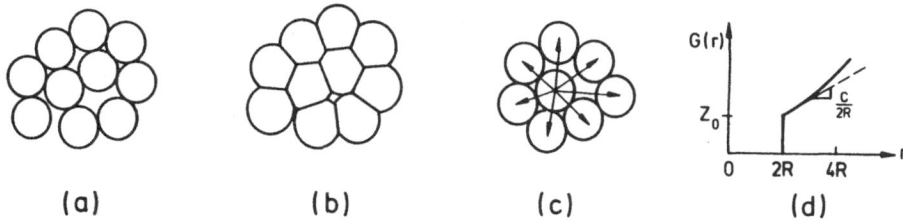

(a)   (b)   (c)   (d)

Figure 1: Compaction of spherical particles by plastic deformation at contact points (a,b). New contacts are created at a rate depending on the radial distribution function of the particle packing (c,d).

As far as the forces active in compaction are concerned, a powder compact can be viewed as a network of lines linking the particle centers. An isostatic compaction pressure will tend to shorten all these links. Each link resists shortening with a force which depends on the flow stress of the material, σ_f , and the size of the contact area, a, in a manner which can be modelled on the known force law of a spherical indenter entering a flat plate,

$$f = 3 \, \sigma_f \, a \tag{5}$$

If there are Z contacts per particle, the equilibrium between the applied pressure and the forces opposing it can be written

$$P = \frac{3}{4\pi R^2} \, \sigma_f \, D \, a \, Z \qquad \text{(R = particle radius)} \tag{6}$$

The quantities a, Z, and σ_f are all functions of the density achieved. In particular, the number of contacts per particle, Z(D), increases steadily as the densification progresses. This, and the average size of the contact flats, a(D), can both be calculated from the RDF of the powder packing (13,14). To calculate σ_f, the average strain in the contact region - which, with given RDF, can again be expressed as a function of density - is inserted into the strain hardening law of the particle material (13).

In stage (iii), an additional term is introduced to model the contribution of entirely surrounded pores to the densification resistance, using weight factors f_h and f_s for the volume fractions of "hard" regions (where the pores are closed) and "soft" regions (where densification still proceeds by flattening of non-contiguous particle contacts). These volume fractions, too, are functions of the packing geometry and the instantaneous density.

The complete compaction formula which results from this treatment,

$$P = \sigma_f(D) \left[f_s (D) \frac{3}{4\pi R^2} (aZ)_D \, D + f_h (D) \, 2\ell n \frac{(aZ)_{D=1}}{(aZ)_{D=1} - (aZ)_D} \right] \tag{7}$$

contains only material constants in the term $\sigma_f(D)$ and only geometrical quantities - all emanating from the RDF - in the brackets. This allows a unified description of the pressure - density relation for spherical powders of plastically deformable materials which is in good agreement with experimental data, without recourse to arbitrarily adjustable parameters.

The model has been derived for monosize spheres. In real powders, the particles have a size distribution, allowing smaller particles to be stacked in the interstices between the larger ones. This will increase the average number of particle contacts, but at the same time, the average size of the contact flats will decrease. Measurements by quantitative metallography on compacts made of powders with a very narrow and a normal size dispersion have shown (13) that the product $(a \cdot z)$, which is the essential variable in equations 6 and 7, is very little affected by the size dispersion. The main effect of the size dispersion is on D_0, the density of the uncompacted powder. In fact, it has been found (13), (16), that equation 7 satisfactorily describes the compaction of moderately non-spherical powders as well. Systematic deviations occur during the phase of transitional restacking, figure 2, and their extent correlates clearly with the deviation from sphericity. In the context of rapidly solidified materials, the model is certainly applicable to melt droplets solidified in free flight, and to powders obtained by mechanical comminution of embrittled ribbons.

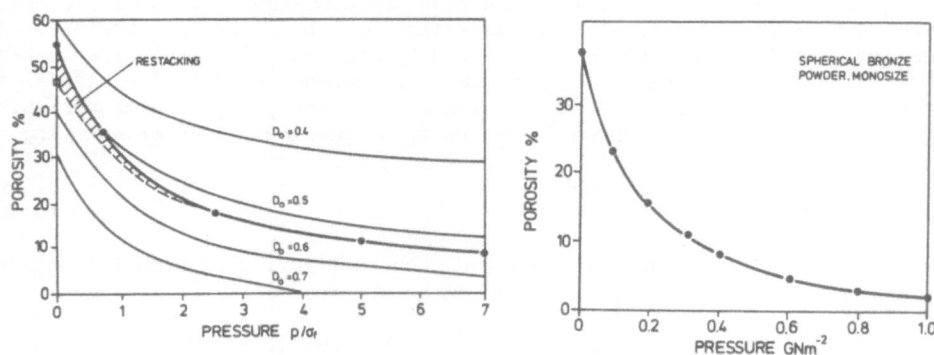

Figure 2: (a) The compaction of spherical bronze powders conforms to equation 7. (b) Non-spherical powders have different initial densities D_0. In a net corresponding to equation 7, with D_0 as a parameter, the compaction of non-spherical zinc powder is seen to conform to equation 7 except in the phase of restacking.

In equation 7, the first term in brackets reflects the fact that in stage (ii), the compaction force is being opposed by an increasing number of particle contacts, with increasing contact areas. This effect has been termed "geometrical hardening" of the powder compact (13). The second term, which comes into play when the relative density exceeds 92 %, describes the rapid increase of the system's resistance to compression as more contact flats impinge on each other, sealing off the pore they surround. While in stage (ii), the flow of material from the contact region into the void space is essentially unconstrained, strong plastic constraints come into play in stage (iii), which therefore has been termed the stage of constraint hardening of the compact. As a result, the resistance to compression mounts rapidly and, in fact, becomes infinite as the pore size and thus the void fraction goes towards zero. This is reflected by the logarithmic term in equation 7. If only the final stage is considered, equation 4 would provide an alternative and formally similar description. Both descriptions allow only an asymptotic approach to full density, which is in full agreement with practical experience.

Arzt, Ashby and Easterling (17) have considered densification by thermally assisted mechanisms, i.e. power law creep and diffusion, with regard to hot isostatic pressing. In the present context, power law creep is of greater interest. Based on a solution for the creep rate of a thick spherical shell surrounding a central void, the following densification rate is obtained for the final stage:

$$\dot{D} = \frac{3}{2} \dot{\varepsilon}_0 \frac{D(1-D)}{[1-(1-D)^{1/n}]^n} \left[\frac{3}{2n\sigma_0}\left(P + \frac{2\gamma}{r}\right)\right]^n \qquad (8)$$

with $\dot{\varepsilon}_0$, σ_0, n = power law creep parameters; γ = surface energy; r = pore radius.

For property-relevant porosities, where the pores are still large enough to allow the surface energy term to be neglected, this becomes

$$\dot{D} = \text{const} \cdot V \cdot P^n. \qquad (9)$$

indicating that the pressure required to uphold a desired densification rate rises with the nth power of inverse void fraction (n=4 is a common value). Complete elimination of pores, therefore, is only possible by sintering with surface energy as the driving force.

3. DYNAMIC COMPACTION

The early development of dynamic powder compaction methods has been reviewed by Clyens and Johnson (19). Important early variants were the Dynapak, a pneumatic high-speed press; the petroforge, a piston machine actuated by ignition of a fuel-air-mixture; various devices for magnetic compaction with strong current pulses, and explosive-driven fluid-dynamic and ballistic presses. To these might be added conventional equipment for moderate-rate metal forming, like forging hammers and extrusion presses. A new regime of much higher compaction rates was opened up by direct explosive compaction, with explosive in direct contact with the powder (e.g., in the form of concentric tubes). This technique is widely practiced (18)-(21), in particular for brittle materials like ceramics (22),(23). In this case, its success is at least partly due to the generation of a high dislocation density in the powder, which makes the material more amenable to densification by sintering. However, the method has also been successfully applied to amorphous metals (24), where subsequent sintering plays no role. The latest development is a ballistic gas gun which projects a flier plate at high velocity against the powder, achieving compaction rates similar to those of direct explosive compaction, while allowing closer control of the energy transfer to the powder (25) to (27).

Table I compares characteristic compaction rates.

Typical platen speeds in the Dynapak and Petroforge units were of the order of 20 m/s; with the Raybould gun, flyer plate velocities up to 2500 m/s have been achieved (29); effective pressures in the powder mass are typically in the range from 2 to 5 GPa, but experiments up to 18 GPa

Table I: Compaction Rates in Various Processes

```
-------------------------------------------------------------------
Extrusion press                                          50/s
Forging hammer                                          100/s
Dynapak, petroforge                                   10-100/s
Explosively pressurized fluid-dynamic press           10-100/s
Magnetic compaction                                   20-100/s
Explosive ballistic compaction                        200-500/s
Direct explosive compaction                        1000-10.000/s
Two-stage air gun                                  1000-10.000/s
-------------------------------------------------------------------
```

have been reported (26). For comparison, typical compaction pressures in powder metallurgy production are of the order of 0.5 GPa. - The range of typical pressures and shock velocities obtained with these guns is similar to those which occur in direct explosive compaction of powders (20)-(22). The shock waves induced in the powder by such high-velocity impact produce effects which cannot occur at the modest speeds of the earlier "dynamic" devices, and their utilization has opened a new era of dynamic powder compaction.

At high rates of deformation, the flow stress of metallic materials increases drastically (30). Strain rates of $10^7 - 10^8$ s^{-1} have been claimed for dynamic compaction experiments with a stainless steel powder (31). Extrapolating from what is known from static compaction experiments (12), this should lend much greater importance to the stage of restacking by particle sliding. Even in the regime of conventional static pressures, faster compaction rates have been found to improve interparticle bonding, as reflected by electrical conductivity and strength of the unsintered compacts (19); also it has been observed that interparticle friction in powders increases significantly under dynamic loading (19). These effects all point to more efficient break-up of surface contamination as the particles slide quickly past each other.

At the very high speeds now coming into use, strong heating at the particle surfaces is observed, while the particle interiors stay relatively cool. Figure 3, reproduced from the work of Morris (32) on a 0.5 % carbon steel powder, demonstrates this convincingly by the etch-resistant martensite zones formed at the periphery of the particles as the material cools down from the temperature attained during shock compaction. The particle interior acts as a heat sink, removing the heat fast enough to quench the periphery from the austenitic to the martensitic state. As the shock pressure is raised beyond 4 GPa, regions of rapidly-solidified melt appear at the particle junctions, indicating a still higher degree of surface heating. Similar melt zones are claimed to occur in explosive welding (33). For steel powders compacted with a shock velocity of 1 km/s (shock pressure 6 GPa), Morris (34) has calculated a surface power density of 10^{11} W/cm^2, more than enough to produce melting at the particle surfaces. Interparticle melting has been observed with powders of widely different materials (35) such as lead (25), tool steels (28). Ni-base superalloys (24), (32), (34), stainless steel (31), aluminium alloys (34) and metallic glasses (35), (36). The structure of the interparticle melt region is either a microcellular solidification structure, or it may be amorphous. A cooling rate of 10^{10} K/s has been calculated from the surface heat input, assuming ideal heat transfer to the particle interior, and rates in the range $10^5 - 10^{10}$ K/s have been inferred from the cell dimensions (28), (34), (36).

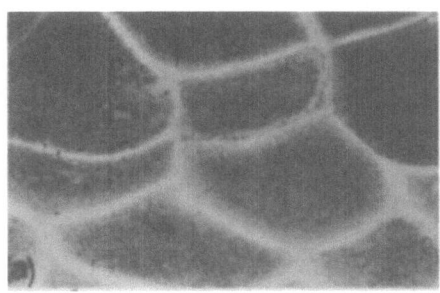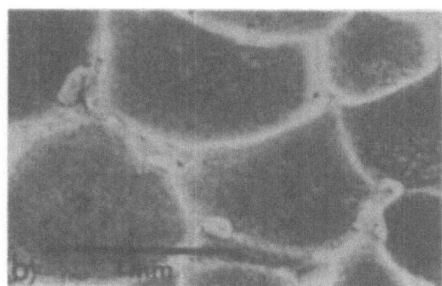

Figure 3: White-etching martensite outlines zones of adiabatic heating in tool steel powder (a) 3 GPa, (b) 5.5 GPa, with melt pool between particles 1 and 2. Micrographs from Morris (32).

Several reasons may account for the concentration of adiabatic heating at the particle surfaces: sliding friction in the first stage of the compaction process; the plastic work done in deforming the contact zones in the second stage; finally, it has been pointed out (24) that when a particle is traversed by the shock front, immediate lateral release occurs at the particle surfaces, transforming the shock wave into a plastic wave at the particle surfaces and slowing it down. This effect is proportional to the square root of the work hardening rate, which may account for material-dependent differences in consolidation behaviour. Judging from the deformation mechanism maps for high strain rates developed by Sargent and Ashby (37), deformation of the particle surfaces in dynamic compaction must occur by adiabatic shear, which would help to explain the localization of heating to the contact zones.

In the light of the densification models discussed above, a strictly local heating of the particle surfaces is the most economical way of using heat to assist the compaction of metastable materials, because it produces softening only where it is required, without superfluous heat input that might jeopardize the metastable structure in the interior of the particles. The effect of adiabatic surface heating on densification is demonstrated very clearly in figure 4, from the work of Raybould (31). Later measurements with less scatter (35) indicate that the elimination of the last fraction of porosity may be asymptotic also in dynamic compaction, but at least a practically satisfactory density of, say, 99 %, is achieved more easily than in the static mode.

144

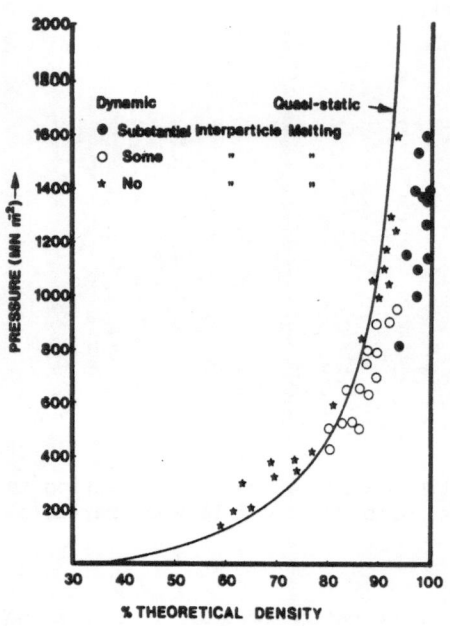

Figure 4: Pressure-density relationship for stainless steel powders,
for static and dynamic compaction.
Adiabatic heating of particle surfaces facilitates the approach to
full density (31).

4. BONDING

In ordinary powder metallurgy, bonding is initiated by cold welding at
deformed particle contacts and completed by diffusional mass transport
during sintering. If metastable materials are to be consolidated, ordinary
sintering treatments are excluded because less atomic transport is necessary for crystallization than for bonding. Thus, bonding has to be achieved,
and perfected, at the moment of consolidation.

Obstacles to bonding are oxide films and other kinds of non-metallic
contamination at the particle surfaces. Recent studies of early stages of
wear (38) have shown that metallic bonds are formed immediately when two
sputter-cleaned metal surfaces are put in touch with each other in an ultra
high vacuum; similar observation have been reported for such dissimilar
material pairs as copper on alumina (39). The classical work of Bowden and
Tabor (40) has made it clear that metallic adhesion at oxidised surfaces is
favoured when a brittle oxide is present on an easily deformable substrate,
and it has been amply demonstrated that sliding or shear deformation will
disrupt oxide films and produce cold welding where mere unidirectional
pressure will not. Frictional heating at rubbing asperities facilitates the
deformation of the substrate, and may in fact cause surface melting. Since
oxides mostly remain brittle up to the melting temperatur of their substrate metal, the thermal softening of the substrate will greatly facilitate the disruption of the film. Once metallic bonding has been achieved at
points, it is assumed that the oxide is stripped and rolled up or otherwise
transformed into a shape in which it does not severely weaken the bond. The
fact that friction welding is fully established as an industrial production
process demonstrates the reliability with which bonding obstacles can be
overcome by surface shear.

In the field of superalloys, unsatisfactory strength of the interparticle bonds has long been a problem - and this in spite of the fact that ample
time and heat for diffusional bonding is available here. In this case the

problem lies in the segregation of impurity elements and in the precipitation of carbides at the particle boundaries during the thermal cycle connected with the consolidation process. For a long time, therefore, there was a tendency to avoid using HIP ped material without additional hot working which was meant to improve bonding by shear deformation. In reality, the shear deformation only produces a more pleasing microstructure, by destroying the particle outlines; to some degree, it may also facilitate the collapse of residual pores. While opinions in the industry are still divided, many technologists now agree that fully satisfactory properties can be obtained by HIP treatment without subsequent forging, provided the particle surfaces are kept clean of external oxide, segregated impurities from the interior, and carbides precipitated during the HIP treatment. This is possible by good hygiene during atomization and handling, cleanliness of the melt, and proper temperature-time-pressure profiles for the HIP operation. In the context of dynamic compaction, we can first of all discount diffusional reactions (segregation, precipitation) at the particle surfaces as obstacles to bonding. The destruction of oxide films by sliding will depend on their thickness and mechanical stability, and will be assisted, as discussed above, by heating at the particle contacts.

Of the many reports on dynamic compaction of powders, only a few have assessed the degree of bonding by mechanical measurements. Ultimate tensile strength and elongation at fracture are sometimes stated (29), (31), (41), but they are sensitive to residual porosity, which makes them poor indicators of bonding. Bonding per se is much better reflected by fracture toughness, especially when combined with SEM fractography to assess the frequency of interparticle fracture. Fracture toughness data have been reported by Morris (35), (42), (43) for a number of materials, including two metallic glasses. The most complete set of data is available for a soft annealed high speed steel powder (figure 5). For this material (which is not amorphous), both density and fracture toughness are seen to approach satisfactory levels only at shock pressures where sizeable amounts of interparticle melt are formed. For his amorphous materials, Morris only states a "pressure for good bonding", which is at least equal to but often in excess (by 20-200 %) of that required to reach 99 % density. Most observers have associated "good bonding" with interparticle melting. The association is sometimes backed up by a purported analogy with explosive welding of plates, where liquid jets are supposed to clean the mating surfaces. It should be pointed out, however, that causality has not been proven: the simultaneous occurrence of melt and bonding might in truth be independent effects of the particle strain rates induced. The authors of a paper on dynamic compaction of a low alloy, 0.1 % carbon steel powder (29) state emphatically that interparticle melting is required for good bonding (44), yet the microstructures reproduced show no evidence of melting, only a general homogenization at higher shock presssures. Very recently, Morris (41) has published results of warm compaction tests with a devitrified Ni-Si-B glass in which slow uniaxial hot pressing was combined with some shear deformation, and of forging tests involving a certain shear component. The densities and strengths obtained were superior to those of dynamically compacted specimens. Bonding is attributed to sufficient interparticle shear for the disruption of oxide films. The shear-pressing experiments were carried out under a protective atmosphere; no data are given for the forging tests.

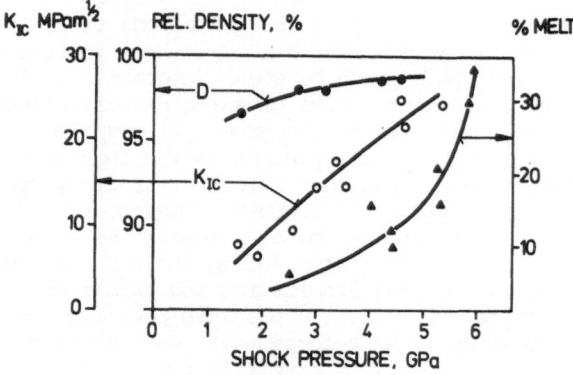

Figure 5: Density, fracture toughness and volume fraction of melt in dynamically-compacted tool steel powder
of type M2, vs. shock pressure *)(data from Morris, ref.
35).

It is not a priori certain that a melt must produce an optimum bond
between two solid surfaces: bonding defects are a frequent, sad experience
both in soldered joints and in fusion welds; solid-solid sliding, on the
other hand, demonstrates its power for bonding painfully in many a wear
couple. Morris (35) states that the molten material which is squeezed into
the pores "often does not bond well". At present, there is simply too
little information to identify the actual bonding mechanism acting in
dynamic compaction.

Surprisingly, no attention seems to have been given to all-inert-gas-
processing of glassy metals; such processing routes are used for, e. g.,
superalloys to ensure particle cleanliness of a degree which allows perfect
bonding. At least it would be worth while to find out whether the surface
oxide on amorphous metal particles can be so reduced as to facilitate
bonding at the more easily controlled strain rates of warm forming.

Another possibiltiy of consolidating amorphous metal particles would be
sintering with a low-temperature liquid phase which perhaps could be absorbed in the amorphous material after it had promoted densification and
bonding. Such transient liquid phase sintering is used, e. g., in the
consolidation of dental amalgams at ambient temperatures, and can be combined with the forging of powders (45).

---

*) The data for volume fraction of melt in figure 5 may not be truly
representative: with comparable energy input (= equalized temperature of
the compact after the shot), all other materials studied by Morris uniformly show about 10% less melt than the tool steel. With this material, it is
not trivial to distinguish rapidly-solidified melt from quenched solid
austenite; alternatively, special behaviour might be caused by the material's pronounced work hardening.

5. PROCESSING WINDOWS FOR DYNAMIC COMPACTION

A window for the dynamic compaction of amorphous metals will be limited by the following considerations (35,44):

(i)   sufficient shock pressure to produce high density (>99%);

(ii)   sufficient shock pressure to produce reliable bonding;

(iii)   the interparticle melt must have time to solidify and become strong before it is reached by the descending flank of the shock pulse (which acts as a rarefaction wave);

(iv)   reflection of the shock wave into the compact must be avoided;

(v)   after temperature equalization on the scale of a particle diameter, the temperature must be low enough to avoid crystallization or other kinds of property degradation of the amorphous material.

Each of these points has its problems.

"Full density" is reported in many papers on dynamic compaction, but the reliable achievement of close-to-zero porosity throughout each specimen in a series of equal shots still seems to be problematic as can be seen from figure 4; elsewhere, density variations of ca. 2 % were stated for "fully dense" dynamically compacted material (29).   It must be remembered that porosity of the order of 1 % generally has a strongly adverse effect on ductility (fracture strain), impact and fatigue strength of consolidated powder bodies.

The topic of bonding will not be resumed here.

The duration of the shock pulse can be controlled via the thickness of the flier plate; reflection of the shock wave can be prevented by momentum traps, e.g. a loosely attached spall plate or an absorber piston at the base of the target assembly (29),(46).   In direct explosive compaction with the commonly preferred, concentric-cylindrical assembly, shock wave inter- action at the center is less easily controlled and can cause blow-out of the core.

Schwarz   et   al. (44) have presented the following estimates for the compaction of a low alloy steel powder:

Table II: Typical time parameters in dynamic compaction.

Duration of shocked state	2 μs
Densification and interparticle melting	40 ns
Solidification of interparticle melt	200 ns
Temperature equalization between particle surface and interior	10 μs
Relaxation to ambient temperature	5 ms

(Times greater by two orders of magnitude have been stated by Morris (28) for (larger?) specimens of dynamically compacted tool steel).

Equating the bonding criterion (ii) with the production of 10 v/o inter- particle melt, and combining it with the shock duration criterion (iii), Schwarz et al. (44) have constructed the consolidation map shown in figure 6, where the normalized coordinates are energy input: $\qquad$ $\varepsilon = (Pv_p/2)/(\bar{c}_p(T_m-T_o))$ $\qquad$ (10)

and shock duration $\qquad$ $\tau = 16\ D_m t/R^2$ $\qquad$ (11)

(P = shock pressure, v_p = volume of pores in unit mass of powder, $\bar{c}_p$ = average specific heat, T_m = melting temperature, T_o = initial temperature, D_m = thermal diffusivity of particle material, R = particle radius).

It will be noted that the latent heat of melting is neglected in the denominator of equation 10 which should be the total specific heat content of the particle material at the melting temperature. No attention is paid to property degradation during cooling after the shot.

Morris (35) has studied the criterion for property degradation (v). In small specimens (25 mm diameter x 10 mm high) of compacted Metglas, crystallization was observed at prior particle boundaries when the specimen temperature after the shot exceeded 450°C. His graphs show that this temperature was reached when as little as 1-2 % of interparticle melt had formed during compaction, and he concludes that "the window for compaction of highly metastable materials is a very narrow one". In figure 6, the upper energy limit (ε=1) is for total melting of the compact. For degradable materials, the limit would be much lower, and it would move further to lower energies with increasing compact size because of slower heat extraction after the shot. Morris' observation of degradation at 1-2 % melt would make the window in figure 6 negative. It is doubtful whether a positive window exists for compacts beyond the button-size which is characteristic of the present state of the art.

Experiments in dynamic compaction mostly seem to have been limited to densification studies, and there is hardly any detailed information on the state of the compacted specimen. Only Roman et al. (36) have reported radial distribution functions which show some sharpening of the peaks at flier plate velocities above 900 m/s. This is interpreted as indication of "some ordering of the structure" under the influence of the thermal effects of the shock.

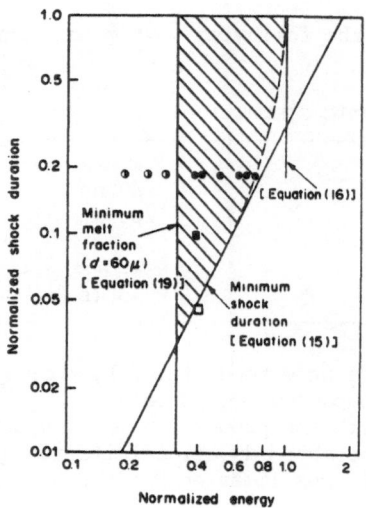

Figure 6: Dynamic consolidation map, after Schwarz et al., (44). Data points for low alloy steel powder - solid circles: compacts with good strength; half filled circles: poor strength; open square: poor strength due to microcracks. The asymptotic limit at ε = 1 corresponds to total melting of the powder. Successful consolidation of this powder is expected within the shaded area.

6. PROCESSING WINDOWS FOR SLOW WARM CONSOLIDATION

By most accounts, the reliable attainment of homogeneous high density in dynamic pressing is still problematic, and the process is limited to very simple shapes. Slower methods like extrusion or uniaxial or isostatic pressing would appear to be more easily controllable and more versatile. Is it possible to find a processing window where full densification can be achieved without devitrification? Published experiments are few, but they offer some encouragement.

The "warm" range is limited, on the one hand, by the transition at T_p from shear band to homogeneous deformation, where the materials begin to show some ductility and on the other hand, by the glass transition temperature T_g. For the popular alloy type $Fe_{40}-Ni_{40}-B_{20}$, T_p = 470 K, T_g = 715 K, but for many other amorphous alloys, T_p is of the order of 550 K (47). Thus "warm" will often mean ca. 280-450°C.

Liebermann (47) obtained fully dense specimens with fair interparticle adhesion by warm extrusion of $Fe_{81.5}-B_{14.5}-Si_4$ at 640 K; uniaxial pressing at 660 K of a similar alloy resulted in 96 pct density at a pressure of 2GPa, but poor adhesion. Full densification is also reported for $Cu_{60}-Zr_{40}$ after extrusion close to the glass temperature (48). Uniaxial compaction of $Ni_{80}-P_{20}$ at constant pressure and heating rate has been studied by Bruson and Maloufi (49) who achieved densities as high as 90 % at only 200°C and 0.8 GPa; the experiments were not extended to higher temperatures, although a tendency for improvement is clearly suggested. The same tendency was observed in warm pressing experiments with $Cu_{60}-Zr_{40}$ (50). A review paper by Miller (51) mentions several examples of successful warm consolidation of amorphous powders, especially at temperatures close to the glass transition (48),(52),(53).

Some property degradation probably occurred in most or all of these experiments, but little information is given. In addition to incomplete densification and bonding, in any warm-slow process one would have to beware of embrittlement and partial crystallization. Magnetic properties of warm consolidated materials generally indicate structural relaxation (51), but in one instance, high temperature forming has been reported to produce no significant loss in magnetic properties (54).

Data from Morris (41) for $Ni_{70}-B_{20}-Si_{10}$ compacted by a variety of techniques at temperatures far above the truly "warm" range are reproduced in figure 7. The data are shown here to illustrate the difficulty of full densification even at temperatures where the material must be expected to fully devitrify during the compaction operation. Dynamic compaction of preheated material gave high density but poor strength, because of insufficient adhesion or particle cracking. Uniaxial pressing with some lateral flow to cause interparticle shear produced good densification and fair adhesion at moderate pressures, but only at temperatures far in excess of typical T_g values (47) for this material. Similar temperatures were necessary to achieve satifactory density by slow forging (isothermal upsetting) of preforms warm pressed at 773 K.

No systematic attempt seems to have been made to indentify optimum "warm" consolidation conditions. Such an approach could be based on the hot consolidation maps developed by Arzt et al. (17). Such maps are shown for Metglas 2826 in figure 8 and 9 although it must be emphasized that available input data are too scanty to allow more than a demonstration of the principle.

The temperature dependence of flow stress of this alloy has been measured by Krenitsky and Ast (55), creep data have been determined by Gibeling and Nix (56), and incubation times for crystallization during creep deformation by Anderson and Lord (57). Unfortunately, the data pertain to

150

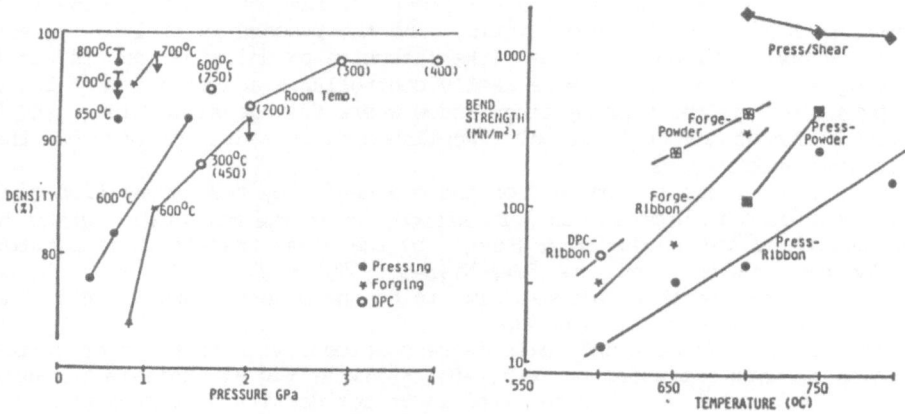

Figure 7: Density and strength achieved by various methods of hot compaction of originally amorphous $Ni_{70}-Si_{10}-B_{20}$ (41). Symbols not explained in figure inserts: DPC = dynamic compaction; bars above ● = higher density obtained by pressing with shear flow; arrows below ● = lower densities obtained when pressing powder instead of ribbons.

different temperatures, stresses and strains, necessitating considerable extrapolation in constructing a map. In particular, the creep and yield data were determined below T_g (which for this alloy is at 643 K (58)), while the crystallization data were measured above T_g. The creep measurements at 523 - 573 K can be described in terms of a power law; their extrapolation beyond T_g is at best a lower bound for the rate of time-dependent deformation in the upper region of the map shown in figure 9. Despite these reservations, the maps seem to reflect reality to a fair degree. Pressing ribbons of $Fe_{81.5}-B_{14.5}-Si_4$ at 600 K with .69 GPa for 600 s, Liebermann (47) reports a density of 75 % while the maps would predict 80 % for a spherical powder of $Fe_{40}-Ni_{40}-P_{14}-B_6$; similar agreement is found for ribbons of $Fe_{40}-Ni_{40}-B_{20}$.

At the right in figure 8, the map shows a field where the stresses are high enough to enforce densification by instantaneous yield. At lower stresses, to the left, only time-dependent deformation can effect densification. Density-pressure curves are shown for various times. At the temperature for which the diagram is constructed, the incubation time for crystallization is 3000 s. Times beyond that, i.e. the left part of the field, must be avoided.

At higher temperatures (figure 9), less pressure is required to produce full density by creep deformation, but crystallization will occur sooner. The disadvantage of shorter incubation times far outweighs the gain due to faster creep. Thus for essentially slow consolidation processes, like warm pressing or HIP, one should chose the lowest temperature compatible with a reasonable cycle time. This may appear surprising, but it is inevitable in the temperature range below the nose of the time-temperature-transformation plot. Empirical data are not available for higher temperatures, but calculations by Bergmann and Fritsch (59) indicate that the nose may be situated around 1000 K.

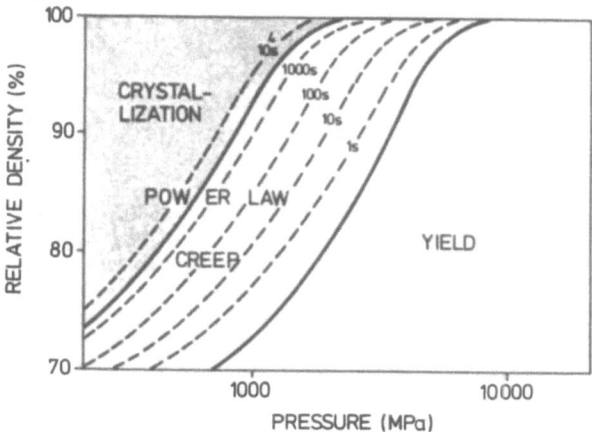

Figure 8: (Semi-hypothetical) map showing regions of instantaneous yield and "power law creep" for the compaction of Metglas 2826 (Fe_{40}-Ni_{40}-P_{14}-B_6) at 633 K. Also shown is the region where crystallization starts before full density has been reached.

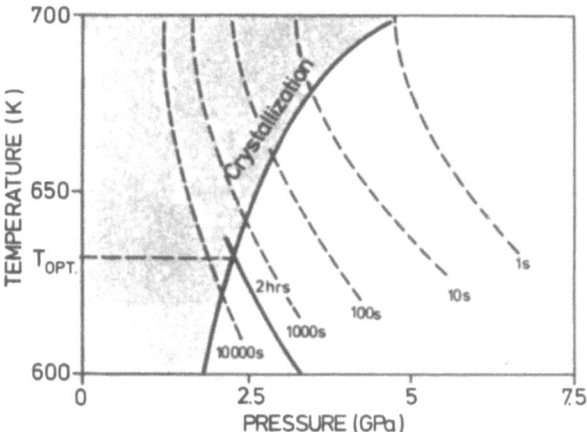

Figure 9: Minimum pressure to reach full density within stated times in warm compaction of Metglas (cf. figure 8), and region where crystallization intervenes before full densification. For a reasonable practical warm pressing time of 2 hrs, the optimum temperature (requiring minimum pressure) is 630 K.

Taken at face value, the maps in figures 8 and 9 indicate that most of the warm pressing experiments made so far used insufficient pressures and times. Pressures of the order of 2.5 GPa, the minimum indicated by figure 9, are beyond the reach of ordinary powder metallurgy dies, but they could easily be achieved in very simple belt-type apparatus. Relaxing the density requirement from 100 to 95 pct would reduce the pressure by about one-half, making it just manageable for conventional powder pressing dies (though still far beyond the present limit of HIP technology at about 0.2 GPa).

Let us now consider extrusion in the light of the maps. Since densification is fast, it has little to gain from thermally assisted deformation, and the required pressures must always be expected to be essentially higher than for slow pressing at equal temperature. Full densification has been reported at extrusion pressures of only 0.78 - 0.98 GPa at nominal temperatures of 640 - 660 K (47). In the light of figure 9, this would indicate substantially accelerated densification, which might have been due to adiabatic heating or to a beneficial effect of shear deformation. However, experience from powder forging (60) suggests that shear is beneficial mostly for bonding, not for densification. While shear may accelerate densification at low levels, it does not usually reduce the pressure requirement for reaching full density.

7. CONCLUDING REMARKS

Dynamic compaction with high velocity impact (gas gun or explosive) seems a promising but still somewhat problematic possibility for producing compact pieces of metastable materials. Open questions are:

- accurate control of energy input;
- reliable attainment of negligible porosity and good bonding, which both are mandatory for good mechanical properties;
- the narrow limits for heat input without property degradation;
- the feasibility of compacting pieces of large size and even moderately complex shapes.

Warm pressing is expected to require pressures which are clearly above the range of conventional die tooling, but manageable with simple high-pressure tooling. Bonding would be a problem in simple pressing, but probably not in isothermal forging which involves sufficient shear flow. The bonding problem could be greatly alleviated by fully inert processing and/or particle surface cleaning, as in present-day superalloy technology.

The warm consolidation maps presented in this paper suggest that all experiments made so far were considerably outside the optimum processing window.

Warm extrusion has already been proven feasible. The discussion in this paper suggests that experiments made so far may have been outside the optimum processing window, so that improvements could be expected.

To establish processing windows for warm forming operations, experimental work should be directed at

- determination of complete crystallization-time-temperature diagrams;
- the possible influence of simultaneous deformation on crystallization kinetics;
- "yield" data at high temperatures;
- "creep" data at high temperatures.

Such work should be worthwhile because the evidence suggests that the optimal processing parameters for warm-forming compact pieces of amorphous metallic alloys should be technically attainable, though requiring modifications of present-day forming equipment and powder handling methods.

8. REFERENCES

1. Bockstiegel G, Hewing J, Arch. Eisenhüttenw.,36, 1965, 751.
2. James PJ, Powder Metall. Internat., 4, 1972, 82, 145 and 193.
3. Kawakita K, Ludde KH, Powder Technol., 4, 1970/71, 61.
4. Arzt E, Fischmeister H, Mem. Sci. Rev. Metall.,76, 1979, 573.
5. Shapiro I, Kolthoff M, J. Phys. Colloid Chem., 51, 1947, 483.
6. Konopicky K, Radex-Rundsch.,3, 1948, 141.
7. Torre C, Berg- u. Hüttenm. Mh., 93, 1948, 62.
8. Heckel RW, Trans. AIME, 221, 1961, 1001.
9. Heliwell N, James PJ, Powder Metall. Internat., 7, 1973, 25.
10. James PJ, Powder Metall. Internat., 20, 1977, 21 and 199.
11. Sundström BO, Fischmeister H, Powder Metall. Internat.5, 1973, 171.
12. Fischmeister H, Arzt E, Olsson LR, Powder Metall., 21, 1978, 179.
13. Fischmeister H, Arzt E, Powder Metall., 26, 1983, 82.
14. Arzt E, Acta metall., 30, 1982, 1883.
15. Scott GD, Nature, 194, 1962, 956.
16. Fischmeister H, Arzt E, 7th Intern. Powder Metall. Conf., Dresden 1981, Vol. 1, 105.
17. Arzt E, Ashby MF, Easterling KE, Metall. Trans., 14A, 1983, 211.
18. Balzerowiak HP, Bock-Nussbaum F, Prümmer R, High Temp.-High Press., 3, 1971, 517.
19. Clyens S, Johnson W, Mater. Sci. Eng., 30, 1977, 121.
20. Lennon CRA, Bhalla AK, Williams JD, Powder Metall. 21, 1978, 29.
21. Gourdin WH, Mater. Sci. Eng., 67, 1984, 179.
22. Davis RF and Palmour H III, Journal of Materials Education (JEMMSE),5, 1, 1983, 151.
23. Davis RF, Horie Y, Scattergood RO, Palmour H III: Advances in Ceramics, Kingery WD, ed., Amer. Ceram. Soc., Columbus, 1984, 157.
24. Meyers MA, Gupta BB, Murr LE, J. Metals, 33, 1981, 21.
25. Raybould D, Morris DG, Cooper GA, J. Mater. Sci., 14, 1979, 2523.
26. Deutsches Patent No. 27 376 74.9.
27. Raybould D. Proc. 15th Intern. Machine Tool Res. and Design Conf., Tobias SA, Koenigsberger F. (Eds) MacMillan, London 1975, 627.
28. Morris DG, Met. Sci., 15, 1981, 116.
29. Kasiraj P, Vreeland T Jr, Schwarz RB, Ahrens TJ, Acta metall.,32,1984, 1235.
30. Campbell JD, Mater. Sci. Eng., 12, 1973, 3.
31. Raybould D, J. Mater. Sci., 16, 1981, 589.
32. Morris DG, Mater. Sci. Eng., 57, 1983, 187.
33. Crossland B, Williams JD, Metall. Rev., 15, 1970, 79.
34. Morris DG, Metal Science, 16, 1982, 457.
35. Morris DG, The Dynamic Compaction of Metal Powders, in "Metastable Crystalline Materials", MRS Boston 1983.
36. Roman OV, Bogdanov AP, Voloshin YuN, Gorobtsov VG, Pikus IM, Termich. Obrab. Metall., 10, 1983, 57.
37. Sargent PM, Ashby MF, Report CUED/C/MATS/TR. 98, March 1983, Cambridge University.
38. Buckley DH, Proc. 2nd Int. Conf. Sci. Hard Mater., Rhodes 1984, The Institute of Physics, Bristol, in press.
39. Pepper SV, J. Appl. Phys., 47, 1976, 801.

40. Bowden FP, Tabor D, Friction and Lubrication of Solids, Oxford University Press, London, 1950.
41. Morris DG: Compaction of Amorphous Ribbons and Powders. RQ5, 1984, Wurzburg.
42. Morris DG, Metal Sci., 14, 1980, 215.
43. Morris DG, J. Material Sci., 17, 1982, 1789.
44. Schwarz RB, Kasiraj P, Vreeland T Jr, Ahrens TJ, Acta metall., 32, 1984, 1243.
45. Fischmeister HF, Larsson LE, Powder Metall., 17, 1974, 227.
46. Morris DG, Metal Powder Report, 1983, p. 405.
47. Liebermann HH, Mater. Sci. Eng., 46, 1980, 241.
48. Pourahimi S, Thesis MS, Northeastern Univ., Boston, 1980, quoted in ref. 51.
49. Bruson A, Maloufi N, Mater. Sci. Eng., 64, 1984, L13.
50. Miller SA, Murphy RJ, in "Proc. 2nd Conf. Rapid Solid. Process.", Mehrabian R, Kear BH, Cohen M (Eds), Claitors, Baton Rouge, 1980,385.
51. Miller SA in "Amorphous Metallic Alloys", Luborsky FE (Ed), Butterworths, London, 1983, 506.
52. Stempin JL, Wexell DR, US Pat. 4 298 382, 1981, quoted in ref. 51.
53. Smith JS, Perepezko JH, Rasmussen DH, Loper Cr Jr, US Pat. 4 282 034, 1981.
54. Gibbs MRJ, Evetts JE, Shah NJ, J. Appl. Phys., 50, 1979, 7642.
55. Krenitsky DJ, Ast DG, J. Mater. Sci., 14, 1979, 275.
56. Gibeling JC, Nix WD, Scripta Met., 12, 1978, 919.
57. Anderson PM III, Lord AE, Jr., Mater. Sci. Eng., 44, 1980, 279.
58. Ast DG, Krenitsky DJ, Mater. Sci. Eng., 23, 1976, 241.
59. Bergmann HW, Fritsch HU, Metal Sci., 16, 1982, 197.
60. Fischmeister HF, Ann. Rev. Mater. Sci., 5, 1975, 151.

ABSTRACTED DISCUSSION OF THE PAPER BY H. FISCHMEISTER:

Participants: C.M. Adam, J. Ågren, L. Arnberg, H.W. Bergmann, R.W. Cahn, M.E. Glicksman, N.J. Grant, H. Jones, J.H.Perepezko and T. Sheppard

The discussion centered on mainly two topics:
- compactability of amorphous powders,
- phenomena connected with adiabatic heating.

Compactability of amorphous powder: naturally, compactability of amorphous powders is a very important question concerning applicability of rapid solidification techniques to the synthesis of new alloys with new properties and of manufactured machine parts. In this area relatively little knowledge has so far been amassed. Nevertheless, it has been conclusively shown that amorphous materials with relatively high glass temperatures can be compacted into parts of modest volume as long as the compacting temperature is not raised above the glass temperature. One of the most promising technologies to be utilized presently in this respect appears to be explosive forming.

Besides amorphous powders, crystalline alloy powders with metastable phases are also of interest. The problems here are quite similar. Therefore information was sought on the possibility of generating a transient liquid phase during compaction between the powder particles (i.e. in the grain boundaries), or else, the introduction of very locally heated regimes (again at powder particles boundaries) were discussed for this purpose. This led naturally to the question of metastability in the original microstructure.

Adiabatic heating: adiabatic heating, a phenomenon to be considered from two aspects, namely, ... how to heat up in local spots ...and ...how to remove the locally generated heat into the generally surrounding matrix ... were the questions of concern. It appears that adiabatic heating can be certainly achieved by utilizing a rapid compaction method, for example, by utilizing explosive forming, very high gradient extrusion including the possibility of quenching the extruded strand upon exiting the die, with other possibilities. The existence of glassy phases in grain boundary or powder particle boundary regions appears to be without question. This was shown by pair correlation function measurements of these microstructural regions.

In connection with adiabatic heating the point was made that employing pressure gradients would not only allow local rapid heating, but also, for example, would shift the nose in TTT-diagrams to such an extent that new possibilities for materials syntheses may open up. An interesting thought mentioned was also to utilize a sudden pressure release to create relatively a very highly undercooled melt condition.

PRODUCTION AND CHARACTERIZATION OF RAPIDLY-SOLIDIFIED PARTICULATES

H. JONES
University of Sheffield
Dept. of Metallurgy
Sheffield S1 3JD, U.K.

ABSTRACT
   The contribution aims to give a perspective of the ever-expanding range
and   variety of methods of manufacturing rapidly-solidified (RS) particulates and the fundamental characteristics of their products. Particular
attention will be given to more recent developments and to possible criteria for assessing their efficacy. These might include rating of the following capabilities:

(i)   to generate a required particle size or size distribution and
       particle morphology.

(ii)   to achieve a required structural state within the particulate.

(iii) to successfully process materials of different characteristics
       such as fusibility and reactivity.

(iv)   to accommodate minimum levels of production volume and rate.

Criteria (i) and (ii) are closely linked and rather basic and will receive
particular attention in the contribution. It is concluded that understanding of the mechanisms by which the various methods form and solidify the
particulate is still largely in its infancy and that relationships between
operating variables and RS structure remain to be determined for many of
these production processes. In view of uncertainties concerning applicable
cooling and solidification conditions for particular processes, there is
some merit in ranking them in terms of their ability to produce a given
microstructure or constitution at a particular particle size. Evaluation of
properties as a function of RS microstructure or constitution and of variables in any subsequent consolidation remain fruitful areas of further
study.

1. INTRODUCTION
   Rapid formation of solid from parent liquid involves sustaining a high
solidification front velocity which can be hundreds of mm/s. This high
front velocity results from or generates a large undercooling which can be
hundreds of degrees ahead of or at the advancing solidification front. Such
large undercoolings result in the large departures from equilibrium constitution in size-refined and even segregation-free microstructures associated
with rapid solidification processing (RSP) for very extended ranges of
material composition. Such high front velocities can be achieved at steady
state only by particular conditions of directed energy processing of surfaces for example by means of scanning laser, electron or plasma beams.
Large undercoolings prior to solidification are generally difficult to
achieve by slow cooling of undivided large volumes. Hence the almost uni-

versal practice of achieving the conditions for RSP by rapid formation of a small elementary section dimension (typically < 0.1 mm) of liquid material in good contact with an effective heat sink. These conditions create a high cooling rate ($\simeq 10^6$ K/s) during solidification which in turn leads to a high solidification front velocity and undercooling.

A considerable variety of techniques are now available for achieving such rapid solidification. These are conveniently divided into those which involve or produce particulate, the subject of the present contribution, those which directly produce continuous wire, filament, fibre, ribbon or strip and those which involve localized melting at or of a surface. The range of techniques has been reviewed previously by the author in 1978 (1) and, more briefly, in 1982 (2), and most recently by Savage and Froes (3) in 1984. Particulate methods and the general characteristics of their products have been reviewed most recently by Maringer (4) in 1980 (for aluminium alloys), by Lawley (5) in 1981, by Grant (6) in 1982 and (in relation to production of amorphous metal powders) by Miller (7) in 1983. The present purpose will be to summarize the present state-of-the art of production of rapidly-solidified (RS) particulate. Particular attention will be given to more recent developments and to possible ways of assessing their efficacy as methods of achieving rapid solidification.

2. PRINCIPLES OF METHODS

The generation of rapidly-solidified (RS) particulate involves essentially two stages (i) formation of the particulate geometry (ii) rapid solidification and cooling. The first stage is a matter of fluid dynamics while the second is one of heat flow. In some processes the two stages are distinct whereas in others they overlap. Separation of the two stages may be an important consideration in designing the ultimate forms of the various processes since ideal conditions for forming the geometry may not be ideal for achieving rapid solidification.

Methods for generating RS particulate are most readily classified in terms of the physical force employed in the first instance to generate the particulate. For instance an emerging column or pendant of melt destabilizes at low velocities directly into spherical droplets under simple capillary action if left to its own devices (figure 1). The condition for this breakdown has been studied repeatedly, e.g. refs. (8-13), usually with a view to producing a stable jet, but also with the object of generating well-controlled metal droplets that cool and rapidly-freeze in free-fall (14-18). Employment of a hollow cylindrical jet allows the generation of RS hollow spheres (microballoons) for fusion technology and other applications (19-21). This process of jet breakdown can be assisted by the superimposition of subsonic (14, 22) or ultrasonic (23, 24) vibrations most effectively in conjunction with impact of a high-velocity jet of another fluid which can be a high pressure gas (25), vapour (26) or liquid (27). Such impact of a pressurized fluid which normally acts both as particulate former and as quenchant is the basis of most of the established commercial processes for generating metallic particulates from their melts. Widmer (26) and, more recently, Miller and Murphy (28) have attempted to separate the formation and quenching stages by employing pressurized vapour or gas ostensibly as the atomizing fluid followed by water jets as the quenching fluid. The efficiency of the 'open' type of impinging jet atomization used traditionally can be improved by moving the point of atomization into coincidence with the point of emergence of the melt i.e. the close-coupled form, figure 2. Examples include the updraught process reported (29) to be operated e.g. by Alcoa, Alcan and Metalloys, the horizontal version of Alpoco (30) and the downdraught variant of Valimet (29, 31) also being studied at GE (32) and Brown Boveri (33). An alternative is to transfer, into a reduced pressure, melt that has been supercharged with gas by the application of high pressure. The resulting release from solution and expansion of the gas disintegrates the melt into droplets in the soluble gas (vacuum atomization) process (34). A further alternative is to replace the atomizing action of a high velocity fluid with direct impingement of the melt column, melt pendant or derived droplets into the roll gap between rotating twin rolls (figure 3a), as developed at Swansea (35), MIT (36) and at Tohoku (37) or onto a rotating drum with a serrated surface (38, 39)(figure 3b), on to a toothed ceramic disc (4) or inside a rotating drum (41). These can be regarded as derivatives of the Degussa process developed during World War II (e.g. see ref. 42) and tend to generate a flake rather than a powder product. Variants employ a liquid layer held centrifugally within a rotating drum or cup to both form and quench a spherical particulate (rotating water atomization = RWA (43-4), rapid spinning cup = RSC (45)) or continuous cylindrical product rather than flake (figure 4a,b), or employ gas jets to pulverize a liquid film generated by melt stream impingement on to a plasma-coated rotating steel drum (46).

The effectiveness of atomizing a melt stream by feeding it on to a rotating disc or cup so that droplets form by destabilization of the resulting liquid film at the disc or cup periphery is as well-established as twin fluid atomization. The introduction of suitable perforations in the cup results in emergence of the liquid through them as a stream of shot particles (rotating perforated cup = RPC) (47). Pratt and Whitney essentially employ an unperforated cup as rotary atomizer in their RSR rapid solidification rate) process adding helium jets normal to the droplet flight path with the aim of enhancing the quench rate compared with that resulting from free flight alone (48). The need to process reactive metals and alloys without contact with incompatible containment materials has resulted in various forms of the rotating electrode process (REP)

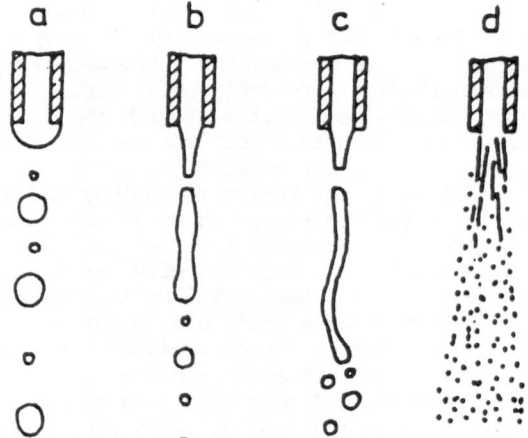

Figure 1: Illustration of transition from coarse direct drop formation (a) to fine atomization (d) with increase of velocity of an emergent column or pendant of liquid (14).

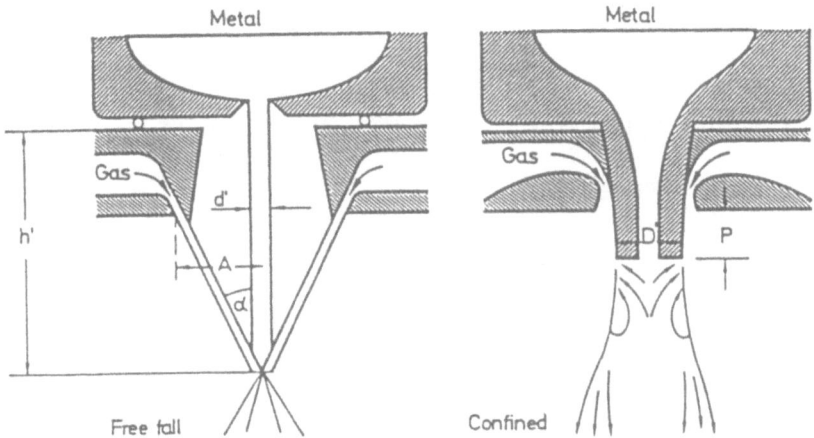

Figure 2: Comparison between free fall and close-coupled down-draught pressurized gas atomization (from H.Schmitt, Powd. Met. Internat., 1979,11,17-21).

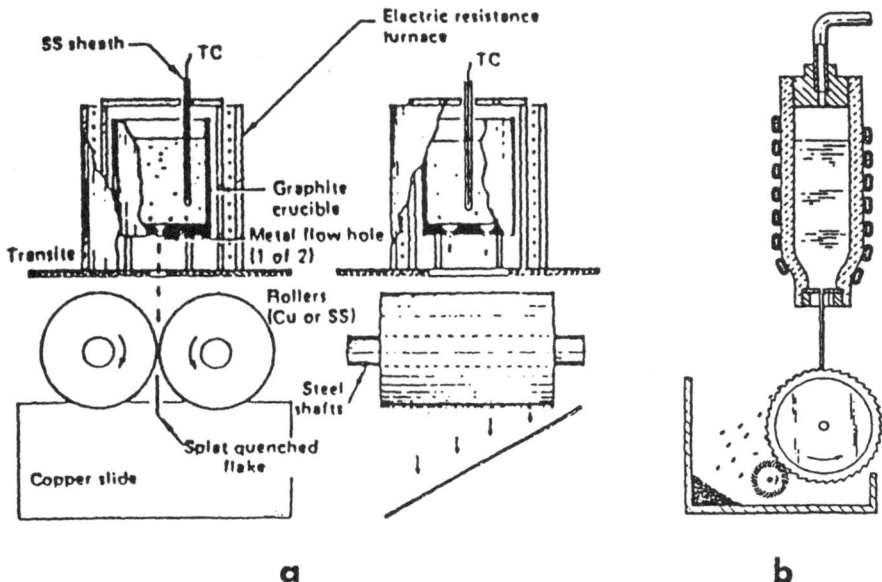

a   b

Figure 3: (a) Twin roller atomization (36) (b) serrated surface rotating drum impact atomization (38).

introduced by NMI two decades ago (50). The principle (figure 5a), is to strike an electric arc at a point on the consumable electrode so that resulting liquid is centrifugally expelled as a stream of droplets. The electric arc can be replaced by a laser (laser (melt) spin atomization = L(M)SA), (50,51) electron beam (Pulvérisation sous vide = PSV) (52) or, (figure 5b), plasma beam (plasma rotating electrode process = PREP) (53) and

160

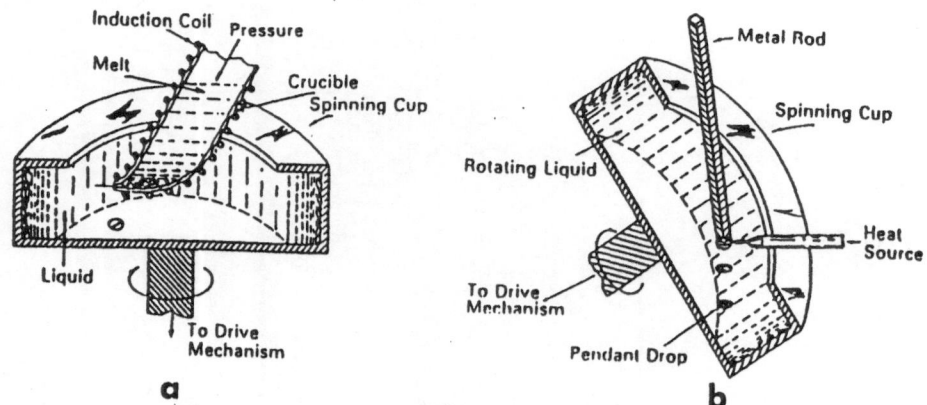

Figure 4: (a) Rapid spinning cup (RSC) atomization (45) (b) pendant drop
rapid spinning cup atomization (45).

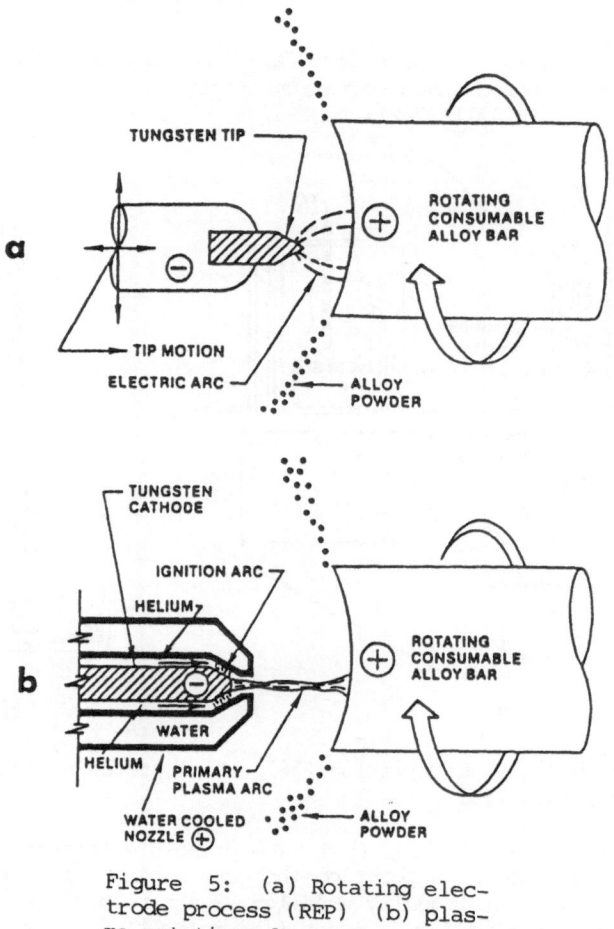

Figure 5: (a) Rotating electrode process (REP) (b) plasma rotating electrode process
(PREP) (53a).

versions have been developed in which, the product particulate is formed by drip feeding the periphery of a rotating cup or disc from a stationary electrode situated above it as in centrifugal shot casting (CSC) (54) and in electron beam splat quenching (EBSQ) (50). All of these methods lend themselves to solidification of resulting droplets in free flight typically to produce essentially spherical particulate on impact with an interposed chill surface to produce splat particulate.

The emergent column or pendant melt , as an alternative to capillarity, vibration , jet or substrate impact or centrifugal action, can generate a stream of droplets under the action of a sufficient electric field as in electrohydrodynamic atomization (EHDA) (56-7) (figure 6a). A pendant drop or crucible of melt can generate RS particulate in the form of staple fibres by contact with a serrated rotating chill disc in the pendant drop (PDME) or crucible (CME) melt-extraction processes (57). Continuous ribbon or wide strip produced by chill-block melt spinning (CBMS) or planar flow casting (PFC) can also be pulverized in a ball, rod, cr hammer mill (58) or mechanically chopped (59) to make RS particulate. Embrittlement by heat treatment (60) or by hydrogen-charging (61) prior to milling aids pulverization when the as-produced ribbon or strip is not already brittle. Spark erosion in a dielectric fluid is a further alternative that produces spherical particulate up to some 40 μm in size (62-63) (figure 6b).

Emulsification of melts with a suitable carrier fluid is another approach with the particular advantage of allowing rapid solidification to be triggered at specific levels of prior undercooling (64-65). Certain nucleation-resistant or easy glass-forming compositions can be readily undercooled in comparatively large volumes in a crucible (66,67) or on a plaque (68) of amorphous material, or with interposed slag layer (66). Sputtering (69), vapour deposition, electroless or electrolytic deposition from salt

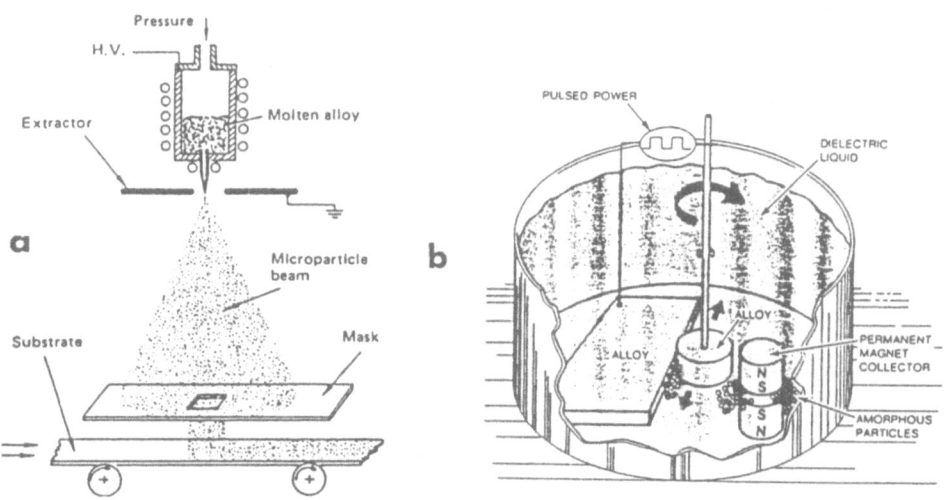

Figure 6: (a) Electrohydrodynamic atomization (EHDA) (Figure supplied by J.F. Mahoney 1985). Reprinted with permission of DATAMATION® magazine, © copyright by Technical Publishing Company, A Dun and Bradstreet Company (1982) - all rights reserved.) (b) generation of amorphous particulate by spark erosion (62b).

solution, mechanical alloying (70-71) or simple chemical reaction (71-75) can also be employed to make products corresponding to those of rapid solidification from the melt.

While for some purposes it is desirable to retain RS particulate in its particulate form, for other purposes it must subsequently be consolidated. One alternative to a separate consolidation step is to progressively build-up a multilayer deposit by incremental solidififcation of spray droplets on a chill surface (figure 7a, 7b). Gas jet (76-77) and plasma jet (83-86) spraying have been used with this objective in view simultaneously with peening in a recent variant aimed at achieving full density in the deposit.

Thus, an impressive range and combination of technologies is available for making RS particulate. It is of interest to pose the question how effective are the various alternatives when measured against particular criteria? Such criteria might include rating of the following capabilities:

- to generate a certain particle size or size distribution (e.g. coarse, medium or fine depending on the end use) and particle morphology (spherical, cylindrical or flake with particular degrees of regularity, again depending on the end use) i. e. effectiveness as a particulate making process.

- to achieve a specific structural state within the particulate at a certain size and shape i. e. effectiveness as a rapid solidification process.

- to successfully process materials of different characteristics such as fusibility and reactivity.

- to accommodate specified levels of production volume and rate.

The first two of these criteria are linked and rather basic and will be particularly featured in what follows. The general characteristics of the various processes have been amply reviewed in refs. 1-7 and so will not receive special attention here.

3. EFFECT OF VARIABLES ON EFFICACY OF PROCESSING METHODS

There is no standard method of assessing the efficacy of a rapid solidification technique. Early practice was reviewed by the author in 1975 (88). These included and include thermal analysis which, where practicable, can yield data on cooling rate, undercooling and solidification front kinetics. An effective system for continuously recording the time-dependence of temperature of travelling droplets has yet to be reported, though use has been made of calorimetry to determine temperature at particular distances from the point of generation (89,90). Measurement of particulate dimensions especially diameter or thickness allows an upper limit to the cooling rate to be estimated corresponding to ideal cooling but prediction of the actual cooling rate requires a value for the operative heat transfer coefficient which can be difficult to estimate for the complex conditions applicable in, for example, many of the atomization processes. Microstructural parameters such as dendrite cell size or eutectic interphase spacing provide a further indication of the actual solidification conditions though prediction of the applicable relationships in particular cases awaits further advances in theoretical understanding. Nonequilibrium constitution provided one of the earliest test of efficacy of new rapid solidification techniques while physical properties give an indirect indication of microstructure and constitution.

3.1 Thermal analysis

Calorimetric determination of temperature as a function of distance travelled by gas atomized aluminium droplets showed a notable increase in percent liquid with increase of superheat and melt orifice diameter and decrease in atomizing pressure for a given distance travelled as expected from the increase in heat content at higher superheat offsetting smaller average droplet size and larger average droplet size obtained respectively for larger melt orifice diameter and lower pressures (89). Photographic studies at Sheffield confirm expectation that droplet velocity exhibits a maximum at a point some distance from the point of atomization, the maximum velocity increasing with increase of atomizing pressure (figure 8a,8b). Such variations together with the decrease in velocity of atomizing gas with increasing distance from the point of atomization and its dependence

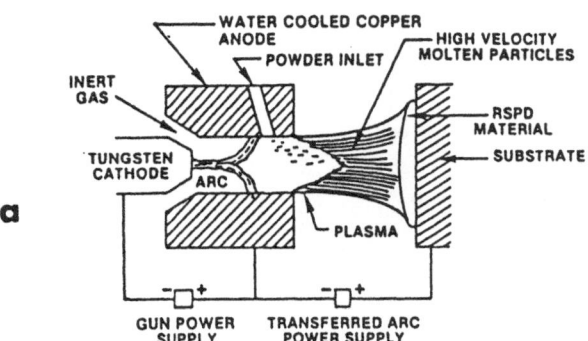

Figure 7: (a) Rapid solidification plasma deposition (RSPD) (86) (b) spray peening (87).Fig. 7a is reprinted with permission from JOURNAL OF METALS, Vol. 33, No. 11, pp. 23-26, a publication of the Metallurgical Society of AIME, Warrendale, Pennsylvania, USA.

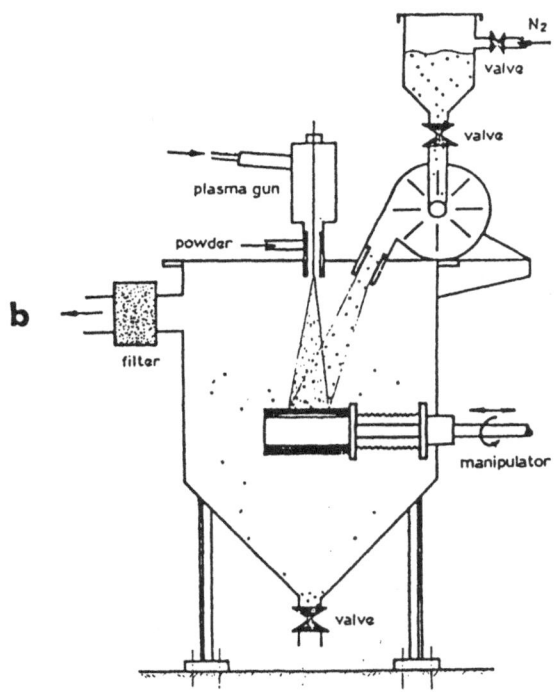

164

on atomizing pressure will result in corresponding variations in heat
transfer coefficient which must be fully taken into account in any realistic model of cooling and solidification history (77b). Interpretation of
calorimetric measurements by Pond and Winter (90) of exit temperatures of
melt-spun ribbon and melt-extracted fibres of tin, lead and zinc are com-
plicated by differences in dwell time on, and velocity of, the chill block,
and differences in product thickness. These differences may partly account
for the factor of 20 increase in quench rate indicated by the results for
the melt-spun material.

3.2 Size and size distribution

Early work involving RS atomized powders only rarely included more than
minimal information concerning particle size or size distribution. Effects
of process (refs. 5, 6, 22-3, 25, 27, 32, 35, 40, 43-5, 77a, 91-123)

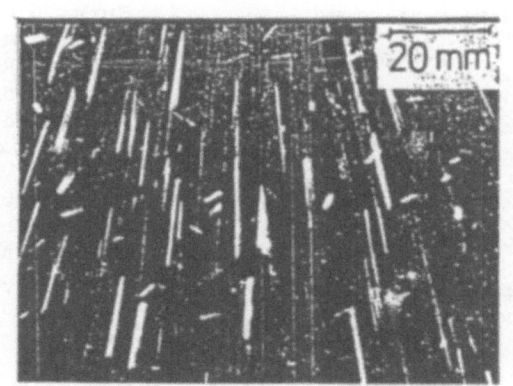

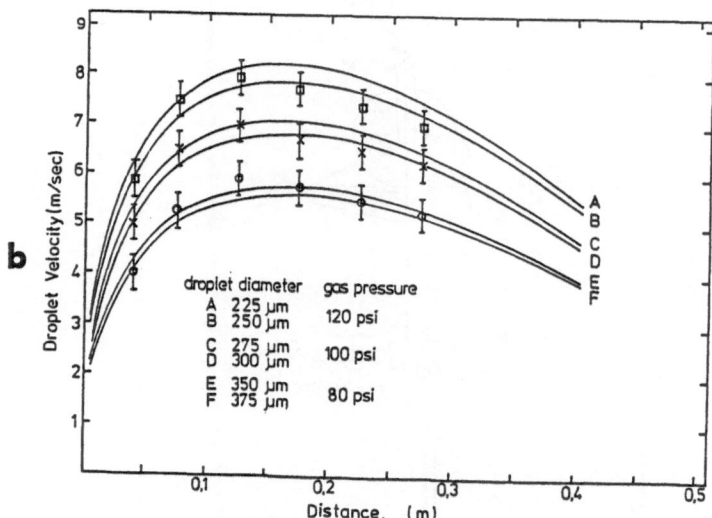

Figure 8: Photographic traces of travelling droplets
and (b) resulting velocity as a function of distance
of travel and atomizing pressure for Al-2wt% Cr
(77a).

(figure 9 a,b) or material (refs. 22-3, 32-3, 37a, 53a, 89, 94, 105, 112-3, 118, 124-9) (figure 10 a,b) variables on mean size distribution have now been reported for many of the processes that manufacture RS particulate. For the relatively well-defined situation of single-step direct drop formation at the rim of a rotating disc at low feed rates, droplet size is expected to be simply determined by the balance between the generating effect of centrifugal force ($= V\rho\omega^2 \Delta/2$) and the restraining effect of surface tension force ($\alpha\gamma d$) where respectively V and d are droplet volume ($\pi d^3/6$) and diameter, ω and Δ are disc speed of rotation and diameter and ρ and γ are liquid density and surface tension, giving (130):

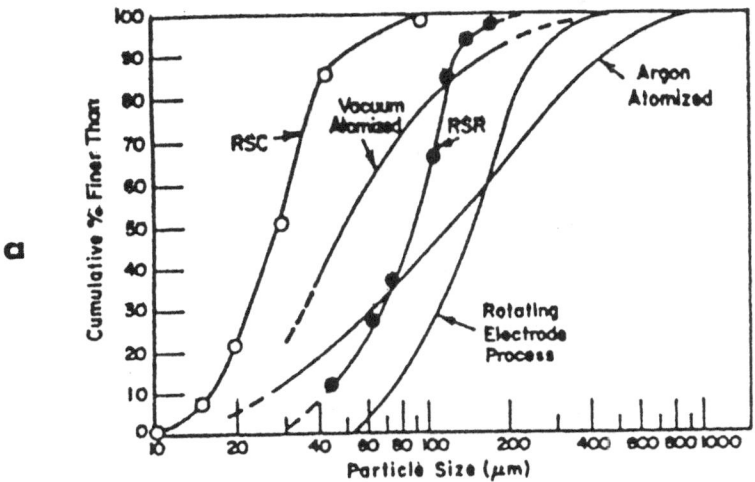

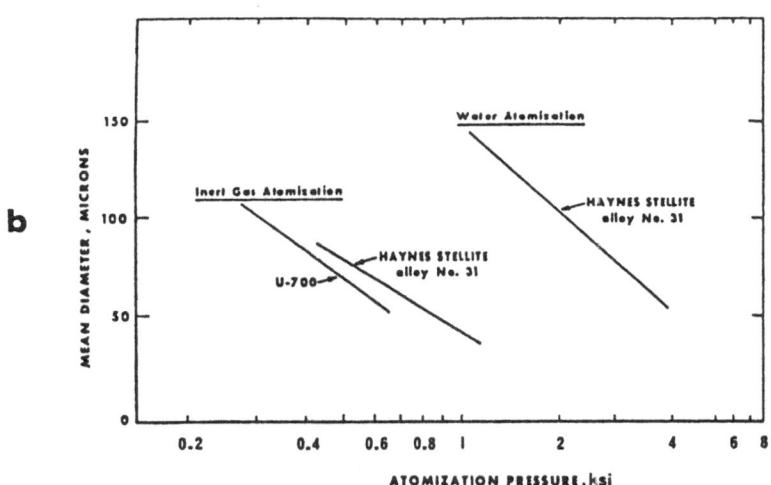

Figure 9: (a) Particle size distribution of superalloy atomized powders as a function of atomizing method (45a,b) (b) mean particle diameter of two superalloy powders as a function of atomizing pressure (98).

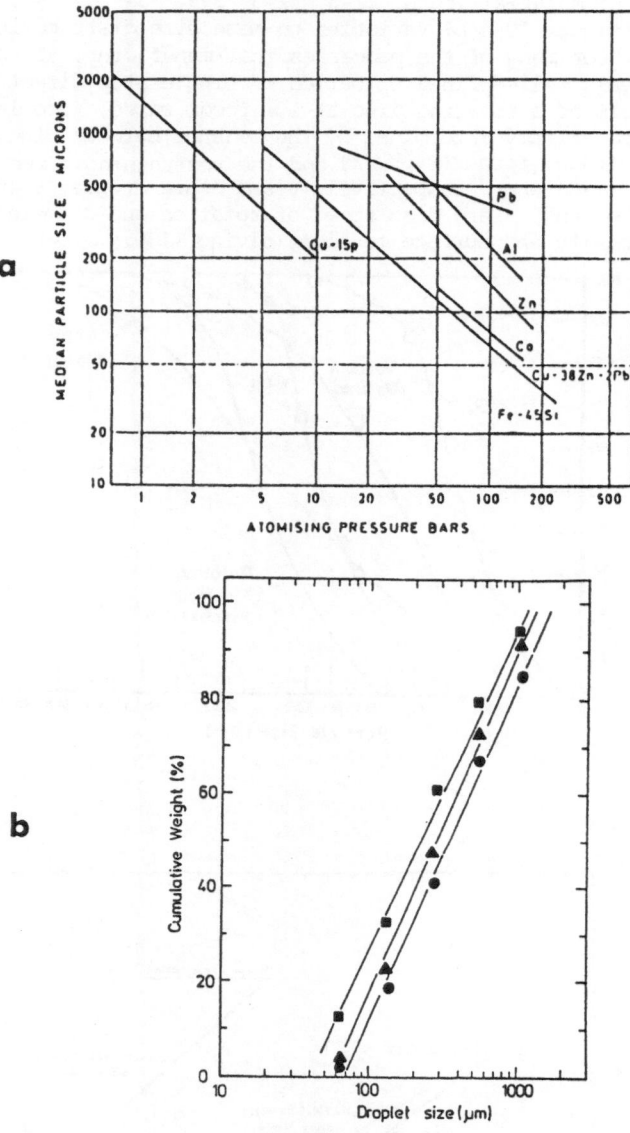

Figure 10: (a) Mean particle diameter as a function of water atomizing pressure for different metallic melts (107) (b) particle size distribution as a function of alloying level for argon atomization of Al-Fe alloys (129). (Key: ■ 0.8 ▲ 2.3 ● 4.0 at%Fe)

$$d/\Delta = A_1(We)^{-n_1} \tag{1}$$

where $We = v^2\rho\Delta/\gamma$ with $v = \omega\Delta/2$, $n_1 = 0.5$ and A_1 is expected to be close to 2 as confirmed experimentally for a variety of liquids including some liquid metals (130-3). Increased feed rate gives first ligament then sheet (film) formation prior to breakdown into droplets (figure 10a-c) (134). A recent study of droplet formation by the rotating electrode process (110a-

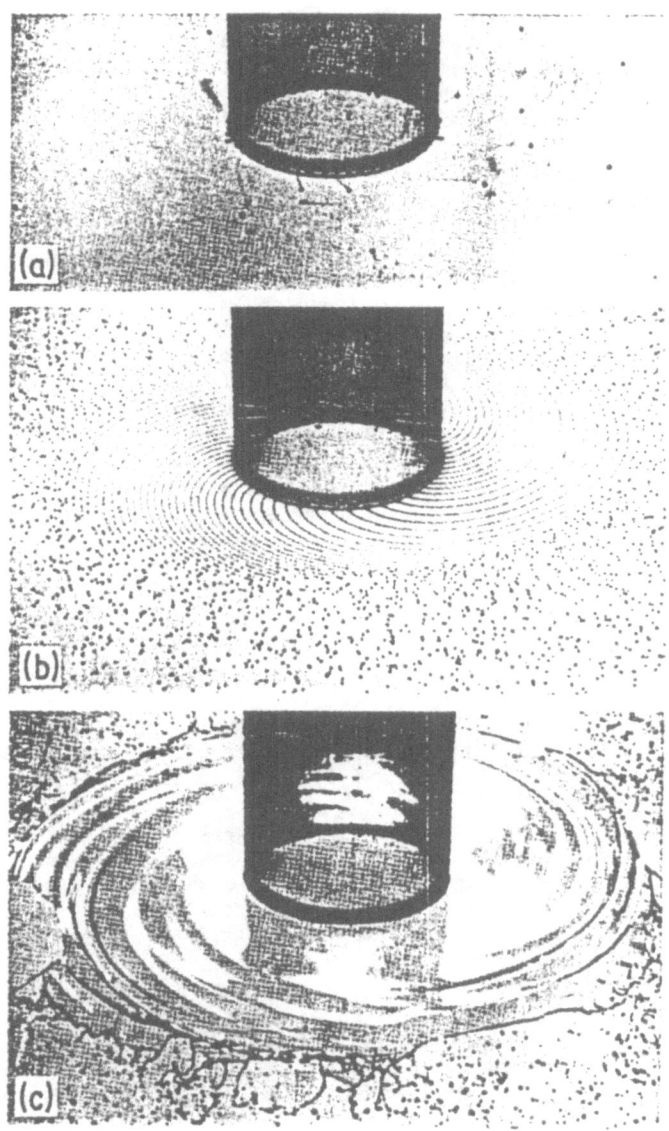

Figure 11: Showing transition from (a) direct drop formation (DDF) to (b) ligament disintegration (LD) to (c) film disintegration (FD) with increasing feed rate (1 to 8 to 45 kg/h) in rotary atomization (134).

c) has established that direct drop formation (DDF) produces a smaller satellite (secondary) droplet (figure 12a,b) by spherodization of the thin thread of liquid that links the primary droplet to the parent melt before separation finally occurs. This results in a bimodal size distribution of spherical particles, the proportion of secondary droplets increasing with increasing feed rate Q (figure 12c). The ligament disintegration mode then occurring produced particles that were more elliptical than by DDF. The conditions for transition from DDF to ligament disintegration (LD) and film disintegration (FD) are represented in figure 12d as a plot of operating parameter $Q\omega^{0.6}/\Delta^{0.7}$ against material parameter $\gamma^{0.9}/\eta^{0.2}\rho^{0.7}$ where η is melt viscosity. The corresponding transitions form DDF to LD and FD with increasing Q have been identified (e.g. ref. 105) for pressurized fluid atomization though theoretical modelling (e.g. ref. 136)* has not yet displaced comparison of results with empirical correlations such as that of Lubanska (135) (figure 13a) for gas atomization of liquid metals and wax:

$$\bar{d}/\Phi = A_2[\alpha(1 + \beta)/We]^{n_2} \qquad (2)+$$

where Φ is melt orifice or stream diameter, α and β are ratios of kinematic viscosity and mass flow rate, respectively of the melt to those of the atomizing fluid and We is the Weber number $v^2\rho\Phi/\gamma$ in which v is relative velocity of atomizing fluid at its point of impact with the melt stream with $A_2 \simeq 40$ to 50 and $n_2 \approx 0.5$ for the conditions explored by Lubanska (e.g. $6 < \Phi < 22$ mm, angle of atomizing gas jets to horizontal 65 to 78°). Rao and Mehrotra (113) subsequently found smaller values of $A_2(\simeq 10)$ and n_2 ($\simeq 0.3$) for a smaller stream diameter $\simeq 2$ mm and atomizing angle ($\simeq 25°$) for Sn, Pb and Pb-Sn alloy, and even smaller values of $A_2(0.8)$ and $n_2(0.2)$ were obtained for transverse atomization of Pb (120). Our results (77a) obtained at Sheffield for downwards cone-jet argon atomization of Al-2.1wt%Cr (figure 13b) give values of a d about half of those predicted by equation (2) with $A_2 = 40$ and $n_2 = 0.5$. The atomizing angle (60°) was similar to that used by Lubanska but the stream diameter (1 mm) was much smaller which might possibly account for the discrepancy which is however still just within the lower bound of scatter of the data used to obtain equation (2) with $A_2 = 40$ and $n_2 = 0.5$. The smaller stream diameter we employed places our data at the top end of the range $0.007 < d/\Phi < 0.3$ explored by Lubanska. Such correlations at least serve to indicate the wide range of mean particle size that can be produced by appropriate control of the operating variables. Larger values of d, e.g. in the range 0.5 to 5 mm, tend to be associated with melt-drop methods (14-16, 20-1), twin fluid atomization at low pressures (25-6, 107, 138-40) or methods of rotary atomization at lower rotational speeds (35, 37b, 54, 78, 139-40). Values of d $\simeq 0.1$ mm approach the lower limit of most rotational methods and of twin fluid atomization as normally operated, but a value as low as 28 μm has been reported (45) for Mar M246 superalloy particulate made (conditions unspecified) by the RSC process (figure 9a) which is comparable with values between 18 and 25 μm reported (32-3, 141-4) for gas atomization of aluminium alloys under (pressumably) close-coupled or ultrasonic (33, 143-4) conditions. A high yield of very small particles is characteristic of spark erosion (62a,b) and EHD processing (55a-d) for which an average particle size of 0.01 μm has been reported (55d) for Al-1 wt%Mn, though sizes of 0.1 to 1 μm (145-6) are perhaps more typical and sizes up to 150 μm have been produced (55c) in Al-

---

* for a recent review see ref. 137

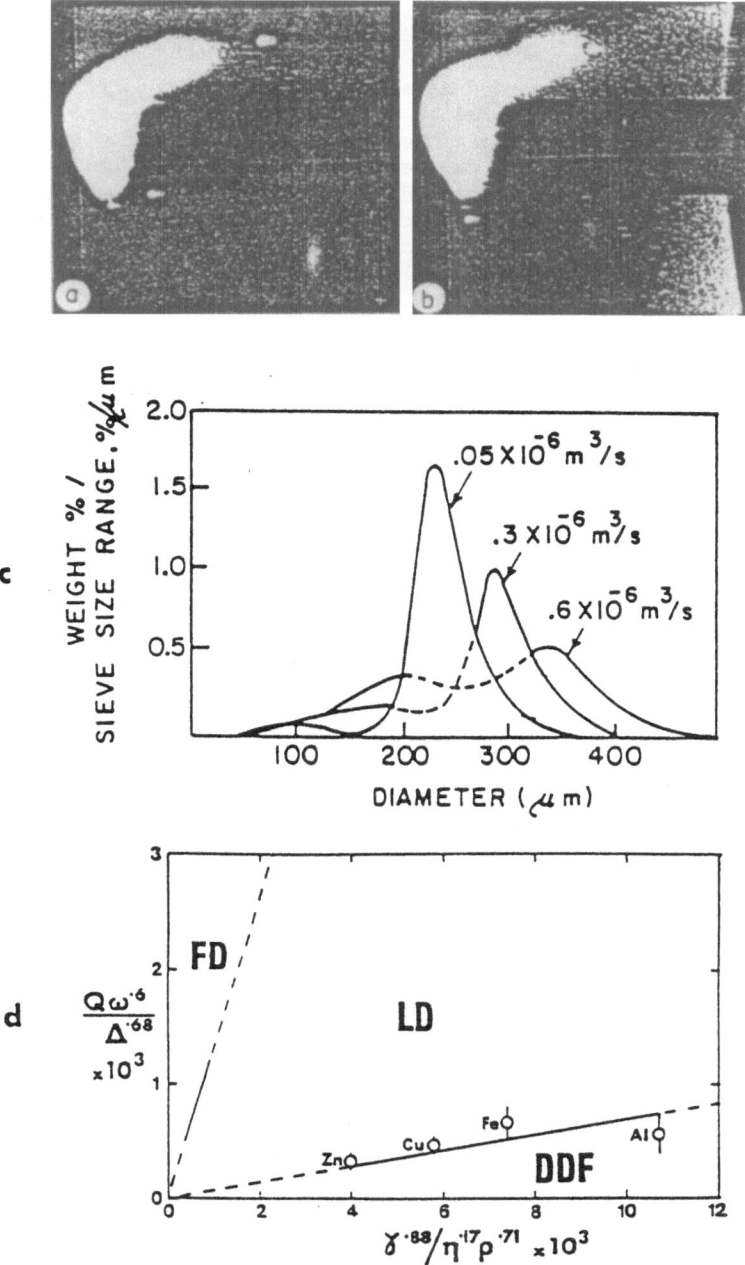

Figure 12: Showing for REP atomization (a,b) formation of main
and secondary particle during DDF mode (SAE 1090 steel, (b)
being taken 0.007s after (a)) (c) effect of increased melting
rate on particle size distribution for SAE 1090 steel (d)
domains of operation of the three atomization modes based on
observations for Zn, Cu, Fe and Al (110c).

5.5wt% Cu. Size distributions differ widely even among the standard processes, for example normal gas atomization giving a much broader distribution than for, say, the rotating electrode process (5) (figure 9a).

3.3 Scale of microstructure

If particle size is regarded as an indicator of the upper limit to the cooling rate that could have applied during solidification, the microstructure itself reflects the conditions that actually applied. The most commonly recorded microstructural parameters are the dendrite cell size in the case of microcrystalline products and percentage of non-crystalline phase in the case of glass-forming systems. Mainly recent data (30, 109, 142b,c, 147-8) on cell size λ as a function of gas atomized powder particle size d for some high strength aluminium alloy compositions are plotted in figure 14a according to the power relation:

$$\lambda = A_3 d^{n_3} \qquad (3)$$

The data are well correlated over the range $0.4 < \lambda < 6$ µm and $0.7 < d < 300$ µm by equation (3) with $A_3 = 0.13$ µm$^{1-n_3}$ and $n_3 = 0.67$. No significant differences are evident between the most recent sets of data though somewhat larger values of λ are given by Lyle and Cebulak (142b,c) at least for particle sizes up to 50 µm. Lyle and Cebulak (142b,c) found no systematic effect of choice of atomizing gas on λ at given d, while the recent data of Wang and Grant (148) showed a reduction in λ by a factor $\simeq 0.6$ when helium

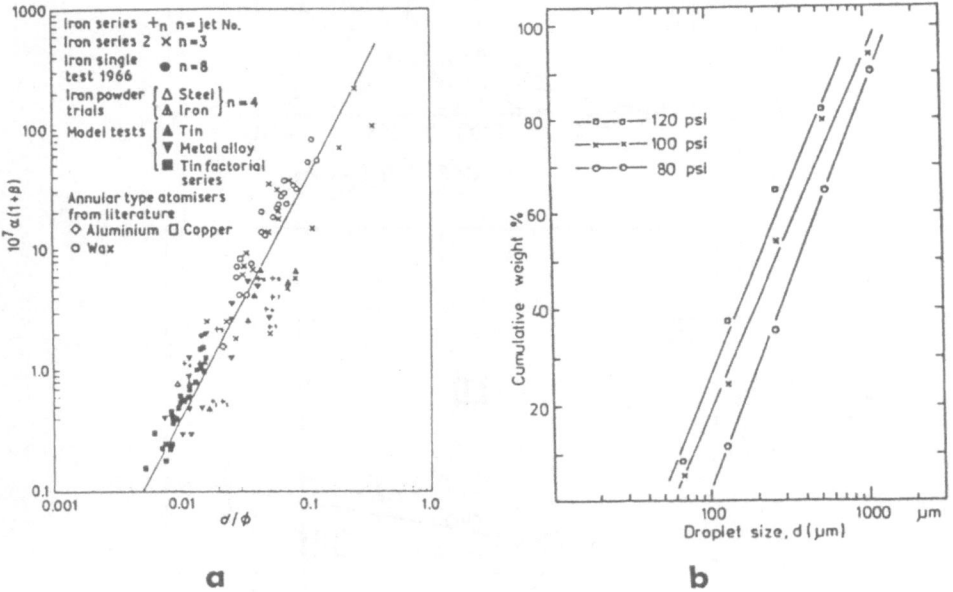

a   b

Figure 13: Showing (a) $10^7 \alpha (1 + \beta)$ as a function of d/ϕ for gas-jet atomization of liquid metals and wax (135) (b) particle size distribution as a function of argon atomizing pressure for Al-2wt% Cr (77a). Figure 13a is reprinted with permission from JOURNAL OF METALS, Vol. 22, No. 2, pp. 45-49, a publication of the Metallurgical Society of AIME, Warrendale, Pennsylvania, USA.

was used rather than argon. Plots corresponding to figure 14a show $n_3 \approx$ 0.2 for coarse-atomized maraging 300 alloy steel powder 140, $n_3 \simeq$ 0.4 to 0.5 for gas (149) or water (150) atomized high speed tool steel powders and $n_3 \simeq$ 0.6 for rotary, argon gas and vacuum atomization of C101 and IN100 super-alloys (115, 151-2). Dendrite spacing at given d has been shown to be dependent on process method (115, 138, 140, 152), atomizing gas (153) and alloy composition (149,151). For example,the effect of proceeding from vacuum to argon to rotary He-cooled atomization of IN100/C101 (figure 14b) is to reduce λ for the same d by some 15 to 20% at each step (115) corres-ponding to a factor of $\simeq$ 2 in cooling rate for each step. Dendrite cell size in continuous cooling situations is usually related to cooling rate $\dot{T}$ by the empirical equation:

$$\lambda = A_4 \dot{T}^{-n_4} \qquad (4)$$

where typically 0.25 < n_4 < 0.5 while cooling rate is expected to be linked to particle size by a relation of the form

$$\dot{T} = A_5 d^{-n_5} \qquad (5)$$

The value of n_5 in equation (5) depends both on the method of atomization and on the mechanism of heat transfer.However, eliminating T between equations (4) and (5) gives the form of equation (3) with $A_3 = A_4/A_5$ and $n_3 = n_4 n_5$. A representative value $n_4 = 1/3$ for aluminium alloys and $n_3 = 2/3$ from figure 14a thus implies $n_5 = 2$ for the conditions applicable which

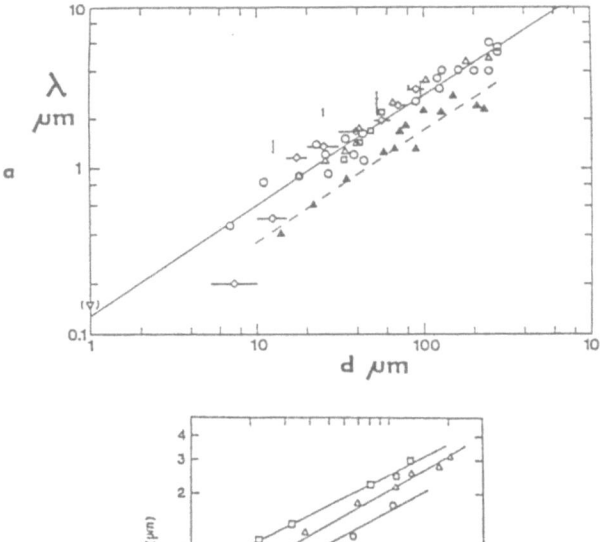

Figure 14: Dendrite arm spacing as a function of powder particle diameter for (a) high strength alu-minium alloys

| Ar, He, air, N_2 updraught atomization (142c)
◇ N ultrasonic (147)
○ Ar ultrasonic atomization (109)
△ ▲ Ar, He ultrasonic (148)
□ Ar updraught atomiza-tion (30)
▽ electrohydrodynamic(145)
(b) nickel-base superalloys C101 and IN100 (RSR = gas-cooled rotary-atomized, AA = argon atomized, VA = vacuum atomized). From ref. 115, Advances in Powder Technology, G.Y. Chin, Ed., American Society for Metals, 1982. With per-mission.

would require some dependence * on particle size d of particle velocity relative to that of the coolant gas (140a) to be fully consistent with forced convective cooling. Data on dendrite-spacing-derived cooling rate plotted as a function of d for water atomized M2 high speed steel particulate (156) also gives $n_5 \simeq 2$ (figure 15a) while a corresponding plot (figure 15b) for rotary-atomized He-cooled IN100 superalloy particulate gave somewhat smaller $n_5 \simeq 1.5$ (48). Interesting recent results (157) show that grain size of RS-Ti6A14V exhibits power relations with REP powder particle and with splat flake or ribbon thickness of the form of equation (3) with $n_3 \simeq 0.9$, the flake/ribbon having the smaller grain size at given particle radius or flake thickness (figure 16a). Prediction of cooling rate as a function of REP particle size via equations 5a and 5b and of flake/ribbon thickness via equation (6)+ with $h = 0.1$ W/mm^2 K typical of chill substrate rapid solidification gives a dependence of grain size on cooling rate (figure 16b) of the form of equation (4) with $n_4 = 0.9$ identical to the value derived theoretically by Boswell and Chadwick (158) for aluminium, with the difference that grain sizes are about an order of magnitude smaller for Ti6A14V.

---

* Convective cooling rate of a spherical droplet of diameter d travelling at velocity v relative to a coolant gas is

$$\varepsilon = 6h(T-T_A)/(c'\rho'd) \qquad (5a)$$

where c', ρ', T are specific heat, density and temperature of the droplet, T_A is temperature of the gas and h is heat transfer coefficient given by:

$$h = 2(K/d) + 0.6(K^4\rho^3c^2/\eta)^{1/6}(v/d)^{\frac{1}{2}} \qquad (5b)$$

in which K, ρ, c and η are conductivity, density, specific heat and viscosity of the gas. Use in equation 5a of h from equation 5b would give equation 5 with $n_5 = 2$ if $v \propto d^{-1/2}$ is a possible solution to the governing momentum equation (154)

$$a = \frac{1}{2} C_D(\rho/\rho')(v^2/d) \qquad (5c)$$

where a is acceleration of droplet by the gas and C_D is drag coefficient (itself dependent on Reynolds number $\rho v d/\eta$) (155).

+ Cooling rate of a flake or ribbon of thickness z dominated by a heat transfer coefficient h operative on the chill substrate side only is

$$\varepsilon = h(T-T_A)/c'\rho'z \qquad (6)$$

where T_A the chill substrate temperature, and c',ρ' and T are as for equation (5a) except that they apply to the solidifying ribbon or flake.

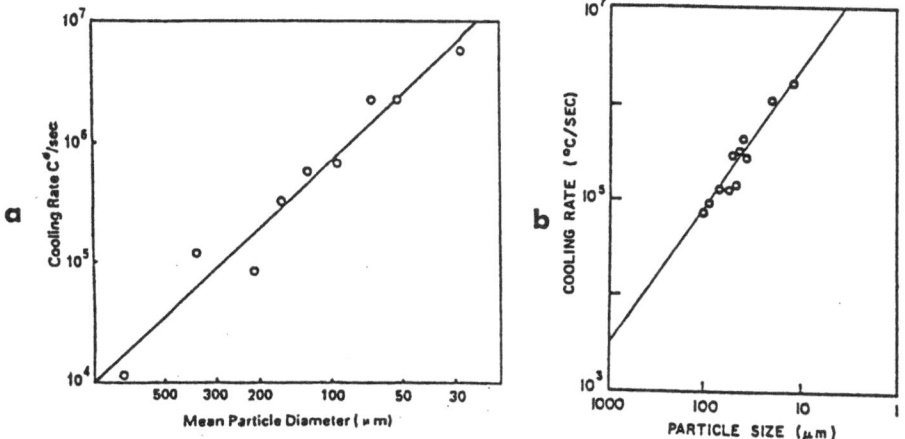

Figure 15: Cooling rate as a function of powder size for
(a) water-atomized M2 high speed tool steel (156),
(b) rotary-atomized helium-cooled IN100 nickel-base superalloy (adapted
from refs. 48 published and for sale by Claitor's Law Books and Publishing
Division, P.O. Box 3333, Baton Rouge, LA 70821, USA, and 104).

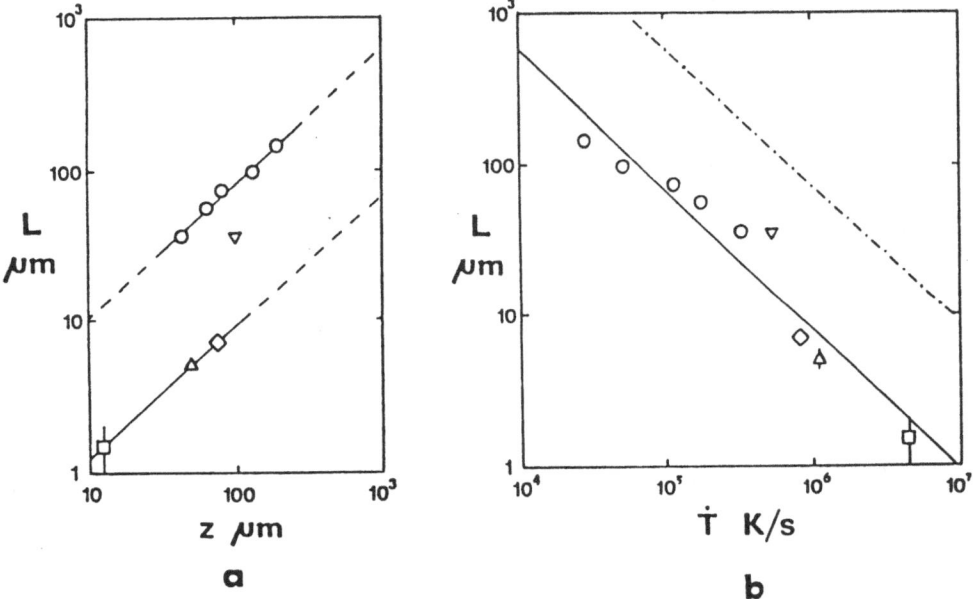

Figure 16: Grain size L as a function of (a) particle radius z and (b)
estimated cooling rate $\dot{T}$ for Ti-6wt% Al-4wt% V particulates made by different methods (157).
Key: O REP powder   ∇ PDME fibre   ◇ EBSQ flake
       △ PDME ribbon   □ Hammer and anvil splat   –·–·–·–Ref. 158 for Al

3.4 Fraction of nonequilibrium phase

Dendrite cell size is restricted as an indicator of the efficacy of a rapid solidification process to the range of conditions that produce dendrites. A sufficiently high solidification front velocity or prior undercooling is expected to result instead in formation of segregation-free solid which may be an extended crystalline solid solution, a new crystalline phase or a noncrystalline (amorphous) phase. Because the distribution of nucleation catalysts and cooling conditions differ among particles even when they are identical in size, a proportion of particles of a given size may attain the condition for segregation-free solidification while the remainder may not. This proportion f however is a useful indicator of the efficacy of a rapid solidification process at a given particle size for a particular composition. Figure 17a shows the expected decrease of f with increase of d for droplets of two Pd-Si compositions made by graphite atomization and solidified during free fall down a drop tube containing He-20%H_2 gas, a higher level of glass-forming addition (Si) significantly increasing the fraction of particles that were amorphous at a given particle size (159). The corresponding plot for two Fe-Si-B compositions made by gas-water atomization showed a much smaller difference between the compositions in that case (160). The effect of melt superheat on this plot for twin-roll atomization of an Ni-Si-B alloy into a water-quench bath showed systematic decreases in f at given d with decrease of superheat, then increasing again for the lowest superheat (37b). This was attributed to completion of solidification before reaching the water bath for the lowest superheat. Levi and Mehrabian (145) similarly reported for EHD-atomized Al-Si alloy particulate an increase in the proportion of segregation-free material with decrease in d, the proportion being higher at given d for the lower silicon content of the two studied (figure 17b). Safai and Herman (161) reported that the stable α-phase in alumina plasma-sprayed into water was progressively displaced by an increasing fraction of metastable phases with decreasing particle size over the range 80 to 20 μm. Similarly a progressive increase in the fraction of atomized particles of VM300 maraging steel exhibiting 'non-dendritic' undercooled structure with decreasing d was reported by Kattamis and Mehrabian (139). Corresponding effects are present in spray-deposits formed by incremental solidification as observed, for example in work by Kim (77b, 129) at Sheffield (figure 17c), in which the proportion of optically-featureless 'zone A' material decreased progressively from an initial level of 100% with increasing distance into an Al-3wt%Fe alloy spray-deposit some 15 mm in total thickness. This effect was consistent with a progressive build-up of thermal resistances in series throughout the growing deposit, associated particularly with porosity trapped between successive impact events. This results in establishment of a temperature gradient within the deposit during deposition, the measured temperature reaching 300°C at 7 mm into the growing water-cooled deposit. The observation that intermittent spraying tends to restore the initial situation at each resumption of spraying suggests that the initial 100% zone A condition could be maintained by sufficiently reducing the deposition rate. It would be of interest to determine for an appropriate alloy composition the extent to which simultaneous spray peening (87) would allow the minimum deposition rate that maintains the initial condition to be increased.

3.5 Physical properties

Differences in cooling rate, undercooling and/or solidification front velocity in rapid solidification give rise to differences in properties and response of the structures and microstructures they produce. The simple bend test employed for glass-forming alloys to give a first indication of

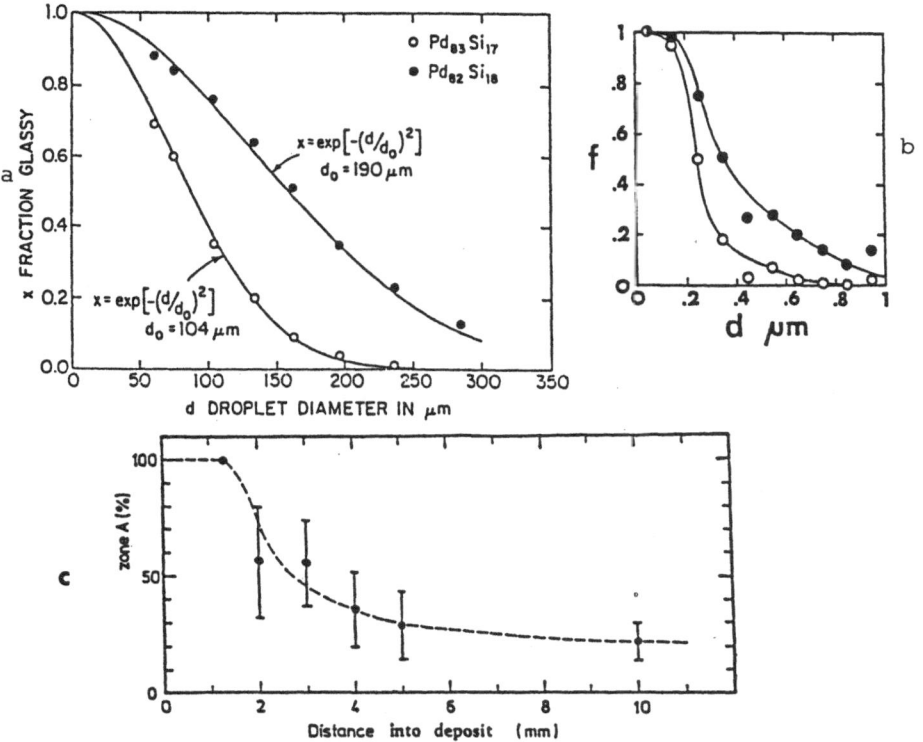

Figure 17: (a) Fraction amorphous as a function of droplet diameter for two glass-forming Pd-Si alloys: reprinted with permission from Scripta Metallurgica, Vol. 15 No. 3, Drehmann A.J and Turnbull D, Solidification Behavior of Undercooled $Pd_{83}Si_{17}$ and $Pd_{82}Si_{18}$ Liquid Droplets, Copyright 1981, Pergamon Press (159), (b) fraction f segregation-free as a function of particle diameter d for two compositions (● 3wt% Si ○ 6wt% Si) of Al-Si made by EHD atomization (145) (c) percentage featureless 'zone A' material as a function of distance into deposit for spray-formed Al-3wt% Fe (77b).

whether or not a melt-spun ribbon is amorphous (bend-ductile) or microcrystalline (bend-brittle) is a familiar example. While this method is not readily applicable to particulate products of rapid solidification, Chaudri and Hutchings (162) propose that compression testing might be used to assess the strength of single particles which in turn reflects their structure. Magnetic properties of ferromagnetic RS particulates have been shown in several studies to depend on particle size as well as on production method (62a;b, 119, 160a,b, 163-4). Fe-Si-B alloy particulate made by gas-water atomization or spark erosion in liquid argon showed increasing levels of remanent magnetisation above the amorphous phase Curie temperature with increase of particle size d from < 10 μm to 20 to 30 μm indicating an increasing fraction of crystalline material (163). No such remanent magnetisation was observed over the same size range for particulate made by spark-erosion in dodecane indicating that this was all fully amorphous. The Curie transition however then broadened increasingly with decrease of particle size over this range presumably reflecting less and less ideal structural states. Variations of magnetic properties with production method and particulate size for corresponding microcrystalline Mn-Al-C alloy RS-particulates was however attributed to differences in composition rather than of cooling rate in this case since dendrite arm spacings did not vary significantly (164). Particle size dependent isothermal crystallization kinetics have been reported (119) for rotary-atomized gas-quenched amorphous Ni-B-Si alloy RS-particulate, 10 to 15 μm sized particulate requiring up to twice as long to crystallize as 55 to 65 μm sized material. A similar result was obtained from magnetisation measurements on spark-eroded amorphous Fe-Si-B (160b) for which crystallization temperature at a heating rate of 40 K/min decreased with decreasing particle size for samples made by spark erosion in dodecane (62b). One essential engineering requirement of RSP is that beneficial effects of rapid solidification identifiable in RS-particulate should survive consolidation into bulkier forms more suitable for a range of engineering applications. In that sense properties during or resulting from consolidation are a valid test of the efficacy of RS processing. Not infrequently comparisons are made between properties in the as-slowly-solidified state and those of RS-material in a wrought state resulting from consolidation. Any differences found then partly depend on the conditions of working of the RS material and the differences may increase, decrease or even, for certain compositions and treatments, disappear if the slowly and rapidly solidified materials are compared following identical thermomechanical treatments subsequent to solidification. Nominally identical thermomechanical treatments do not, even so, provide a definitive comparison since the optimum treatment for an ingot material will not in general be identical with that for rapidly-solidified material (165). The significance of any such reported differences thus needs careful consideration and may not be immediately evident. Thus the doubling of UTS reported (166) for extrusions of splat-quenched compared with atomized Al-8wt% Fe (figure 18a) was undoubtedly in part due to differences in extrusion conditions. Recent reports of increased UTS and increased elongation to fracture for the same extrusion conditions of RS Al-4.8wt% Fe-3.3wt% Ni-6.7wt% Co (167) and Al-14wt% Fe-2wt%V (168) pulverized melt-spun ribbon, when softer less rapidly solidified material was excluded from the consolidate (figure 18b), are reassuring. For a different alloy, Lyle and Cebulak (142a) reported that grain size and interparticle spacing as-extruded decreased with decreasing RS powder particle size in parallel with the dendrite cell size of the unconsolidated powder (figure 19), though consequential effects on properties were not detailed. The effect of atomized powder particle size fraction on extrudability and extrusion properties has been reported by Sheppard et al for aluminium (169), different aluminium alloys (170-1) and a copper alloy (172) powder. Extrusion pressure increased notably with decreasing particle size below ≃150 μm for aluminium (169) and Al-3.6wt% Mn (171), reflected in increased proof strength and decreased elongation for the aluminium and, at 400 °C, for the Al-Mn alloy. An increase in proof strength with decrease in particle size was also found for Al-2.5wt% Fe (170a) together with the decrease in elongation at 400 °C found for Al-Mn. The effect of powder particle size was otherwise not large for the range of compositions and conditions investigated. Large property differences in general require sufficiently large differences in solidification conditions, conditions above compared with those below a threshold, and susceptible alloy compositions as well as susceptible properties. For example, figure 20a, b shows hardness and UTS as a function of Mn-content in Al-Mn alloys for chill-cast and splat-flake material extruded under identical conditions of billet size, temperature and extrusion ratio to give double the UTS and up to five times the tensile elongation to fracture for the RS material (173). Any difference in solidification conditions between atomized and splat-quenched material was evidently, for these alloys, insufficient to give a detectable effect on these properties for the conditions employed, unlike the situation referred to earlier for Al-Fe based compositions where the conditions of rapid solidification are evidently much more critical in determining the levels of final properties. Hildemann et al (117), however, reported 14 percent (30 MPa) increases in yield strength of Al-3.7wt%Ni-1.5wt%Fe alloy particulate forgings in proceeding from centrifugally-atomized powder ($\bar{d} > 50\mu m$) to air-atomized fine powder (d = 11.7) to splat particulate, attributed to the decrease in dendrite arm spacing from 1.0 to 0.5 to <0.5 µm because this governs the spacing of the Al_9FeNi strengthening dispersoid in the forgings. A much larger decrease in dendrite arm spacing from 7 µm to 0.7 µm for splat particulate compared with air atomized powder gave no greater (in fact the same) increase in yield strength in 2024-T4 aluminium alloy extrusions simply because this alloy contains a relatively small fraction of dispersoid to contribute to its strength.

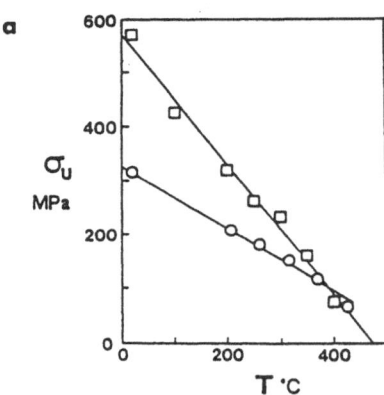

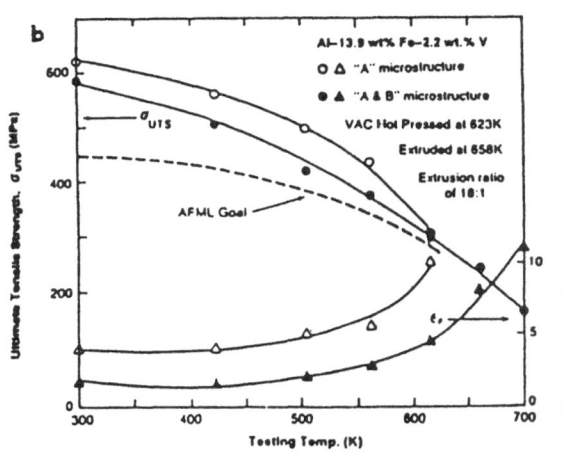

Figure 18: (a) Strength σ_U as a function of test temperature T for Al-8wt%Fe consolidated by compaction and extrusion from ○ air-atomized (141) and ☐ splat-quenched (166) feedstock materials;
(b) Strength and strain to fracture as a function of test temperature and initial microstructure for melt-spun, pulverized, hot-pressed and extruded Al-14wt%Fe-2wt%V: reprinted with permission from Scripta Metallurgica, Volume 18, No. 9, Skinner DJ and Okazaki K, High Strength Al-Fe-V Alloys at Elevated Temperatures produced by Rapid Quenching from the Melt, Copyright 1984, Pergamon Press, (168).

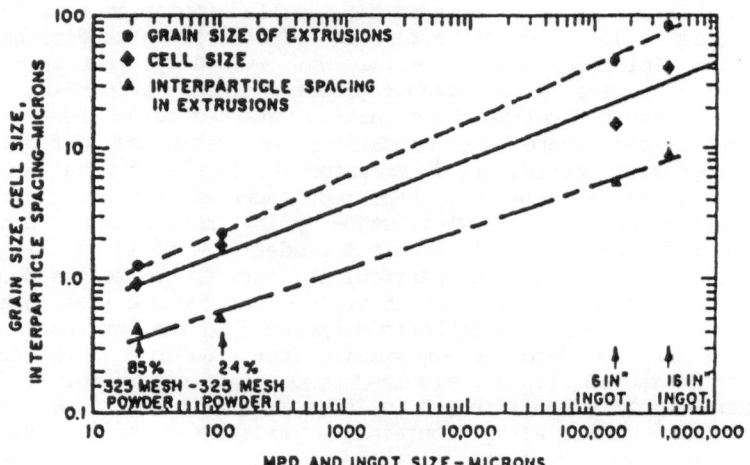

Figure 19: Dendrite cell size as-solidified, grain size after extrusion and segregate interparticle spacing after extrusion as a function of powder mean particle size (MPD) or ingot size for Al-5.9Zn-2.6Mg-1.8Cu-0.8Fe-0.8Ni alloy (in wt%). From ref. 142, Powder Metallurgy for High Performance Application, edited by John J. Burke and Volker Weiss. Proceedings of the 18th Sagamore Army Materials Research Conference. Syracuse, N. Y.: Syracuse University Press, 1972.

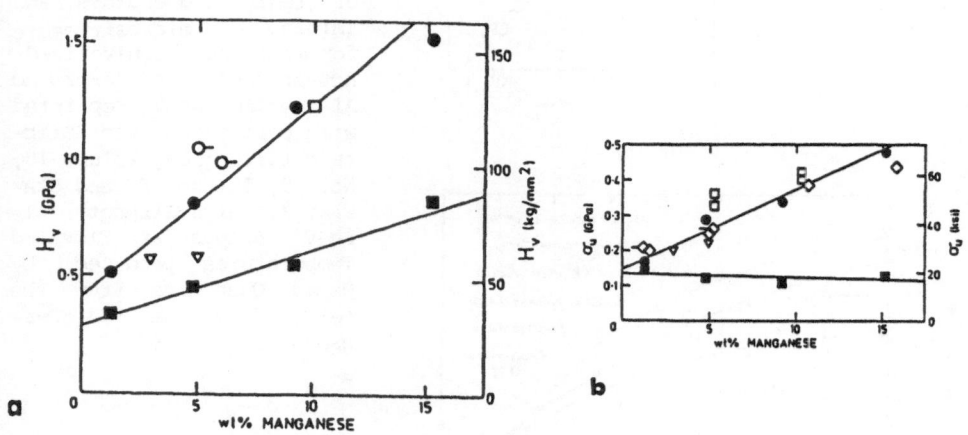

Figure 20: Hardness H $_v$ and ultimate strength σ_u as a function of Mn-content of Al-Mn alloys. Filled points: from bar 16mm diameter extruded at 300 °C from billet 73mm in diameter of ● canned splat-quenched flake ■ cast ingot. Open points: literature data for ◇-◇▽ extrusions from atomized pre-alloyed powder, ○ forgings of atomized pre-alloyed powder, □ rolled from spray-deposited material. From ref. 173.

4. SUMMARY AND CONCLUSIONS

Progress has been made in exploring the possibilities of a wide variety of methods for generating rapidly-solidified particulate. Understanding of the mechanisms by which the various methods form and solidify the particulate is as yet very much in its infancy and the relationship between operating variables and rapidly-solidified structure remains to be determined for many of these processes. In view of uncertainties concerning applicable cooling and solidification conditions for particular processes, there is some merit in ranking methods of making RS particulate in terms of their ability to produce a given microstructure or constitution at a given particle size. Evaluation of properties is a function of RS microstructure or constitution and of variables in any subsequent consolidation remain fruitful areas of further study.

5. REFERENCES

1. Jones H, in "Ultrarapid Quenching of Liquid Alloys",ed. H.Herman, Treatise on Materials Science and Technology, Vol. 20, ,Academic Press, New York, 1981, 1-71.
2. Jones H: Rapid Solidification of Metals and Alloys, Monograph No. 8, The Institution of Metallurgists, London, 1982, Ch.2.
3. Savage SJ and Froes FH, J. Metals, 1984,36 (4), 20-23.
4. Maringer RE, SAMPE Quart., 1980, 11 (4), 30-34.
5. Lawley A, J. Metals, 1981, 33 (1), 13-18.
6. Grant NJ;(a) in "Advances in Powder Metallurgy", ed. Chin GY, American Society for Metals, Metals Park, Ohio, 1982, 1-21; (b) in "High Strength Aluminium Powder Alloys", ed. Koczak MJ and Hildeman GJ, The Met. Soc. of AIME, Warrendale, Pa, 1982, 3-18 and J. Metals, 1983, 35 (1), 20-27.
7. Miller SA; in "Amorphous Metallic Alloys", ed. Luborsky FE, Butterworths, London, 1983, 506-520.
8. Butler IG, Kurz W, Gillot J and Lux B, Fibre Sci. and Techn., 1972, 5, 243-262.
9. Hubert JC, Mollard F and Lux B, Z. Metallkunde, 1973, 64, 835-843.
10. Manfré G and Servi G, Wire Industry, 1975, 42, 281-286.
11. Anthony TR and Cline HE, (a) J. Appl. Phys., 1978, 49, 829-837, (b) Cline HE and Anthony TR, G.E. Corporate R & D Report, 78CRD066, April 1978.
12. Ohnaka I and Fukusako T, J. Jap. Inst. Light Metals, 1978, 42, 415-124.
13. Liebermann H H, J. Appl. Phys., 1979, 50, 6773-8.
14. Aldinger F, Linck E and Clausen N, in "Modern Developments in Powder Metallurgy", Vol 9, Ed. Hausner H H and Taubenblat P V, MPIF/APMI, Princeton, NJ, 1977, 141-150.
15. Lacy L L et al, (a) J. Cryst. Growth, 1981, 51, 47-60; (b) U.S. Pat. 4278083, 3rd Feb. 1981, (c) J. Appl. Phys., 1982, 53, 682-9, (d) in "Materials Processing in the Reduced Gravity Environment of Space", ed. Rindone G E, Elsevier North Holland, N.Y., 1982, 87-94.
16. Steinberg J, Lord A E, Lacy L L and Johnson J, Appl. Phys. Lett., 1981, 38, 135-7.
17. Kendall J M et al, J. Vac. Sci. Technol., 1982, 20, 1091-3.
18. Lee M C, Kendall J M and Johnson W L, Appl. Phys. Lett., 1982, 40, 382-4.
19. Lee M C et al, as ref. 17, 95-104 and 105-113.
20. Johnson W C and Lee M C, J. Vac. Sci. Technol. A, 1983, A1, 1568-70.

21. Lee M C, (a) SAMPE Journal, Nov./Dec. 1983, 7-11; (b) in Rapidly
     Quenched Metals, ed. Steeb S and Warlimont H, Elsevier North Holland,
     N.Y., 1985, 119-122.
22. Matei G, Biczak W, Huppmann W J and Claussen N, as ref. 14, 153-9.
23. Lierke E G and Griesshammer G, Ultrasonics, 1967, 5, 224-8
24. Ruthardt R and Lierke E G, in "Modern Developments in Powder Metallur-
     gy", Vol. 12, ed. Hausner H H et al, APMI/MPIF, Princeton, N.J., 1981,
     105-111.
25. Klar E and Shafer W M, in "Powder Metallurgy for High Performance
     Applications" ed. Burke J J and Weiss V, Syracuse University Press,
     Syracuse, N.Y. 1972, 57-68.
26. Widmer R, ibid., 69-84.
27. Gummerson P U, ibid., 27-55.
28. Miller S A and Murphy R J, Scripta Met., 1979, 23, 673-6.
29. Liddiard P D, in "P/M Aerospace Materials", Vol. 1, M.P.R. Publishing
     Services, Shrewsbury, 1984, paper 26.
30. Smith P, ibid., paper 27, and Metal Powd. Rep., 1985, 40, 159-161.
31. Fortman W K and Ullman S, Metal Powder Report, 1984, 39, 259-261.
32. Miller S A, presented at RQ5-Würzburg, Sept. 1984, paper N38.
33. Couper M J and Singer R F, as ref. 21b, 1737-42.
34. Wentzell J M, J. Vac. Sci. Technol., 1974, 11, 1035-37; S.M.E. Tech.,
     Paper MF 72-503, 1972.
35. Singer A R E and Roche A D, as ref. 14, 127-140; Powder Met., 1980 23,
     81-85.
36. Sankaran K K and Grant N J, Mater. Sci. Eng., 1980, 44, 213-217.
37. Ishii H, Naka M, Masumoto T, (a) Sci. Rep. Res. Inst., Tohoku Univ. A,
     1981, A29, 343-350; and (b) in "Rapidly Quenched Metals", ed. Masu-
     moto T and Suzuki H, The Japan Inst. of Metals, Sendai, 1982, 35-8.
38. Carbonara R S, Raman R V and Clauer A H, as ref. 37b, 141-4.
39. Onoyama T, Ando O and Minakuta T, U.K. Patent Appl., GB 2118080A,
     26th October 1983.
40. Ohnaka I, Yamauchi I, Morimoto M and Fukusako T, J. Jap. Inst. Met.,
     1983, 47, 1010-15.
41. Ray R, U.S. Patents 4221587, 9th Sept. 1980; 4326841, 27th April 1982.
42. Goetzel C G, "Treatise on Powder Metallurgy", Vol. 1, Interscience,
     N.Y. 1049, 43-5.
43. Ohnaka I, Fukusako T and Tsutsumi H., J. Jap. Inst. Met., 1982, 46,
     1095-102.
44. Yamauchi I, Kawamoto S, Ohnaka I and Fukusako T, ibid., 1983, 47,
     1016-21.
45. Raman R V, Patel A N and Carbonara R S, Progr. in Powd. Met., 1982,
     38; 99-105; Metal Powd. Report, 1984, 39, 105-7.
46. Narasimha Rao and Sekhar J A, Mater. Lett., 1984, 2, 407-410.
47. Daugherty T S, (a) Progr. Powder Met., 1963, 19, 146-151; (b) J.
     Metals, 1964, 16, No. 10, 827-830; (c) Powder Met., 1968, 11, 342-357.
48. Cox A R, Moore J B and van Reuth E C, in "Superalloys: Metallurgy and
     Manufacture", Proc. 3rd. Internat. Symp., ed. Kear E H et al, Clai-
     tor's, Baton Rouge, La, 1976, 45-53.
49. Kaufmann A R and Muller W C, in "Beryllium Technology", Vol. 1, ed.
     Schetky L M and Johnson H A, Gordon and Breach, N.Y., 1966, 629-646,
     Kaufmann A R, U.S. Patent, 3099041, 30th July 1963.
50. Sastry S M L, Peng T C, Meschter P J and O'Neal J E, J. Metals, 1983,
     35, No. 9, 21-28; Metal Powder Report, 1984, 39, 537-8.
51. Konitzer D G, Walters K W, Heiser E L and Fraser H L, Met. Trans. B,
     1984, 15B, 149-153.

52. Devillard and Herteman J P, in "Powder Metallurgy of Titanium Alloys", ed. Froes F H and Smugeresky J E, The Met. Soc. AIME, Warrendale, Pa, 1980, 59-70.
53. Loewenstein P, (a) Progr. Powder Met., 1981, 37, 9-21; (b) Metal Powder Report, 1981, 36, 59-64.
54. Hodkin D J, Sutcliffe P W, Mardon P G and Russell L E, Powder Met., 1973, 16, 277-313.
55. Perel J et al, (a) in "Rapid Solidification Processing: Principles and Technologies", ed. Mehrabian R et al, Claitor's, Baton Rouge, La, 1978, 258-269, (b) ibid. II, 1980, 287-293, (c) in "Advances in Metal Processing", ed. Burke J J et al, Plenum, N.Y., 1981, 79-89, (d) in "Rapidly Solidified Amorphous and Crystalline Alloys", ed. Kear B H et al, Elsevier North Holland, N.Y., 1982, 131-6.
56. Clampitt R et al, in "Rapidly Quenched Metals III", Vol. 1, ed. Cantor B, The Metals Society, 1978, 57-62.
57. Maringer R E and Mobley C E, J. Vac. Sci. Technol., 1974, 11, 1067-71, Maringer R E, Rudnick A and Mobley C E, U.S. Pat. 3904344, 9th Sept. 1975.
58. Ray R et al, (a) Met. Progr., 1982, 121, No.7, 29-31; (b) as ref. 38, 1515-19; (c) J. Metals, 1983, 35(6), 30-35; (d) Metal Powder Report, 1984, 39, 287-9; (e) U.S. Patents 4347076, 31st Aug. 1982, 4359352, 16th Nov. 1982; 4379720, 12th April 1983; 4402745, 6th Sept. 1983; 4400212, 23rd Aug. 1983; 4405368, 20th Sept. 1983; 4403115, 7th Febr. 1984.
59. Whang S H and Giessen B C, Mater. Letters, 1984, 2, 230-1.
60. Ray R, U.S. Patent 4290808, 22nd Sept. 1981.
61. Maeland A J and Libowitz G G, Mater. Letters, 1982, 1, 3-5.
62. Berkowitz A E and Walter J L, (a) in "Rapid Solidification Processing: Principles and Technologies II", ed. Mehrabian R et al, Claitor's Baton Rouge, La., 1980, 294-305; (b) Mater. Sci. Eng., 1982, 55, 275-87.
63. Enokizono M and Narita M, Jap. J. Appl. Phys., 1981, 20, 2423-4.
64. Rasmussen D H, Perepezko J H and Loper C R, in "Rapidly Quenched Metals", ed. Grant N J and Giessen B C, MIT Press, Cambridge, Ma., 1976, 51-58.
65. Perepezko J H, Mater. Sci. Eng., 1984, 65, 125-135.
66. Kattamis T Z and Mehrabian R, J. Vac. Sci. Technol., 1974, 11, 1118-22.
67. Lux B, Haour G and Mollard F, as ref. 62a, 429-439.
68. Drehman A J, Greer A L and Turnbull D, Appl. Phys. Lett., 1982, 41, 716-17.
69. Ohnuma S., Nakanouchi Y and Masumoto T, as ref. 21 (b), 1117-24.
70. Koch C C, Cavin O B, McKamey C G and Scarborough J O, Appl. Phys. Lett., 1983, 43, 1017-19.
71. Schwarz R B, presented at RQ5-Würzburg, Sept. 194, paper K71.
72. Yeh X L et al, Appl. Phys. Lett., 1983, 42, 242-244.
73. Schwarz R B and Johnson W L, Phys. Rev. Lett., 1983, 51, 415-418.
74. Van Rossum M, Nicolet M A and Johnson W L, Phys. Rev. B., 1984, 20, 5598-5503.
75. Atzmon M, Verhoeven J D, Gibson E D and Johnson W L, Appl. Phys. Lett., 1984, 45, 1052-53.
76. Singer A R E, (a) Light Metal Age, 1974, 32 (9,10), 5-8, (b) as ref. 55a, 154-64, (c) Proc. Agard Conf. 256 "Advanced Fabrication Techniques", Florence 1978, Publ. 1979, Paper 19; (d) Powder Met. 1980, 23, 172-5.

77. Kim M H and Jones H (a) as ref. 37b, 85-88; (b) as ref. 21b, 139-142.
78. Singer A R E and Kisakurek S E, Metals Technol., 1976, 3, 565-570.
79. Singer A R E, Hodkin D J, Sutcliffe P W and Mardon P G, 1983, 10, 105-110.
80. Shingu P H., Shimomura K and Ozaki R, Trans. Jap. Inst. Met., 1979, 20, 33-5.
81. Miura H, Isa S, Omura K and Tanigumi N, Trans. Jap. Inst. Met., 1981, 22 597-606; as ref. 37(b) 43-6.
82. Warlimont H and Kunzmann P, as ref. 64, 197-204.
83. Moss M, Acta Mat., 1968, 68, 321-6; Moss M and Schuster D M, Trans. Amer. Soc. Metals, 1969, 62, 201-5.
84. Krishnanand K D and Cahn R W, as ref. 64, 67-75; Cahn R W, as ref. 55a, 129-139.
85. Giessen B C, Madhara N M, Murphy R J, Ray R and Surette J, Met. Trans. A, 1977, 8A, 364-6.
86. Jackson M R, Rairden J R, Smith J S and Smith R W, J. Metals, 1981, 33 (11), 23-26.
87. Singer A R E, Metals Technology, 1984, 11, 99-104; U.S. Patent 4224356, 23rd Sept. 1980.
88. Jones H, as ref. 64, 1-27.
89. Singer A R E, Coombs J S and Leatham A G, in "Modern Developments in Powder Metallurgy", Vol. 8, ed. Hausner H H and Smith W E, Plenum, N.Y. 1974, 263-280.
90. Pond R B and Winter J M, Mater. Sci. Eng., 1976, 23, 87-89.
91. Thompson J S, J. Inst. Met., 1948, 74, 101-132.
92. Tamura K and Takada T, J. Jap. Soc. Powd. Met., 1963, 10, 153-9; Trans. Nat. Res. Inst. Metals, 1963, 5, 82-86.
93. Hirata T, J. Jap. Soc. Powd. Met., 1964, 11, 29-32.
94. Silaev A F, Soc. Powd. Met. Met. Ceramics, 1967, 5(53), 350-353.
95. Nichiporenko O S, ibid, 1967, 60, 947-949.
96. Balasubramanian M S N and Tendolkar, Indian J. Technol., 1968, 6, 205-212.
97. Tamura K and Wanikawa S, Trans. Nat. Res. Inst. Metals, 1968, 10, 196-7.
98. Small S and Bruce T J, Internat. J. Powd. Met., 1968, 4(3), 7-17.
99. Rao P et al, (a) J. Vac. Sci.Technol, 1970, 7, 5132-6, (b) in "Chemica '70", Butterworths, 1971, 1-16.
100. Domsa A and Berkovits S, in "Modern Developments in Powder Metallurgy" Vol. 4, ed. Hausner H H, Plenum, N.Y., 1971, 63-74
101. Grandzol R J and Tallmadge J A, (a) AIChEJ., 1973, 19, 1149-58; (b) Internat. J. Powd. Met. and Powd. Technol., 1975, 11, 103-114.
102. Sundaresan R, Krishnan R V and Raghuram A C, PMAI News Letter, 1977 3(2), 4-9.
103. Glickstein M R, Patterson R J and Shockey N E, in ref. 62, 46-62.
104. Lawley A, Ann. Rev. Mater. Sci., 1978, 8, 49-71.
105. See J B and Johnston G H, Powder Technol., 1978, 21, 119-133.
106. Tallmadge J A, in "Powder Metallurgy Processing: New Techniques and Analyses", ed. Kuhn H A and Lawley A, Academic, N.Y., 1978, 1-32.
107. Dunkley J J, Wire Industry, 1978, 48, 365-371.
108. Vanstone R H, Rizzo F J and Radavich J F, as ref. 62a, 260-272.
109. Anand A, Kaufman A J and Grant N J, as ref. 62a, 273-285.
110. Champagne B and Angers R, (a) Internat. J. Powd. Met. and Powd. Technol., 1980, 16, 259-67; (b) as ref 24, 83-104; (c) Powd. Met. Internat., 1984, 16, 125-8.
111. Roberts P R and Loewenstein P, as ref. 52, 21-35.

112. Mehrotra S P and Khedkar P Y, Trans. Ind. Inst. Met., 1980, <u>33</u>, 361-6.
113. Rao K P and Mehrotra S P, as ref. 24, 113-130.
114. Neubing H C, Powder Met. Internat. 1981, <u>13</u>, 74-78.
115. Tien J K and Howson T E, as ref. 6a, 155-187.
116. Patterson R J, Ledwith D L and Dwyer J C, in "Processing of Metal and Ceramic Powders", ed. German R M and Lay K W, The Met. Soc. of AIME Warrendale, Pa., 1982, 33-47.
117. Hildeman G J, Lege D J and Vasudevan A K, as ref. 6b, 249-76.
118. Uygur E M, Metal Powd. Rep., 1982, <u>37</u>, 229-238.
119. Hugo P E and German R M, Internat. J. Powd. Met. and Powd. Technol., 1982, <u>18</u>, 301-311; as ref. 116, 49-64.
120. Nagarjuna N, Mukherjee A and Mukunda P G, Internat. J. Powd. Met. and Powd. Technol., 1983, <u>19</u>, 91-96.
121. Tsipunov, Temovoi Y F, Kuratchenko S B and Kuimova O M, Sov. Powd. Met. Met. Ceramics, 1983, <u>22</u>, 788-793.
122. Koria S C and Lange K W, Ironmaking and Steelmaking, 1983, <u>10</u>, 160-8.
123. Meschter P J, O'Neal J E and Lederich R J, in "Aluminium-Lithium Alloys II", ed. Starke E A and Sanders T H, The Met. Soc. of AIME, Warrendale, Pa., 1984, 419-432.
124. Brooks R G, Leatham A G and Moore C. in "Powder Metallurgy Super-alloys: Aerospace Materials for the 1980s', Vol. 2, Metal Powder Report, Shrewsbury, 1980.
125. Dunstan G R et al, Progr. in Powder Met., 1981, <u>37</u>, 23-38.
126. Klar E and Fesko J W, ibid., 47-66.
127. Voss D P, in "Modern Developments in Powder Metallurgy", Vol. 13, ed. Hausner H H, MPIF/APMI, Princeton, N.J., 1981, 467-481.
128. Anon., Metal Powder Report, 1983, 38, 563-6.
129. Kim M H, Thesis Ph D, Sheffield, 1982, quoted by Jones H in "Rapidly Solidified Metastable Materials", ed. Kear B H and Giessen B C, Elsevier North-Holland, New York, 1984, 303-315.
130. Walton W H and Prewett W C, Proc. Phys. Soc. 1949, <u>62</u>, 341-350.
131. Fraser R P and Eisenklam E P, Trans. Inst. Chem. Eng., 1956, <u>34</u>, 294-319.
132. Busk R S, Light Metals, 1960, <u>23</u>, 197-200.
133. Dunskii V F and Nikitin N V, J. Eng. Phys., 1965, 9(1), 41-45.
134. Fraser R P, Dombrowski N and Routley J H, J. Inst. Fuel, 1963, <u>36</u>, 316-329.
135. Lubanska H, J. Metals, 1970, 22(2), 45-49.
136. Bradley D, J. Phys. D: Appl. Phys., 1973, <u>6</u>, 1724-36, 2267-72.
137. Mehrotra S P, Powd. Met. Internat., 1981, <u>31</u>, 80-84, 132-5.
138. Joly P A and Mehrabian R, J. Mater. Sci., 1974, <u>9</u>, 1446-55.
139. Kattamis T Z and Mehrabian R, J. Mater. Sci., 1974, <u>9</u>, 1040-3.
140. Acrivos C, J. Mater. Sci., 1976, 11, 1159-60, 1752-3.
141. Towner R J, Met. Progr., 1958, 73(5), 70-76, 176, 178.
142. Lyle J P and Cebulak W S, (a) as ref. 25, 231-254, (b) Met. Trans. A, 1975, <u>6A</u>, 685-9, (c) as ref. 55a, 324-333.
143. Domalavage P K, Grant N J and Gefen Y, Met. Trans. A., 1983, <u>14A</u>, 1599-1606.
144. Clyne T W, Ricks R A and Goodhew P J, as ref. 21b, 903-6.
145. Levi C G and Mehrabian R, Met. Trans. A, 1982, 13A, 13-23.
146. Kaufman M J and Fraser H L, Met. Trans. A, 1983, 14A, 623-33.
147. Couper M J (a) as ref. 29, paper 28, (b) Baumann R and Couper M J, to be published.
148. Wang W and Grant N J, Internat. J. Rapid Solidification, 1984/5, 1(2).

149. Takigawa H, Manto H, Kawai N and Homma K. Powder Met.. 1981. 24.
     196-202.
150. Kato T, Metal Powder Report, 1983, 38, 505-8.
151. Holiday P R, Cox A R and Patterson R J, as ref. 55a, 246-257.
152. Cosandey F, Kissinger R D and Tien J K, as ref. 55d, 173-8.
153. VanStone R H, Rizzo R J and Radavich J F, as ref. 62a, 260-272.
154. Ranger A A and Nicholls J A, AIAA Journal, 1969, 7, 285-290.
155. Kurten H et al, Chem.-Ing.-Tech., 1966, 38, 941-8.
156. Anon., Metal Powder Report, 1983, 38, 513-14.
157. Broderick T F, Jackson A G, Jones H and Froes F H, Met.Trans. A,
     November 1985, 16 (11).
158. Boswell P G and Chadwick G A, Scripta Met., 1977, 11, 459-465.
159. Drehman A J and Turnbull D, Scripta Met., 1981, 15, 543-8.
160. Yamaguchi T and Narita K, (a) IEEE Trans. Magnetics, Vol. MAG-13,
     No. 5, Sept. 1977, 1621-3; (b) Appl. Phys. Lett., 1978, 33, 468-470.
161. Safai S and Herman H, as ref. 1, 183-214.
162. Munawar Chaudri M and Hutchings I M, J. Mater. Sci. Lett., 1984, 3
     79-82.
163. Walter J L and Berkowitz A E, Mater. Sci. Eng., 1984, 67, 169-177.
164. Berkowitz A E, Livingston J D and Walter J L, J. Appl. Phys., 1984,
     55, 2106-8.
165. Paton N E, Bampton C C and Ghosh A K, in Proc. ICSMA6, Vol. 2, ed.
     Gifkins R C, Pergamon, N.Y., 1983, 713-719.
166. Thursfield G et al, (a) Fizika, 1970, 2 Suppl. 2, paper 19, (b) J.
     Mater. Sci., 1974, 9, 1644-60.
167. Dickson J, Okazaki K and Sanders T H, U.S. Pat. 4389258, 21st June
     1983.
168. Skinner D J and Okazaki K, Scripta Met., 1984, 18, 905-9.
169. Sheppard T and Chare P J M, Powder Met., 1972, 15, 17-41.
170. Chare P J M and Sheppard T, (a) Powder Met., 1973, 16, 437-58, (b)
     Internat. J. Powd. Met. and Powd. Technol., 1974, 10, 203-215.
171. Sheppard T, McShane H B, Zaidi M A and Tan G H, J. Mech. Work.Technol.
     1983, 8, 43-70.
172. Sheppard T and Greasley, Powder Met., 1978, 21, 155-162.
173. Savage S J and Jones H, as ref. 37b, 159-162.
174. Grant N J, Proc. 3rd. Nordic High Temperature Symposium, Vol. 1, ed.
     Rasmussen J G, Polyteknisch Forlag, 1973.

ABSTRACTED DISCUSSION OF THE PAPER BY H. JONES

Participants: L. Ajran, O. Arkens, L. Arnberg, H.W. Bergmann, R.W. Cahn, B. Cantor, H.L. Fraser, N.J. Grant, S. Hock, R.E. Lewis, J.H. Perepezko, R.E. Maringer, P.R. Sahm, T. Sheppard and J.V. Wood

The discussion was concerned mainly with the importance of powder particle size, shape and structure in relation to further processability (e.g. degassing, compacting, extrusion/hot working) and final properties.

There was concern that the increased surface area of finer powders having higher cooling rates tends however to introduce into the consolidate increased amounts of any surface film material (e.g. oxide) formed during powder production. There were toxicity and pyrophoricity hazards with sub-10 μm powders and there was a need for consolidation studies to be carried out on 100 per cent sub-10 μm fractions to ascertain property benefits.

The point was made that particle structure is not simply determined by particle size. Two droplets of identical size and the same cooling rate can exhibit markedly different undercoolings depending on the presence or not of uncleants within or at the surface of each droplet so that very different microstructures result on solidification.

More information was needed on the pros and cons of using splat flake or pulverized planar-flow-cast material instead of fine powder particulate since the chill-block quenching involved should be able to maintain a given initial structure to an increased thickness compared with melt atomization. The importance of generating an initial particulate structure that can be readily consolidated was emphasized, preferably with the capability of enhancing properties obtained by rapid solidification (rather than degrading them) during consolidation or by further treatment. Attention was drawn to the importance of control and optimization at every stage of processing in what was evidently a multiparameter process, and there is ample scope for work that will lead to better definition and understanding of processing/structure/property relationships for all stages of processing.

RAPID SOLIDIFICATION TECHNOLOGY: THE POTENTIAL FOR INNOVATIVE ALLOY PROPERTIES

C.M. ADAM
Materials Laboratory, Allied Corporation
Morristown, NJ 07960, USA

ABSTRACT

During the last decade, alloys of novel composition, prone to severe microsegregation when solidified by conventional casting, have been developed for industrial applications. A variety of industrial techniques have been developed for producing rapidly solidified alloys, all of which produce alloy material of dimensions 10-100 μm. In many alloy systems solidification commences in a strongly undercooled metallic droplet or film and the surrounding environment allows adiabatic or quasi-adiabatic solidification. Progress has been made in understanding the factors controlling microstructure during solidification of such alloys at high solid-liquid interfacial growth velocities. This review will cover the principal systems currently producing rapidly solidified alloys in tonnage quantity, appropriate mathematical models of the solidification process, and factors responsible for the evolution of microstructure. Some applications of rapidly solidified alloys will also be reviewed, from amorphous metallic foils for magnetic device applications, to consolidated powder metallurgy components for structural applications.

1. INTRODUCTION

A widespread perception of the importance of rapid solidification can be attributed to the novel engineering properties possible from new combinations of the elements of the periodic table. The commercial development of ferromagnetic metallic glasses for transformer applications (1,2) is one example of a new technology based on amorphous Fe-B-Si alloys, which was unknown a decade ago. The development of new hard magnets based on the rapid solidification of crystalline Fe-Nd-B alloys (3) is another example. The benefits of rapid solidification have been widely reviewed for a large number of conventional crystalline alloy systems (4,5), and there has been a widespread tendency to attempt to apply this technology to the design of new alloys. The early pragmatic belief that the technology, per se, should confer substantial engineering benefits to existing commercial superalloys quickly lead to excursions into new alloy compositions (6).

Modern engineering alloys have evolved during the last hundred years from production practices generally involving large ingot casting, high temperature solute homogenization, thermomechanical reduction to specific cross-sectional shapes, and subsequent fabrication, frequently followed by additional heat treatments to optimize properties. In many instances, therefore, alloy properties reflect compromises in alloy chemistry imposed by fabrication practice restrictions, rather than by the widest trade-offs in desireable properties from broader alloy chemistries available to the alloy designer. The use of Fe-3wt.% Si alloys for 0.3 and 0.25 mm transformer sheet, rather than the lower core-loss alloy Fe-6.5wt.% Si with zero magnetostriction, is one example of an alloy chemistry compromise forced by our inability to hot-roll cast Fe-6.5wt.% Si ingots into 0.3 mm thick sheet of appropriate (110) <001> Goss texture. The extreme difficulty of forging the cast superalloys Astroloy and IN100 into gas-turbine engine discs, is another example of fabrication practices influencing alloy properties, rather than alloy chemistry. In new high speed steels, powder metallurgy has allowed development of alloy chemistries and properties unattainable in conventional ingot high speed steels, as with the Crucible Fe-10V-2C alloys (7). With the removal of the concomitant metallurgical evils of solute segregation and coarse alloy phases, alloy designers should now be free to employ a wider range of alloy chemistries than has hitherto been available.

2. RAPID SOLIDIFICATION METHODS

It is worth devoting some short time to an archival example which provides insight into the development of a superior engineering alloy, and results from work on the production of lead shot from the early part of the last century.

Figure 1 shows the Merchant's Shot Tower in Baltimore, built in 1828 and in continuous operation producing some half-million 25 lb. bags of spherical lead shot each year until 1892. The tower is roughly 230 ft. (70 ˙m) tall. The production of shot involved various lead alloys, and the beneficial influence of several alloying elements was broadly appreciated at that time. Arsenic encouraged the formation of spherical shot, presumably through the modification of surface tension. Antimony and copper produced harder shot.

Relatively simple heat-transfer calculations, using equation 1

$$hd/k = 2 + 0.6 Re^{1/2} Pr^{1/3} \tag{1}$$

(where h = heat transfer coefficient, k = thermal conductivity, d = particle diameter, Re = Reynolds number for velocity v, Pr = Prandtl number for air), show that 2 mm diameter spheres with a velocity of 30 m/s under free-fall conditions over 50 m (Reynolds number $\simeq$ 4000, Prandtl number $\simeq$ 0.9) experience a heat transfer coefficient, h, of 3 kW/m^2 sec K, leading to a droplet cooling rate greater than 10^4 K/s. One can conclude that such lead droplets were then rapidly solidified some 150 years ago. Although the solid solubility of antimony in lead is substantial, 3.5wt.% at 252°C, the solubility of copper is not and has been reported to be less than 0.007wt.% at 326 °C (8). Without further direct evidence, one may speculate that the improved hardening produced by copper was an early example of increased solid solubility produced by rapid solidification.

2.1 Atomization

Conventional gas-atomization of liquid metals is a relatively mature technology, having been used since the 1930's to produce a wide variety of metallic powders for diverse applications. Each Space Shuttle launch, for example, consumes 160,000 kg of atomized aluminum powder (9) as part of the solid fuel propellent mixture. In atomization processes alloy melts are finely divided into roughly spherical particles of 10-50 µm average diameter, although the size distribution may vary from 1 to 200 µm, depending on the particular process.

Figure 1: Merchant's Shot Tower, Baltimore.

In general, gas atomization processes rely on high energy gas streams to fragment a coaxially impinging liquid metal stream. Liquid filaments and ligaments are produced from liquid sheets by two-dimensional capillarity-driven breakdown of the liquid volume. The liquid cylinders thus produced then undergo rapid Rayleigh surface instability driven spheroidization (10) with a velocity $V = (2\pi\sigma/\rho\lambda)^{1/2}$, where and are the liquid surface tension and density, respectively, and is the instability wavelength. A lower bound solution to the spheroidization time may be calculated from the Rayleigh solution $t = (\rho\lambda^3/2\pi\sigma)^{1/2}$. Table 1 shows the calculated spheroidization times for instabilities of various wavelength in liquid aluminum, copper, iron and nickel, for example. An important feature of this calculation is that it shows that spherical powder production will be impossible if the liquid cooling rate is greater than the spheroidization rate; in this case, solidification produces irregular and fragmented powder, frequently containing entrained gas in the convoluted powder mass, and having a higher specific surface area than its spherical counterpart. Depending upon particular applications these differences may assume importance. Catalysts and powder fuels require high specific surface areas, whilst powders destined for use as structural aerospace applications, do not. Air atomized aluminum powders, for example, frequently display the ragged characteristics of solidification before spheroidization, particularly at particle sizes in the 50-100 μm size range. Finer powders in the 10-20 μm size range (which are more typical of air atomization) frequently display more nearly perfect spheroidization in only the finer size fractions.

Table 1: Spheroidization Times for Various Liquid Metals*

Instability Wavelength (μm)	Mg	Al	Fe	Ni	Cu
			(microseconds)		
1	0.02	0.02	0.03	0.03	0.034
10	0.72	0.73	0.85	0.91	1.06
50	8.1	8.2	9.6	10.3	12
100	22	23	27	29	34
200	63	65	76	82	95

* Data for liquid metal densities and surface tensions from "A Handbook of Chemistry and Physics", 1970, CRC Publishing, Cleveland.

Microstructures of air-atomized aluminum alloy powders show greatly reduced microsegregation when compared to direct-chill cast ingot structures, and structural alloys show superior strength and corrosion resistance when consolidated by powder forging (11), not withstanding their non-spherical origin and attendant degassing problems. Measurements of the secondary dendrite arm spacings of such powder alloys yield values of 2 to 5 µm. Research workers in this field have assigned cooling rates of 10^3 - 10^5 K/s to such alloys by extrapolating data, like those of Spear and Gardner (12), by some two to three orders of magnitude beyond the original experimental range. The cooling rates derived from such microstructural data have been used to support the relative merits of various atomization processes; inclusion of data generated by considering different spherical atomized powders has significantly broadened our understanding of rapid solidification phenomena, however as will be shown later, dendritic structures do not represent the ultimate limit of this technology. Grant (13) has recently summarized a wide range of microstructural benefits associated with atomization.

During the 1960's considerable effort was devoted to improving the 1000K strength and high cycle fatigue properties of the nickel-base superalloys Waspaloy (Pratt and Whitney Aircraft) and Rene 41 (General Electric), intended for new gas-turbine engine disks to be placed in service in advanced engines during the 1970's. Allen, Athey and Moore (14) have described the development of forging of gas-atomized IN 100 and Mar M200 powders, employed principally to circumvent extensive microsegregation typical in such alloys when produced as-cast billets prior to forging. Argon atomization was used in the original work to produce spherical powder with dendrite arm spacings of 2-3 µm and vastly reduced microsegregation; after inert consolidation by extrusion and forging, the high cycle fatigue lifetime of IN 100 was increased by virtually an order of magnitude by simply removing from the microstructure coarse microsegregated phases. Using extrapolative techniques for nickel-base superalloys, similar to those employed for aluminum alloys, the cooling rate of 100 µm diameter powder was assumed to be $\simeq 10^4$ K/s. During the following decade significant progress was made by Pratt and Whitney in Florida in developing a more advanced inert gas atomization technique (15), which produced a range of nickel-base superalloy powders of sub-micron dendrite arm spacing. Several of these alloys also produced a very fine grained (1000 Å) equiaxed microstructure with much reduced microsegregation (16), which was subsequently referred to as a "microcrystalline" structure and to which was ascribed a cooling rate of 10^5 - 10^6 K/s. The basic atomization device consisted a rapidly spinning disc which produced powder by centrifugal atomization. This central rotary disc produced a normal distribution of powder droplets ranging from 10-200 µm in diameter, which were subsequently subjected to forced helium convective cooling used to produce droplet cooling rates between 10^5 and 10^7 K/s. This device is currently one of the most sophisticated "clean" atomization processes in production in North America; it represents some fifty years of evolutionary change in powder production, figure 2, producing some 10^9 powder particles per second in a helium environment. This rapid solidification rate (RSR) process has subsequently produced amorphous Ni-B-Si alloy powders (16), in one of the few cases where direct comparison between similar compositions made by melt-spinning and centrifugal atomization can be made. In melt-spinning, some 15-40 µm thick amorphous films of this composition can be produced, whilst in the centrifugal atomization process amorphous powder was produced in particles less than 30 µm diameter. The conclusion can therefore be drawn that the two processes are capable of generating roughly similar solidification rates in the -30 µm size regime. The work of Angers and Champagne (17) on the rotating electrode process (REP) bears some relevance to rotary atomization, clarifying the effect of liquid metal parameters (viscosity, density and surface tension) on the powder size distribution produced by the RSR process.

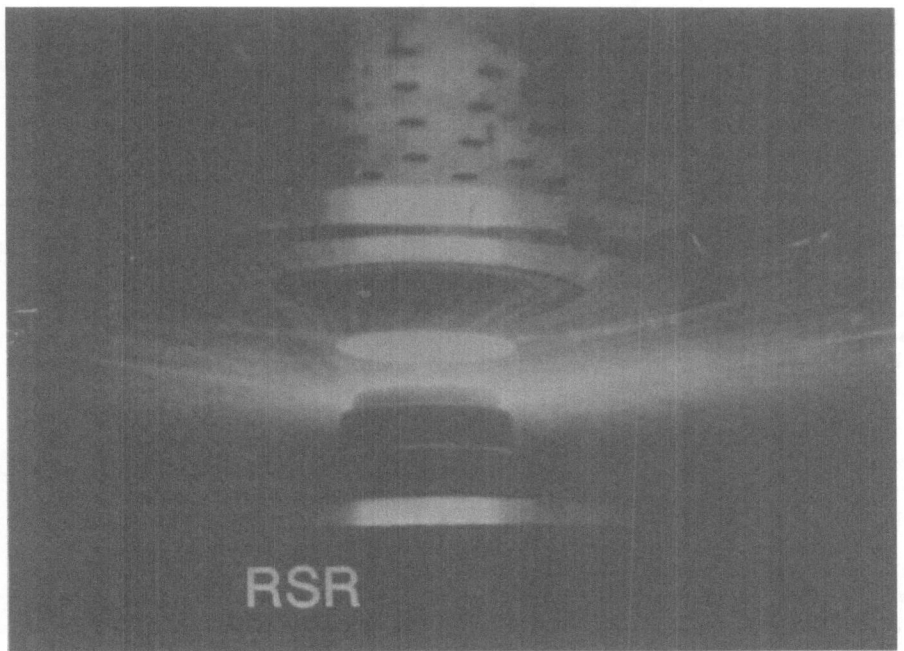

Figure 2: Pratt and Whitney Centrifugal Atomization.

2.2 Ribbon Casting

The development of equipment to produce rapidly solidified continuous filaments commenced with the work of Pond (18) in 1958. This melt spinning process was subsequently used during the 1960's to produce amorphous ribbons of several alloys, e.g. $Pd_{80}Si_{20}$ (19), as reported by Masumoto and Maddin. The process consisted of casting a thin free jet of liquid metal onto the inner surface of a rotating drum, and the prevailing design allowed centrifugal force to maintain adequate heat transfer between the ribbon and the quenching surface. An intrinsic disadvantage of this system was the relatively small quantities of material produced. Work at Allied Corporation commenced in the early 1970's and the development of various processes to produce scalable quantities of high quality amorphous ribbon has recently been reviewed by Davis et al. (20). Early work during this period by Chen and Polk (21) was performed on a double roller-quenching apparatus which simply squeezed liquid metal between two rapidly rotating rolls. Whilst this equipment was probably state-of-the art at that time and produced larger quantities of marginally acceptable amorphous material than had ever been available formerly, there were serious deficiencies preventing scale-up. Control of liquid metal flow into the roll gap and accurate positioning of the stream was difficult; the short contact distance with the roll surfaces produced marginally acceptable quenching, and roll gap deformation of the quenched ribbon frequently produced fracture. This process has recently been reviewed by Miyazawa (22), who analyzed both mass flow and heat transfer aspects of the process for the solidification of aluminum alloys. It is now apparent that the position of the solid-liquid interface during twin roll casting is dynamically unstable, at least during casting of amorphous alloys. During 1973, work at Allied Corporation (23) was successful in jet casting thin liquid streams onto the external surface of a water-cooled copper drum some 100 mm in diameter, rotating with a surface speed of $\simeq$ 30 m/s. This was, to some degree, unexpected, since it had been anticipated that the liquid jet would simply "bounce" off the cooling substrate. This jet casting process demonstration provided significant impetus to Allied's decision to produce metallic glass alloys in large scale and to ultimately offer them for sale commercially, after the further development of its Planar Flow Casting process. The jet casting process was successful in producing a wide range of amorphous alloys; however, the melt-puddle was not particularly stable and, as a result, the process is restricted to the production of ribbon less than about 5 mm wide.

This potentially serious production deficiency was solved by M.C. Narasimhan (24) who invented the Planar Flow Casting process (PFC) and provided the practical basis for what is now a routine production capability of 100 mm and 180 mm wide strip of 20-30 μm thickness. This Planar Flow Casting process (figure 3), inter alia, positions the casting nozzle in close proximity to the rotating chill block and employs a rectangular slot orifice. The molten pool of metal is constrained into a stable shape, as it is now understood, by balancing surface tension forces with the applied liquid metal pressure, to achieve a dynamically stable melt pool on the rotating substrate. Several parameters for stable operation include substrate velocity, liquid metal pressure, nozzle design and nozzle-substrate separation. Both jet casting and planar flow casting have been carried out with suitably protective shielding atmospheres, preventing oxidation of the liquid metal surface and preventing air entrapment under the liquid film. Normally reactive iron and aluminum base alloys have been routinely planar flow cast into strip with very little surface oxidation, and more reactive materials such as magnesium and titanium base alloys have been cast in vacuum. At present it does not appear to be economically feasible to produce large quantities of reactive alloys in large vacuum chambers; however, there appears to be no practical limit to the width of planar-flow cast strip cast in air.

Allied Rapid Solidification Process

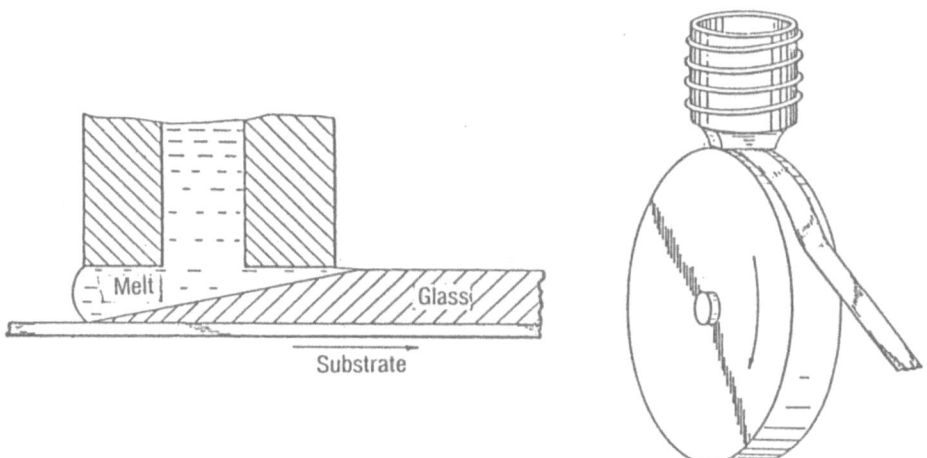

Figure 3: Allied Planar Flow Casting.

3. SOLIDIFICATION MODELS
    It has been considered obvious that liquid cooling rate involved in melt spinning and planar flow casting processes which rely largely on conductive cooling, should be higher than the cooling rates involved in atomization processes which rely predominantly on convective cooling. Microstructures produced by these two processes can show clear differences; the attainment of amorphous structures in a wide range of melt-spun alloys has been used to support this assumption of generally higher heat transfer rates in melt-spinning. However, very few cases exist where similar alloys have been studied when independently produced by the two processes. In alloys which are traditionally difficult to produce in the amorphous state, the dominant mode of solidification at liquid cooling rates of 10^6 - 10^7 K/s is the microcrystalline structure, consisting of a three dimensional assemblage of fine grains from 100 Å in size for some nickel base alloys (14,4,6) to several micrometers for some aluminum alloys (25). The arguments advanced by Boswell (25) for various mathematical treatments of both columnar and equiaxed growth include the critical questions of undercooling and nucleation, which, have been generally ignored in treatments of rapid solidification.
    Perepezko and co-workers (26,27) demonstrated that many of the effects of rapid solidification, viz. supersaturated solid-solutions, metastable phase formation, and microcrystalline structures, are produced during adiabatic solidification of very slowly cooled fine metallic droplets, when levels of undercooling approaching one third of the melting point, T_m, are attained. Following nucleation after substantial undercooling, ΔT, a solidifying interface rejects latent heat both into the supercooled liquid and into the newly formed solid. At the instant of nucleation in a strongly undercooled melt the solid-liquid interfacial growth velocity should ap-

proach the velocity of sound, with the growth velocity decaying rapidly as the supercooled liquid ahead of the interface absorbs latent heat during recalescence. The hypercooling limit, ΔT_H, has been defined as the latent heat per unit volume divided by the specific heat, L/C_p, and represents the maximum possible temperature rise following nucleation in a strongly under-cooled melt in an adiabatic system. During the period of recalescence the temperature of the region of the melt ahead of the solid-liquid may quickly approach the equilibrium melting temperature for the alloy, T_m, provided that the initial undercooling was less than the hypercooling limit. We can define the hypercooling ratio, $\Delta T/\Delta T_H$, as a rough measure of "rapid" adiabatic solidification, and presume that following nucleation and re-calescence the remainder of the available melt solidification in a slower post-recalescence regime. In our consideration of rapid solidification of strongly undercooled melts by either the melt spinning process or gas atomization we should, therefore, assume that undercooling prior to nuclea-tion plays a major role in developing microstructure, rather than the liquid cooling rate which led to that degree of undercooling. Subsequent extraction of heat from the system following nucleation then should deter-mine only the extent of post-recalescence solidification. Both Davies (28) and Boswell and Chadwick (29) have produced time-temperature transformation curves which illustrate how increased cooling rate might be expected to produce increased undercooling before nucleation of crystalline phases. This view of nucleation presumes that the system under investigation is one for which glass-formation is difficult (very low glass transition tempera-ture, T_g (3), associated with a steeply plunging T_0^* vs composition phase diagram (25)) and for which there are various competitive solidification modes in the crystalline metastability regimes. It is likely that competi-tive continuous cooling transformation curves for at least a few different morphological varieties should be considered (figure 4), involving competi-tion between massive partitionless solidification, restricted solute parti-tioning producing microcrystalline structures, and well-defined microcellu-lar and microeutectic solidification involving extensive solute partitio-ning (30), where the microcellular and microeutectic structures could contain different metastable phases. Alloys of composition substantially richer than the T_0-composition curve must solidify with extensive solute partitioning, one example of which is the microeutectic structure in Al-8Fe-2Mo (30, 37) and Al-12Fe-2V (46) alloys, while somewhat more dilute alloys solidify with a microcellular structure, designated Zone A, by Jones (39) in Al-8Fe binary alloys. Solidification closer to the T_0 curve for dilute aluminum alloys produces microcrystalline structures (25). In gene-ral, nucleation appears to be delayed in alloys with higher liquid solute contents, and the effect of increasing solute can be considered to move the liquid transformation curves toward longer times. For an appropriate coo-ling rate of $10^6 - 10^7$ K/s either amorphous structures or massively solidi-fying partitionless solidification structures appear to be produced. The former situation appears to be the case for aluminum-iron-silicon alloys for which amorphous structures are produced only at compositions near β - $Al_9Fe_2Si_2$, and for which Suzuki et al. (31) estimated a required undercoo-ling of 450 K and a reduced glass transition temperature of 0.47 (= T_g/T_m) attained by cooling at $\simeq 5 \times 10^6$ K/s. The icosohedral phase reported in Al-14Mn alloys (32), and the 0 phase reported in Al-12Fe-2V alloys (47) are examples of massive solidification following homogeneous nucleation.

-----

*T_0 is defined as the temperature at which solid and liquid of a given composition have the same free-energy; thus, it represents the upper limit temperature for composition-invariant solidification.

Structural Effect of Cooling Rate

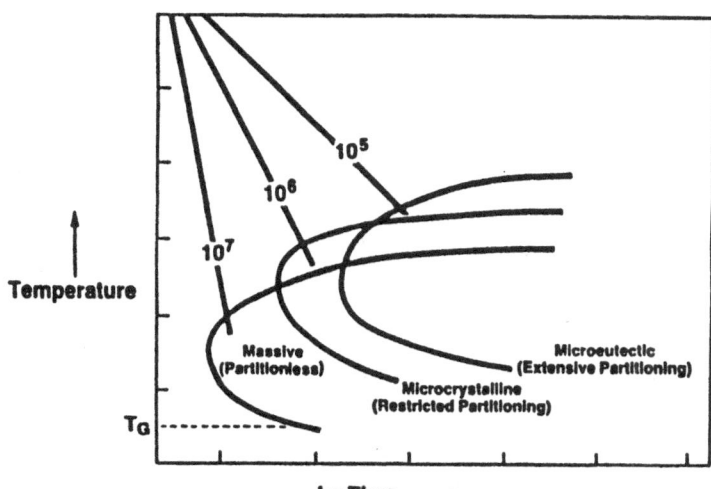

Figure 4: Liquid cooling transformation diagrams for nucleation and growth with various degrees of solute partitioning.

3.1 Atomization

Early heat transfer studies of the convective cooling of metallic droplets concentrated predominantly on the external heat transfer to the cooling gas, and presumed that the internal solidification process was largely controlled by the rate of heat removed from the droplet surface. As recently as 1980 (33), the influence of droplet undercooling was ignored in formal mathematical analyses of the solidification of crystalline metallic droplets. In considering the influence of undercooling on solidification velocity Gill et al. (34) showed that solidification consisted of two regimes. The first was the recalescence or rapid solidification stage where the undercooled liquid absorbed latent heat released at the advancing solid-liquid interface, with heat loss to the surrounding environment playing a minor role in the overall thermal balance. The second stage was one of somewhat slower pseudo-isothermal growth controlled by external heat transfer to the environment. The relative amounts of material solidified under the two different regimes depended on particle size and droplet undercooling, so that small particles of aluminum alloys of radius 5 µm, which were undercooled a few hundred degrees, solidified almost entirely in the very rapid recalescence regime, whilst particles of radius 100 µm, undercooled some fifty degrees, would be expected to solidify predominantly under conditions dictated by external heat transfer. Levi and Mehrabian (35) presented a later analysis for surface nucleation in which plots of dimensionless temperature as a function of the fraction of the droplet solidified agree substantially with the earlier analysis of Gill et al. (34). A further analysis of homogeneous nucleation within a liquid sphere combined two well-known heat-transfer situations (36) into a mathematical model for which a simple analytical solution was subsequently found. The results of these calculations show the effect of undercooling on the solid-liquid interfacial growth velocity for an aluminum droplet of 50 µm diameter (figure 5).

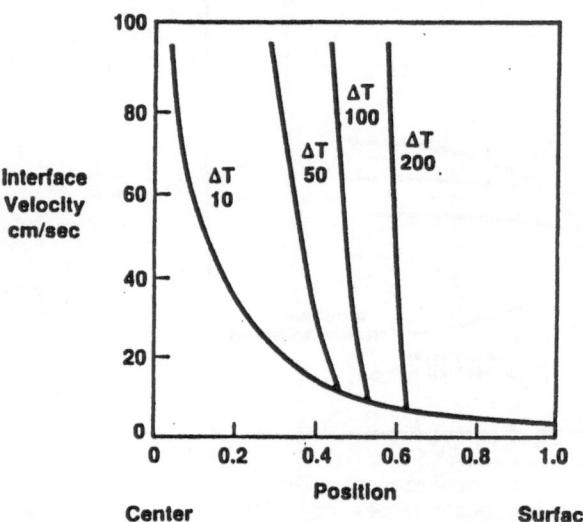

Spherical Solidification Model

50μm diameter, Center Nucleation

h = 2.85 x 10⁴ W/m²K

Figure 5: Solid-liquid interfacial velocity for a 50 μm diameter aluminum sphere; center nucleation with h = 28.5 kW/m² K.

The percentages of the volume solidified during the recalescence period is 2% ($\Delta T = 10K$), 14% ($\Delta T = 50K$), 27% ($\Delta T = 100K$) and 55% ($\Delta T = 200K$) for pure aluminum. Rapid deceleration of the solid-liquid interface follows nucleation, and the steady-state stage of interface velocity (to which the transient recalescence solidification growth velocities are asymptotic) represents growth largely under external heat flow conditions. It is thus apparent that different morphological structures should be observed within the same powder particle, and that major structural differences should be observed from particle to particle (37) if different particles undercool by differing amounts. The range of structural differences possible in very fine submicron powders, where the undercooling and transient recalescence stages dominate solidification, has been convincingly shown by Levi et al. (38). They showed homogeneously solidified Al-6wt% Si alloy powders, and breakdown from a planar to a cellular interface, lending support to the general theory of decreasing solid-liquid interfacial growth velocities after nucleation at substantial undercoolings. Similar microstructural effects were observed in larger 20-50 μm droplets of Al-Fe alloys (30,37) in which microeutectic structures solidified initially at the droplet surface following nucleation, and presumably were the highest solidification velocity zones, but which subsequently were replaced by dendritic structures toward the droplet center as the solidification velocity decreased. In several aluminum alloy systems where regular cellular eutectic structures were formed during droplet solidification, measurements of the eutectic inter-rod spacing yielded solid-liquid interfacial growth velocities varying from a few millimeters per second to several centimeters per second (37). Available experimental data of this type generally confirm the validity of the two major atomization models (33, 34, 36) in predicting solid-liquid interfacial growth velocities. In general, dendritic structures in atomized droplets indicate a "post-recalescence" solidifica-tion structure, and that undercooling has been minimal. For many practical alloy systems, refined dendritic structures produce improved mechanical properties, and may be perfectly adequate. Substantial droplet supercooling appears to be a necessary precursor condition to produce fully "rapidly solidified" structures with their attendant benefits.

3.2 Ribbon Casting

There have been no published heat transfer analyses of the Allied planar flow casting process; however, analyses of splat quenching (39) and jet casting (40) provide some insight into aspects of this process. In both jet casting and planar flow casting, solid product of average thickness 20 μm is produced at velocities of 10-20 m/s. Splat quenching has generally produced temperature gradients of 10^4 -10^6 K/mm, and it has been assumed that both jet casting and planar flow casting produce similar gradients. In planar flow casting, figure 6, for ribbon with thickness t = 25 μm, a liquidus temperature T_l of 1700 K, and a glass-transition temperature T_g of 700 K, the temperature gradient can be considered as $(T_l - T_g)/t$ since t is small, and is thus 4×10^4 K/mm. The solid-liquid interfacial veloci-ty, V_I, can be related to the ribbon velocity, V_χ, by $V_\chi = V_I \sin \theta$ where θ is the angle between the interface growth direction and the direction normal to the ribbon surface.

Planar Flow Casting

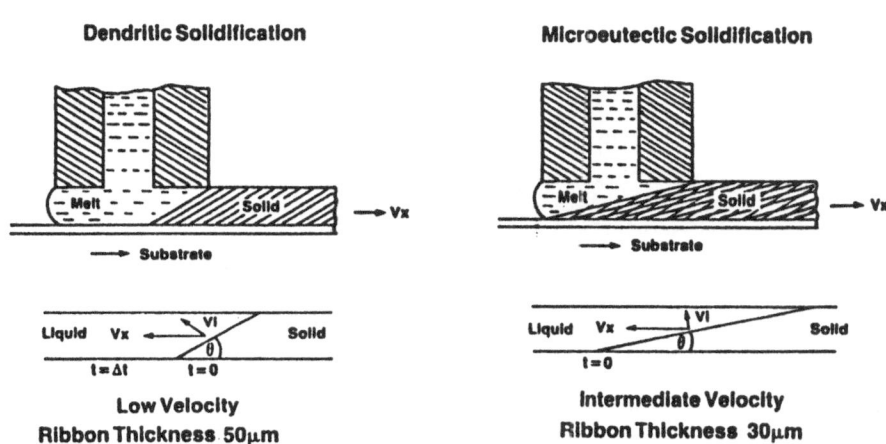

Figure 6: Allied Planar Flow Casting for Crystalline Alloys

For crystalline alloys θ can be established experimentally from the orien-tation of solidified grains within the ribbon; for Fe-6.5wt% Si, for ex-ample, this angle is approximately 10-12 °C (41), within a ribbon that had a casting velocity of 6 m/s. Using these values of θ for crystalline ribbons produces solid-liquid interfacial growth velocities in ex-cess of 1 m/s. Measurements of the contact distance, d, have been carried out by Warrington et al. (42) by filming jet casting on high speed infrared film and subsequently measuring d from the temperature profiles. Since θ is small, $V_I = V_\chi t/d$, and the average cooling rate of the liquid ahead of the solid-liquid interface can therefore be cal-culated in a relatively primitive way from $V_\chi(T_1-T_g)/d$. For jet

casting the contact distance varied up to a few milimeters, confirming
cooling rates of ≃ 10^6 K/s for jet-cast ribbon. For planar-flow casting of
amorphous alloys, Narasimhan (24) considers that the maximum angle θ can
assume is approximately 5° when casting at a velocity of 20 m/s, yielding a
similar interfacial growth velocity of 1 m/s. At higher ribbon velocities,
which produces somewhat thinner ribbon, θ decreases and the interface is
presumed to become almost flat, leading to potentially higher solidification velocities. Crystallization velocities of 1 m/s have been experimentally measured in the "explosive" crystallization of amorphous germanium
films (43) at room temperature, and such velocities should not be considered extreme for solidification during planar flow casting.

A consequence of higher solidification velocities in planar flow casting
of aluminum-transition metal alloys is the abrupt transition from dendritic
solidification structures to a much finer microeutectic structure. The
latter structure, which resembles a branched rod eutectic structure, consists of 10 nm diameter fibrous branching intermetallic rods in an aluminum
matrix, with a rod spacing of 100 nm. Quite substantial differences in
microhardness exist between the lower velocity dendritic structures and the
microeutectic structures, figure 7, leading one to conclude that both
increased solute trapping and reduced intermetallic obstacle spacings are
responsible for the observed hardness differences.

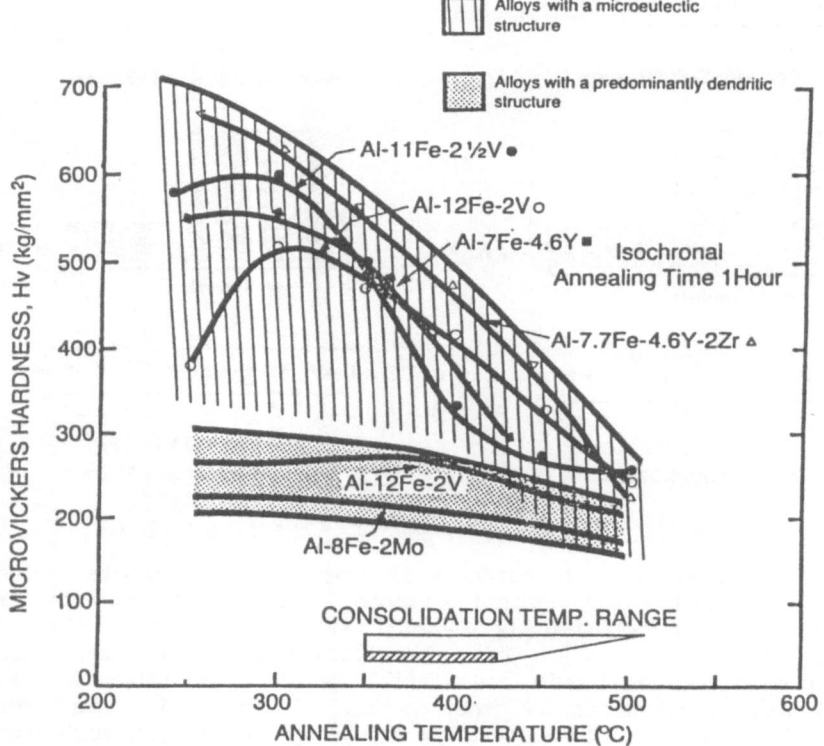

Figure 7: Effect of annealing on the as-cast microhardness of several Al-
Fe-X alloys solidifying with both dendritic and microeutectic structures.

The maximum possible solid-liquid interfacial growth velocity required to produce an amorphous structure during the type of rapid solidification described above can be calculated in an approximate way, by assuming that solute trapping results when the diffusive solute boundary layer thickness, δ, is approximately equal to the mean atomic separation distance in the liquid, a_0. In conventional solidification theory the effective diffusive boundary layer thickness is given by V/D (44), where V is the solidification velocity and D is the diffusivity of solute species in the liquid. The maximum solidification velocity for partitionless solidification is then given by D/a . For a solute species of diffusivity 10^{-9} m^2/s in a liquid metal of atomic mean free spacing 3×10^{10} m, which is typical of transition elements in liquid aluminum (45), the limiting solidification velocity is $\simeq$ 3 m/s. For more sluggishly diffusing species, like silicon in liquid iron, the limiting solidification velocity is $\simeq$ 0.3 m/s, a value which helps to explain the relative ease of glass formation in the Fe-B-Si system. Alternatively, for relatively pure liquid metals, where the diffusivity of vacancies can be considered the controlling species, the maximum solidification velocity needed to produce an amorphous structure could be as high as 50 m/s.

In several aluminum alloys containing substantial amounts of the transition elements, and cast at high ribbon velocities, we have observed a new solidification structure which appears to results from homogeneous nucleation within the ribbon volume. This structure, called 0-phase (46,47), in an Al-12Fe-2V alloy, for example, consists of 100-200 nm diameter spheres of about 500 nm spacing; it results from casting relatively thin ribbon, 15-20 μm thickness, at ribbon velocities greater than 25 m/s. These structures nucleate independently (homogeneously) within the ribbon volume, figure 8, and grow to their final size in less than 100 nanoseconds by what appears to be a massive transformation, growing to a size of 100 nm radius at a solidification velocity somewhat greater than 1 m/s. The transition from a continuously growing interface producing a microeutectic structure, to an independently nucleating phase is not abrupt. Both structures coexist at intermediate ribbon velocities, but higher solute concentrations and faster ribbon velocities produce almost entire 0-phase microstructures. This structure presumably results from nucleation after increased liquid supercooling as a consequence of the faster casting velocity.

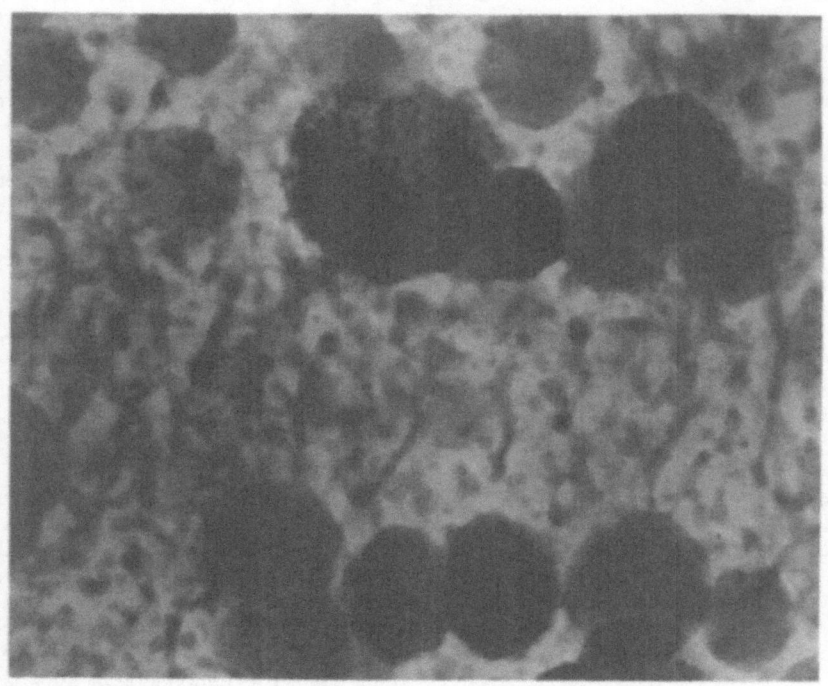

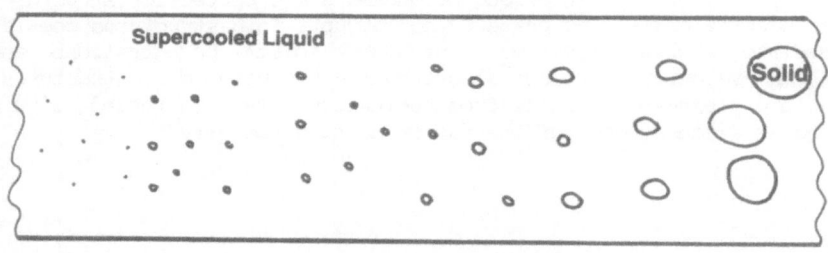

High Velocity – Ribbon Thickness 20μm

Figure 8: Nucleation of 0-Phase in Al-12Fe-2V Alloys.

4. APPLICATIONS

The critical magnetic properties of metallic glasses useful for distribution transformers have been reviewed recently (48); the Allied papers alone now constitute over 250 publications. Allied Corporation has installed continuous casting capability for 10 cm and 18 cm wide strip, with a production capacity of some 20,000 tonnes per year. It is worth summarizing recent developments with two new classes of alloys, nickel-base and aluminum-base alloys, which have evolved from the Allied process of PFC casting followed by mechanical comminution to produce rapidly solidified powder, figure 9, and to review prospects for forming continuous products from such powder.

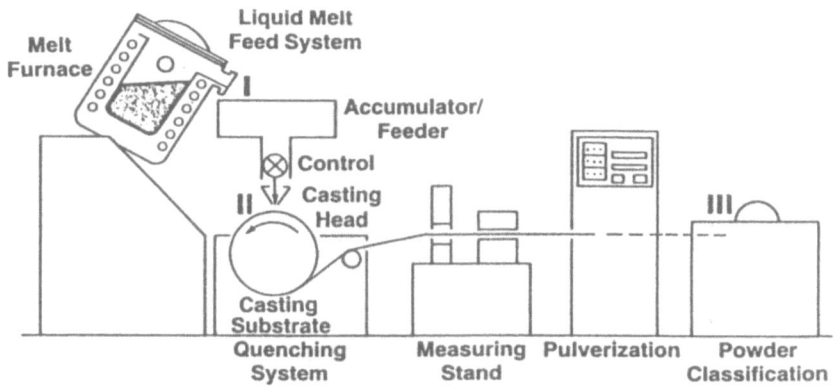

Figure 9: Allied RST Powder Process.

In both Ni-Mo-B and Ni-Mo-Cr-B alloys the as-cast microstructure and comminuted powder can be amorphous under appropriate PFC conditions. Differential scanning calorimetry studies have shown that the amorphous structure first transforms on heating to an fcc supersaturated solid solution, which then precipitates borides and intermetallic compounds with increasing temperature (49). During powder consolidation, crystallization results firstly in the production of the fcc solid solution, followed by the precipitation of both refractory borides and intermetallic compounds. The precipitation kinetics of the borides and intermetallic compounds are appreciably different, so that different mechanical properties have been produced by varying the consolidation conditions. Das optimized alloy chemistries for both Ni-Mo-B alloys (50) and Ni-Mo-Cr-B (51) to produce alloys with higher hot hardness, impact strength and corrosion resistance than traditional high speed steels intended for hot tooling applications. Some of these applications have recently been reviewed by Raybould (52), and include permanent injection molds for aluminum die-casting, high-speed and mill cutters, extrusion and wire dies, and high temperature valves. These diverse applications require substantially different properties and microstructures, varying from extreme hardness and wear resistance, to high temperature thermal stability and toughness. Such property variations are generally available from Ni-Mo-X-B alloys by selecting consolidation parameters which reflect the degree of microstructural coarsening appropriate for the desired application. Figure 10 summarizes the way in which various consolidation processes have been used for Ni-Mo-X-B alloys to attain specific properties. These results show that rapidly-solidified Ni-Mo-Cr-B amorphous alloys can be fabricated into bulk crystalline solids with good tensile strength, impact strength, hot hardness, and wear and corrosion resistance. Consolidation by extrusion or forging is preferable to HIP ping for obtaining improved toughness, and dynamically consolidated parts combine the best hot hardness and strength properties. This class of alloys has great promise for use as hot tooling in plant applications requiring corrosion and oxidation resistance, as well as higher hot strength and toughness than that of conventional high speed steels.

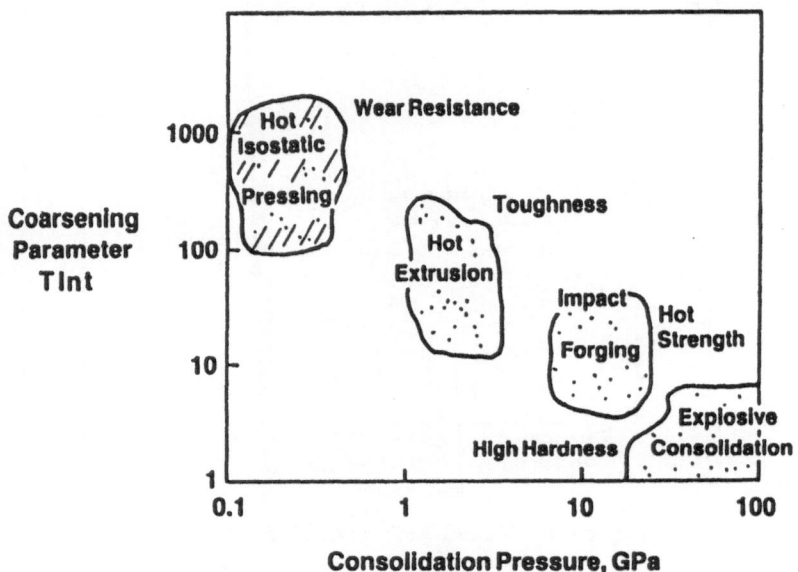

Figure 10: Schematic Powder Consolidation Diagram.

During 1983 Allied Corporation and Amax Specialty Metals concluded a joint agreement to market extruded Ni-Mo-X-B alloy sections for a variety of tooling applications. Such high performance alloys, consolidated by hot extrusion, are now available commercially from Amax.

During the past decade considerable attention has been devoted to development of high temperature aluminum alloys capable of competing with titanium alloys on a specific strength basis over the temperature range of 505 K (450°F) to 620 K (650°F).These alloys are initially destined for a wide range of aerospace applications: gas-turbine engines (low temperature fan and compressor cases, vanes and blades), missile applications (fins, winglets, rocket motor cases) and applications where the improvements in both temperature capability and elastic modulus can be utilized. Such alloys are typically strengthened by transition metal intermetallic compounds, at solute levels considerably higher than is acceptable for conventional direct-chill (DC) cast commercial precipitation hardening alloys. At such solute levels approaching 10-15 wt% transition elements, rapid solidification has allowed development of thermally stable intermetallic compounds in consolidated powders of ternary and quaternary aluminum-iron alloys, viz. Al-8Fe-2Mo (30,53), Al-8Fe-4Ce (54), Al-5Fe-3Ni-6Co (55), and Al-12Fe-2V (47). Figure 11 shows the ultimate tensile strength as a func-

tion of temperature for several such alloys consolidated from powder by
extrusion at temperatures in the range of 600 K (620°F) to 700 K (800°F).

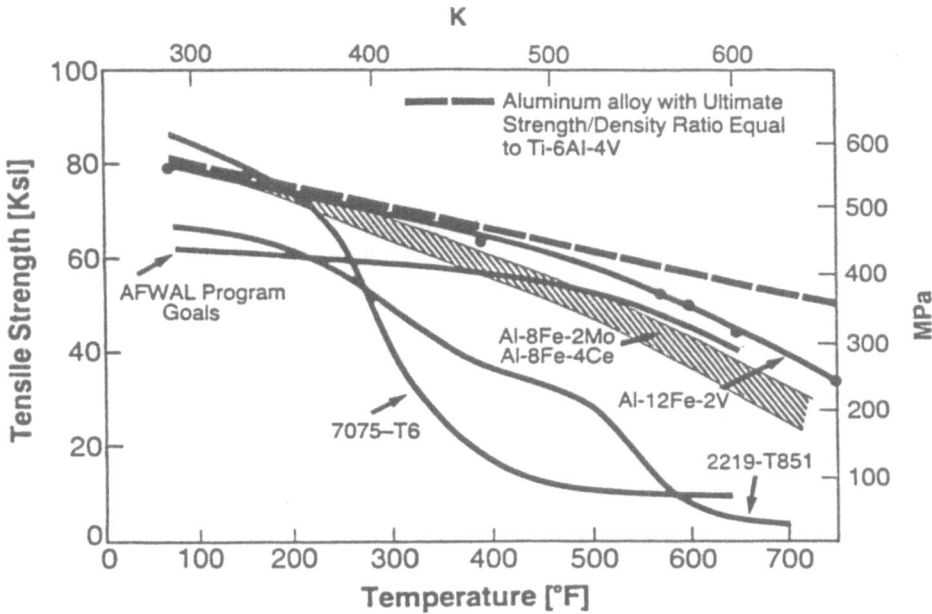

Figure 11: Elevated Temperature Strength of Advanced Aluminum Alloys.

In all practical alloys of the aluminum-transition metal systems, rapid
solidification produces either microeutectic or dendritic solidification
structures with much improved strength properties. An important result of
recent alloy development for stiffness critical applications is the significant improvement in Young's Modulus, table 2, and shear moduli, with a
room temperature modulus of 96 GPa (14 x 10   psi) having been attained for
consolidated P/M alloys. A comparison, figure 11, of the Al-Fe-V-Zr alloys
with other rapidly-solidified aluminum alloy systems, such as Al-8Fe-2Mo
made from rotary atomized powder, or Al-8Fe-4Ce made from splats, clearly
shows the advantage of planar-flow-cast Al-Fe-V-Zr alloys, and in particular Al-12Fe-2V over other alloys.

Table 2: Elastic (Young's) Modulus of Extruded Al-12Fe2V Alloys.

Temperature °C (F°)	Young's Modulus GPa (Msi)
27 (80)	96.5 (14.0)
93 (200)	92.4 (13.4)
149 (300)	88.9 (12.9)
204 (400)	85.5 (12.4)
260 (500)	82.0 (11.9)
316 (600)	77.9 (11.3)
371 (700)	74.5 (10.8)

These high temperature aluminum alloys are especially suitable for aerospace applications for a wide range of modern missiles operating in the hypersonic regime and exposed to saline environments for long periods of time before use. In general, structural aluminum alloys show severe saline pitting corrosion, requiring substantial protection either by an electrolytic anodic film deposition (anodization) or by chemical protection followed by epoxy coating. High temperature/high strength aluminum alloys recently developed at Allied Corporation have shown significantly reduced weight loss in ASTM B-117 Salt Fog testing. Typical performance of several Allied alloys exhibiting outstanding high temperature strength is shown in figure 12, indicating that the best of these alloys is virtually free from pitting corrosion attack after some 70 days exposure. Similar saline corrosion resistance has also been reported for Al-6 to 8 wt% Fe alloys (56) used in desalination plant equipment studies.

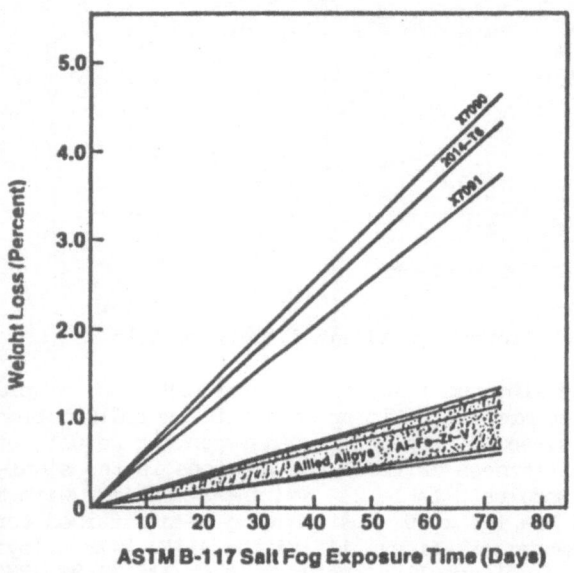

Figure 12: Saline Environment Weight-loss Data for Several Aluminum Alloys.

At least two direct powder rolling processes exist today in North America, consolidating powder directly into sheet. Reynold's Aluminum has operated for several years an in-line mill, figure 13, for conversion of very coarse particulate of 1-3 mm diameter, produced by quenching droplets of molten aluminum in water, into sheet stock for subsequent rolling. Another fabrication company has developed a similar process line, figure 14, for gas atomized powder. Both processes have some relevance for consolidation of aluminum RSP materials into sheet products. Table 3 lists the tensile properties of a high temperature aluminum alloy sheet, with an elastic modulus of 14 x 10[6] psi (94 GPa), fabricated by direct rolling of 1.5 in thick (38 mm) preforged billet into sheet of various thicknesses. Such alloys are intrinsically capable of continuous rolling of the type shown in figure 13 and 14 to produce continuous product.

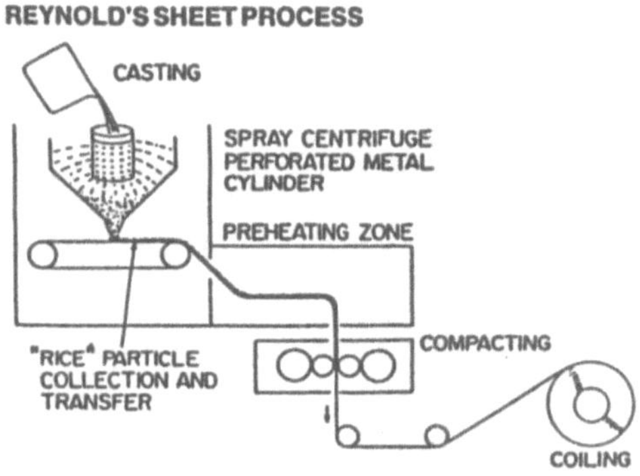

REYNOLD'S SHEET PROCESS

Figure 13: Schematic illustration of the Reynold's Aluminum Sheet Rolling Process.

DIRECT POWDER ROLLING

Figure 14: Schematic illustration of a direct rolling process for air atomized aluminium powder.

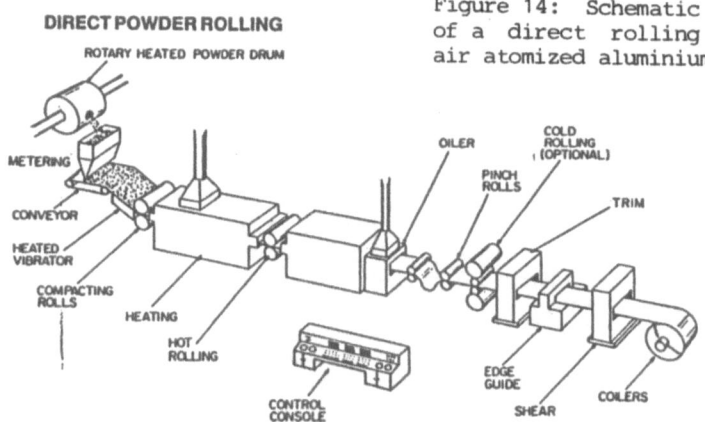

Table 3: Room Temperature Tensile Properties of Al-10Fe-2V Sheet Produced from 38 mm Thick Slab Rolled to Various Thicknesses.

Thickness	Yield Strength Ft_y	Ultimate Strength Ft_u	Elongation %
5.6 mm (0.220")	441 MPa (64 ksi)	496 MPa (72 ksi)	14%
3.8 mm (0.150")	441 MPa (64 ksi)	490 MPa (71 ksi)	14%
2.4 mm (0.095")	470 MPa (68 ksi)	496 MPa (72 ksi)	10%
1.7 mm (0.067")	345 MPa (50 ksi)	393 MPa (57 ksi)	18%

5. SUMMARY

Rapid solidification processing has enabled us to design novel alloy systems for high temperature applications. The examples given above for Ni-Mo-Cr-B and Al-Fe-V-Zr alloys clearly show the advantage of using cooling rate $> 10^6$ K/s in obtaining improved mechanical properties in the final consolidated bulk samples for structural use. The use of the Allied planar flow casting process produces material suitable for mechanical comminution to powder, which can be subsequently consolidated into bulk products as shown for both Ni-base and Al-base alloys. The mechanical and corrosion properties of these new alloys offer significant potential for exploitation in industrial applications.

ACKNOWLEDGEMENTS

The considerable assistance of my colleagues in Corporate Technology, Drs. Brown, Das, Davis, Raybould and Skinner, is gratefully acknowledged.

REFERENCES

1. See, for example, "Structure and Properties of Glassy Metals", Hasegawa R(ed), CRC Press, Boca Raton, Florida, 1983, and "Amorphous Metal Alloys", Luborsky F E(ed), Butterworths, London, 1983.
2. Davis L A and Hasegawa R, Metallurgical Treatises, (Tien J K and Elliott J F, ed), Met. Soc. of AIME, 1982, 301.
3. See, for example, papers presented at Conference on Magnetism and Magnetic Materials, Pittsburgh, 1983, J. Appl. Phys, 1984, 55, 2063-2087; Stadelmaier H H, Elmasry N A and Chang S, Mater. Lett., 1983, 2, 169; and Croat J J, Herbst J F, Lee R W and Pinkerton F E, J. Appl. Phys., 1984, 44, 148.
4. Grant N J: Rapid Solidification Technology, ASM, Metals Park, OH,
5. Tietz T E and Palmer I G: Advances in Powder Technology", ASM, Metals Park, Ohio, 1983, 189-224.
6. Holiday P H, Cox and Patterson R J: Proceedings of 1977 Reston Conference on Rapid Solidification Processing: Principles and Technologies, Mehrabian R (ed), Claitors, Baton Rouge, LA, 1978, 246.
7. Dixon R: Advances in the Development of Wear Resistant Tool Steels, ASM Publication, 1985, 8201-085.
8. Hansen M: Constitution of Binary Alloys, McGraw-Hill, 1958, 609.
9. NASA Fact Sheet KSC 191-80: A Primer on Propellants, November, 1980.
10. Lord Rayleigh, Proc. London Math. Soc. 10, 1879, 4.
11. Lyle J P and Cebulak W S, Metall. Trans., 6A, 1975, 685.
12. Spear R E and Gardner G R, Trans. AFS., 1963, 71, 209.
13. Grant N J, Jnl. Metals, 1983, 35, 20.
14. Allen M M, Athey R L and Moore J B: Source Book on Powder Metallurgy, Bradbury S (ed), A.S.M. Metals Park, Ohio, 1979, 88-98.
15. Mayfield J: Aviation Week and Space Technology, Vol. 118, 1981, 46.
16. Patterson R J: Rapidly Solidified Amorphous and Crystallin Alloys, Kear B H et al. (ed), Elsevier, New York, 1982, 123-129.
17. Champagne B and Angers R, Powder Met. International., 16, 1984, 125.
18. Pond R B, U.S. Patent 2,825,108; 4 March, 1958.
19. Masumoto T and Maddin R, Acta Met., 1971, 19, 725.
20. Davis L A, DeCristofaro N J and Smith C H, Proceedings of Conference on Metallic Glasses: Science and Technology, Hargiti C et al. (ed), Central Research Institute for Physics - Budapest, Vol. 1, 1981, 1.
21. Chen H S and Polk D E, U.S. Patent 3,856,513; 3 Dec. 1974.
22. Miyazawa K and Szekely J, Met. Trans.A, 1981, 12A, 1047-1057.
23. Bedell J, U.S. Patent 3,862,658; 28 Jan. 1975.

24. Narasimhan M C, U.S. Patent 4.142.571; 6 March 1979; also private communication, Feb. 1985.
25. Boswell P G, Metals Forum, 1979, 2, 40-54.
26. Perepezko J H and Boettinger W J: Proc. Materials Research Society Symposium on Alloy Phase Diagrams, Bennett L H, Massalski T B and Giessen B C (ed), Elsevier-North Holland, 1983, 223.
27. LeBeau S E, Perepezko J H, Mueller B A and Hildeman G J, Proc. ASTM Conference on RSP Aluminum, 1984, to be published.
28. Davies H A, in: Rapidly Quenched Metals III, Vol. 1, Cantor B (ed), Metals Society, London 1978, 1.
29. Boswell P G and Chadwick G A, Jnl. Mater. Sci. 1977, 12, 1879.
30. Adam C M in: Rapidly Solidified Amorphous and Crystalline Alloys, Kear B H, Giessen B C and Cohen M (ed), Elsevier, New York, 1982, 411.
31. Suzuki R O, Komatsu Y, Kobayashi K F and Shingu P H, Jnl. Mater. Science, 1983, 18, 1195-1201.
32. Shechtman D, Blech I, Gratis D, Cahn J W, Phys. Rev. Lett. 53, 1984, 1951.
33. Levi C G and Mehrabian R, Metall. Trans. B, 1980, 11B, 21.
34. Gill W N, Jang J W and Mollendorf J C, Chem. Eng. Comm., 1981, 12, 3.
35. Levi C G and Mehrabian R, Metall. Trans. A, 1982, 13A, 221.
36. Gill W N, Jang J W, Mollendorf J C and Adam C M, Jnl. Crystal Growth, 66, 1984, 351-368.
37. Adam C M and Bourdeau R G, Proc. Second Int. Conf. on Rapid Solidification Processing: Principles and Technologies II, Mehrabian R, Kear B and Cohen M (ed), Claitors Publishing Division, Baton Rouge, La., 1980, 246.
38. Levi C G and Mehrabian R, Metall. Trans. A, 1982, 13A, 13.
39. Jones H, Mater. Sci. Eng., 1969/70, 5, 297-299.
40. Katgerman L and van den Brink P J, Proc. 4th Int. Conf. on Rapidly Quenched Metals (Sendai), Masumoto T (ed), 1982, 61-63.
41. Chang C F, Bye R L, Laxmanan V and Das S K, I.E.E.E. Trans. Mag., MAG 20, 1984, 553.
42. Warrington D H, Davies H A and Shohoji N, Proc. 4th Int. Conf. on Rapidly Quenched Metals (Sendai), Masumoto T (ed), 1982, 69-72.
43. Takamori T, Messier and Roy R, Jnl. Mater. Sci., 8, 1973, 1809-1816.
44. Burton J A, Prim R C and Slichter W P, Jnl. Chem. Phys., 21, 1953, 1987.
45. Clyne T W, Metall. Trans. B, 15B, 1984, 369-381.
46. Skinner D J and Okazaki K, Scripta Met., 18, 1984, 905.
47. Skinner D J, Okazaki K and Adam C M in: Proc. ASTM Conference on Rapidly Solidified Powder Aluminum Alloys; ASTM, 1985, in press.
48. Fish G, Can. Elect. Authority, to be published.
49. Das S K, Proc. 3rd Rapid Solidification Processing Conference, NBS Proc. RSP Principles and Technologie, III, Gaithersburg, 1982, 559.
50. Adam C M, Das S K and Davis L A: Advances in Materials Technology for Process Industries, N.A.C.A., 1984, in press.
51. Das S K and Raybould D, in Rapidly Quenched Metals, ed. Steeb S and Warlimont H, Elsevier North Holland, N.Y., 1985, 1787-1790.
52. Raybould D, Metal Powder Report, 39(5), 1984, 282.
53. Adam C M, Simon J W and Langenbeck S: Rapid Solidification Processing, Principles and Technologies, III, Mehrabian R (ed), Nat. Bureau of Standards, Washington, 1983, 629.
54. Hildeman G W, U.S. Patent 4,379,719; April 12, 1983.
55. Dickson J, Okazaki K and Sanders T H, U.S. Patent 4,389,258; June 21, 1983.

56. Faninger G, Merz D and Winter H, Proc. Second International Conference on Rapidly Quenched Metals, MIT, Cambridge, 1976, 483.

ABSTRACTED DISCUSSION OF THE PAPER BY C.M. ADAM

Participants: W.J. Boettinger, B. Cantor, H.L. Fraser, M.E. Glicksman,
              S. Hock, L. Katgerman, R.E. Lewis, E. Vogt, H. Warlimont and
              J.V. Wood

Discussion arose of the differences in high temperature properties expected
for conventionally gas-atomized powders and the Allied materials. It was
generally agreed that the planar-flow casting process allowed access to
higher solute level alloys than various atomization processes, and the
higher elevated temperature strength properties presented in this paper
reflected higher concentrations of fine metastable intermetallic phases in
the Allied alloys. Several features of the finely dispersed spherical phase
(0-phase) described by C.M. Adam were discussed. This phase has variously
been described as quasi-crystals (Cahn), icosohedral phase (NBS,
Boettinger) and 0-phase (Allied). In Allied material the 0-phase is frequently 100 - 200 nm in size, with 200 - 500 nm spacing and has a composition identical with the melt; it may be, therefore, a product of diffusion-
less solidification at high undercoolings. It was stated that at high
undercoolings the driving force for solidification should be very large,
and the very high density of solidification nuclei thus presents an apparently anomalous contradiction between very high nucleation rates and
very high growth rates. M.E. Glicksman expressed the view that although the
nucleation rate should be very large in deeply undercooled melts, the
resulting extremely rapid growth rates expected may be modified by the low
diffusivities expected at the lower temperatures involved, and the solidification process then effectively compromises between high nucleation rates
and diffusion-modified growth rates.

ENGINEERING PROPERTIES AND APPLICATIONS OF RAPIDLY SOLIDIFIED MATERIALS

N.J. GRANT
Massachusetts Institute of Technology
Dept. of Materials Science and Eng.
Cambridge MA 02139
USA

ABSTRACT
Because of more generous funding, several classes of alloys have received much more attention and development than others. In most cases the R&D efforts were large enough to generate useful data from which potential applications become almost self-evident.

Particular emphasis was placed on a fairly wide range of aluminium alloys, with particular interest in lithium-containing alloys for reasons of density and specific properties such as modulus and strength. Similarly, major support of the superalloys is clearly evident among publications. Of interest is the shift in property and temperature emphasis as a result of very large changes in grain size. Such an event is demonstrated by a new application of the RS-PM IN-100 alloy as a disc alloy at 650 °C in contrast to its original use of a coarse-grained casting for use up to 950 °C.

Less well known are the equally interesting developments in high strength, high temperature, high conductivity copper base alloys, of interest as water cooled blades for stationary power turbines, and for fusion reactor applications.

Equally of interest are developments with the stainless steels and heat resistant alloys. The stainless steels include the ferritic grades, the austenitic grades and the $\alpha + \gamma$ microduplex alloys with attractive superplastic behavior. The stainless steels, long viewed as ordinary commodity-type alloys, were considered to be marginally benefited by rapid quenching, but in fact offer broad potential for new alloy classes with highly attractive properties.

Finally, a brief look at super-fine grained alloys produced from the glassy state, of selected compositions, will indicate some exciting avenues for alloy and structure development, bringing along equally interesting potential for improved properties: mechanical, corrosion, magnetic, etc.

1. INTRODUCTION

Production of rapidly-solidified particulates and products, though still quite a young technological field, has nevertheless matured sufficiently to have permitted documentation of a fairly wide range of alloys in a number of alloy systems to provide mechanical, physical, chemical, magnetic, and other properties for an assessment of RS-PM product potential.

It is true that funding of R&D has been much heavier in some alloy systems than in others, so that there is a relative abundance of data to report for aluminium-base alloys and superalloys. The literature shows important commercial successes as well as significant recent progress towards commercialization with stainless austenitic steels, copper-base alloys, ferritic steels, heat and corrosion resistant alloys, and others, suggesting early success there as well.

Production successes can be reported for high speed tool steels, superalloys, brazing alloys, austenitic stainless seamless tubing, some routine ferritic steels, and a small initial effort with aluminium. That's not too bad a performance from what was a laboratory curiosity as recently as 1970, at the time of the First International Conference on Rapid Solidification (1). Several other large markets are simply awaiting the correct moment to make their debut.

The purpose of this presentation is to look at the attractive structures and properties of some of the reported alloy systems in order to demonstrate that superior structures and properties are indeed available at this moment, and that perhaps a few refinements in particulate production and processing are still required, but more importantly, that a little courage and some financial and business ventures are now overdue. It's proper in this instance to let the reported structures and properties speak for themselves.

This report will look at three well defined and still developing alloy systems, as well as at some unusual natural advantages associated with RS products. The alloy systems to be discussed are: 1) aluminium-base alloys, 2) nickel and cobalt-base superalloys, and 3) austenitic stainless steels. Additionally, the attributes of a superfine grain size will be discussed briefly to demonstrate the attractiveness of a particular structural characteristic as the basis for product development.

2. ALUMINIUM-BASE ALLOYS

Aluminium alloys have much to gain from RST: increased solid solubility, including many elements that normally tend to segregate severely and/or form coarse complex constituent particles in ingot practice; a highly refined grain size (typically 1 to 5 μm); and a capability to use a variety of elements basically forbidden to ingot technology, added in major amounts (Li, Ce, Zr, Cr, Fe, Ni, Co, Mn, etc.).

Table 1 lists the tensile properties of RS-PM alloy 7091, which contains about 0.4% Co as a grain stabilizer in an alloy of the 7075 class (alloy 7090 contains 1.4% Co). Some improvement over 7075 is evident, with enhancement of smooth bar fatigue properties in particular, as well as gains in yield and ultimate strength (2,3). The 7091 alloy is the first commercial RS-PM aluminium alloy, with every expectation that other alloys will follow.

TABLE 1. Room Temperature Tensile and Toughness Data for RS-PM 7091*

Aging treatment	UTS ksi	0.2% Y.S ksi	Elong. %	N_{UTS} / Y.S	K_{IC} ksi $\sqrt{in}$
121°C, 6 h + 163°C, 4 h	83.5	77.5	11	--	--
121°C, 6 h + 163°C, 4 h	88.5	80.6	13	--	--
121°C, 24 h + 163°C, 4 h	59.3	42.4	14	--	--
121°C, 50 h	80.0	74.3	15	1.12	--
121°C, 24 h	85.9	77.0	10	--	--
	81.1	70.0	13	1.22	30.7
(Transverse)	82.4	71.4	6	--	11.1

*Alloy 7075 plus 0.4% cobalt.

Tables 2 and 3 show typical room temperature properties for two experimental RS-PM 2024 + Li alloys containing 1.6% Li and 1.0% Li (4). Earlier studies included a 3% Li alloy prepared from RS twin roller foils (5). In general the mechanical test data for the RS foils and the RS powders are comparable in spite of the fact that the RS foils achieved solidification rates up to one order of magnitude greater than the powders. Table 4 records room temperature tensile values for a variety of 2XXX series alloys, including ingot-based 2024, which serves as the reference material.

Compared to IM 2024, the better of the two 2024 + Li alloys (alloy 3) shows by increased yield strength by about 50%, an increased ultimate strength by about 20%, and decreased ductility values by about 50%. Corrected for the decrease in specific gravity due to lithium (about 5% at 1.6% Li and about 3% at 1% Li), the improvements in yield and ultimate strengths are further enhanced as seen in the last column of Table 2. Also associated with the lithium additions are improvements in the elastic modulus ranging up to 20% for a 3% addition of lithium, giving specific modulus improvements up to 30% for a 3% lithium alloy (5): see Table 4 for typical density corrected yield and modulus values for 2024 + Li alloys.

TABLE 2. Room Temperature Tensile and Toughness Data for an RS-PM 2024 + 1.6% Li Alloy*.

A. Tensile Data

Treatment	UTS ksi	0.2% Y.S. ksi	Elong. %	Density Corrected UTS $\times 10^4$ in.
SHT 510°C - T4	77.2	63.8	7.1	81.3
SHT 510°C - T6	75.9	64.7	6.4	79.9
SHT 495°C - T4	70.4	55.4	8.3	74.1
SHT 495°C - T6	70.0	57.6	8.3	73.7

B. Toughness Data

Treatment	Smooth Bar % Elong.	% RA	Notched T.S. ksi	$\dfrac{N_{UTS}}{\sigma_{Y.S.}}$	K_Q ksi $\sqrt{in}$
SHT 510°C - T4	7.1	10.0	77.6	1.2	30.8
SHT 510°C - T6	6.4	5.0	55.1	0.8	--
SHT 495°C - T4	8.3	8.3	75.1	1.4	--
SHT 495°C - T6	8.3	10.0	58.0	1.0	--

*2024 + Li Composition: 3.78 Cu, 1.40 Mg, 1.63 Li, 0.11 Fe, 0.13 Si, 0.18 Ti, nil Mn and Cd, bal. Al: weight percent.

As seen in Table 3, additions of about 1 to 1.5% Li preserve the notch tensile properties of alloy 2024 and result in acceptable K_Q values (4). The better notch tensile and fracture toughness properties are associated with the more ductile alloy conditions. Enhancement of the notched strength and toughness values is vital to the commercialization of lithium-modified aluminium alloys; the alloy structural features associated with toughness are discussed further below.

TABLE 3. Room Temperature Tensile and Toughness Data for an RS-PM 2024 + 1% Li Alloy*.

A. Tensile Data

Treatment	UTS ksi	0.2% Y.S. ksi	Elong. %	Density Corrected UTS $\rho \times 10^{4}$ in.
SHT 510°C – T4	81.4	58.4	11.9	83.9
SHT 510°C – T6	81.1	68.2	9.0	83.6
SHT 510°C – T815	81.5	69.0	7.6	84.0
SHT 495°C – T6	74.0	60.5	10.7	76.3

B. Toughness Data

Treatment	Smooth Bar % Elong.	% RA	Notched T.S. ksi	$\dfrac{N_{UTS}}{\sigma_{Y.S.}}$	K_Q ksi$\sqrt{in}$
SHT 510°C – T4	11.9	17.5	81.4	1.4	33.9
SHT 5106°C – T6	9.0	13.3	71.9	1.1	22.7
SHT 510°C – T815	7.6	10.0	72.2	1.0	--
SHT 495°C – T6	10.7	12.5	75.9	1.3	--

*2024 + Li Composition: 4.16 Cu, 1.80 Mg, 0.96 Li, 0.50 Mn, 0.18 Cd, 0.14 Fe, 0.14 Si, bal. Al: weight percent.

Table 4 summarizes typical room temperature strength values for several 2XXX alloys, comparing ingot with RS-particulate alloys, and including density-corrected yield strength and Young's modulus values. It is apparent that a 3 w/o Li addition to an RS-PM 2024 alloy can easily double the yield strength over that of IM 2024 (even prior to a density correction) while increasing the density-corrected elastic modulus by 30%.

TABLE 4. Room Temperature Tensile Properties of Selected Al-Li Containing Alloys.

Alloy	UTS ksi	MPa	0.2%Y.S. ksi	MPa	Density – corrected YS and E: YS 10^4in	E 10^7in
2024 Ingot	41.5	286	67	465	41.5**	10.6**
2024 + 1 Li – RS Flake*	56.3	388	76	524	58	11.5
2024 + 3 Li – RS Flake*	82.8	571	85	583	91	13.5
01420 Ingot (1)	43.0	279	65	448	46	12.2
Al-Mg-Li RS PM (2)	61.6	425	77	531	66	12.2
X2020 RS PM (3)	90.4	623	94	650	95	12.0

*Twin roller product
**Reference states
(1) Al-6Mg-1.5Li: USSR alloy
(2) Al-6.7Mg-1.6Li
(3) Al-4.4Cu-1.55Li

The 01420 Al-Mg-Li alloy is a Soviet lithium-modified ingot alloy which is used in Soviet aircraft; however, as an ingot product, strength and modulus benefits are apparent only after density corrections (4,6). By contrast, a comparable composition prepared as an RS-PM alloy shows a yield strength improvement of 50%, density uncorrected, and over 60%, density corrected, over the ingot product.

The X2020 alloy, an Al-Cu-Li alloy (no magnesium) is of considerable interest because it is able to develop very high tensile strength values which are, however, associated with low ductility. Tables 4 and 5 show yield and ultimate strength values in the 80 and 90 thousands; however, even when the elongation and reduction of area values for smooth bar tests are relatively high, notched tensile values are often below 1.0, and K_Q values are only 10 to 11, far below acceptable toughness values (7). These very poor notched tensile and K_Q values have been a puzzle, especially in view of the far more acceptable values for the 2024 types of alloy seen in Table 3.

TABLE 5. Room temperature tensile and toughness data for RS-PM Aluminium Alloy X2020

Alloy and Condition		UTS ksi	0.2% YS ksi	Elong. %	R of A %	$\frac{N_{UTS}}{\sigma_{YS}}$	K_Q ksi$\sqrt{in}$
68 - T6		87.9	81.9	5.8	8.9	0.80	10.4
- T7		80.6	74.5	6.3	15.5	0.89	--
- 175°C, 4 h		82.2	76.6	5.5	10.2	0.96	--
- Cold Work + T6		89.4	82.2	8.9	14.2	0.73	--
- Cold Work + T7		85.4	78.4	10.8	16.0	0.83	--
69 - T7		94.3	90.4	5.3	6.0	0.47	11.2
70 - T6		87	78	4.7	3.7	--	--
Compositions	Cu	Li	Mn	Cd	Al		
68	4.6	1.0	0.4	0.10	bal.		
69	4.3	1.5	0.4	0.15	bal.		
70	3.6	3.0	0.5	0.25	bal.		

It is of course not necessary to depend on Li additions to achieve excellent mechanical properties in RS-PM aluminium alloys. Table 6 shows extremely attractive combinations of strength, ductility and toughness values for a 7075 alloy modified by the addition on 1% Ni plus 0.8% Zr (8). Both elements are normally classified as impurities in ingot alloys, undesired beyond small amounts of less than 0.5 and 0.2%, respectively. In the absence of significant segregation as a result of rapid soldification, fine precipitates of Al_3 Zr and Al_3 Ni are formed, producing a dispersion strenghtening effect on top of the normal strengthening due to aging in the 7075 base composition.

Note that yield values up to 95,000 psi and UTS values in excess of 100,000 psi are achieved, coupled with elongation values up to 10% and notch-strengthened tensile values (8). K_Q values however are only fair.

TABLE 6. Room temperature tensile and toughness data for RS-PM aluminium Alloy 7075 + 1% Ni+0.8% Zr

Condition	UTS ksi	0.2% YS ksi	Elong. %	N_{UTS} / σ_{YS}	K_{IC} ksi $\sqrt{in}$
SHT 460°C + 120°C, 24 h	98	91	9.0	1.23	--
SHT 490°C + 120°C, 24 h	99	91	10.0	1.23	21.2
SHT 490°C + 120°C, 66 h	102	95	7.3	1.19	
SHT 460°C + 130°C, 24 h	98	91	8.7	--	--
SHT 490°C + 150°C, 24 h	78	88	9.5	1.32	23.2

Figure 1 is a plot of stress amplitude (rotating beam) versus the number of cycles N, at room temperature, in tests performed in an air atmosphere. Comparisons are shown among IM 2024, IM 01420 (the Soviet Al-Mg-Li alloy with 1.6% Li), RS-PM Al-Mg-Li (1.6% Li), RS-PM 2024 + 1% Li (4) (see Table 2), and RS-foil 2024 + 3% Li (5). All of the RS-particulate alloys are significantly more fatigue resistant than the two reference ingot-based alloys. In particular, the RS-foil 2024 + 3% Li alloy, with a fatigue strength for 10^7 cycles of 42,000 psi, is 70% stronger than its lithium-free IM2024 alloy. Even though this RS-foil 2024 + 3% Li alloy is very much stronger than the IM 2024 alloy, the notched tensile to yield ratio was less than 1.0. The big improvement in fatigue strength is due to the highly refined grain size and microstructure which sharply delayed crack initiation; once cracking was initiated, crack growth rates were high and failure occured relatively quickly.

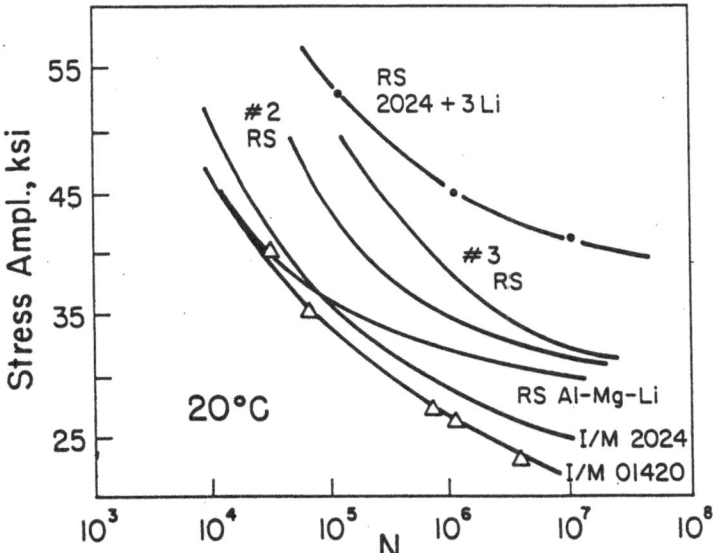

Figure 1: S-N fatigue plot, reversed bending, zero mean stress, for RS aluminium alloys versus ingot products 2024-T4 and 01420, an Al-Mg-1.6 Li alloy. The RS alloys are Al-6.7 Mg-1.6% Li, X2020 with 1% Li and RS 2024 + 3% Li. Tests in air atmosphere at room temperature.

Figure 2 illustrates the unfavorable slope of the fatigue crack growth rates of the RS-foil 2024 + 3% Li alloy compared to that for a RS-foil 2024 plus 1% Li alloy and IM 2024 (5). The steep slope is typical of those found for RS alloys which have received limited hot deformation. The major reason for the high crack growth rates is indicated to be due to the presence of fine oxide stringers in the rapidly solidified and extruded aluminium alloys.

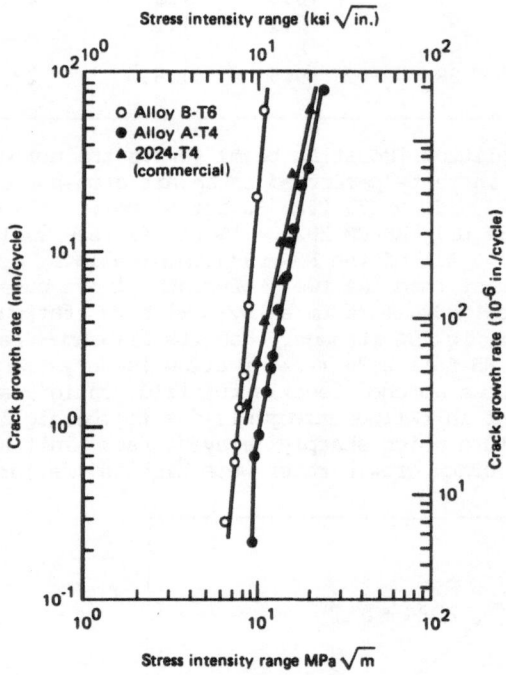

Figure 2: Fatigue crack growth rate data for 2024 alloys: B-T6 is 2024 + 3% Li, T6 condition, made from extruded RS-twin roller foils; AT-4 is 2024 + 1% Li, T4 condition similarly processed, and 2024-T4 is commercial ingot product.

The inevitable surface oxides emerge as stringers in the extrusion direction, being more abundant the finer the powders, increasing with lithium content, and increasing with increasing exposure time in an open air atmosphere prior to canning and hot extrusion. Figure 3 shows typical fine oxides as part of oxide stringers (7). Figure 4 shows fine, isolated regions containing intercrystalline cracks due to cold stretching of 2 to 3% after solution heat treatment in an X2020 alloy containing 1.6% Li (7). Clusters of fine powders result in heavier local oxide enrichment than on average and contribute to early cracking during cold deformation.

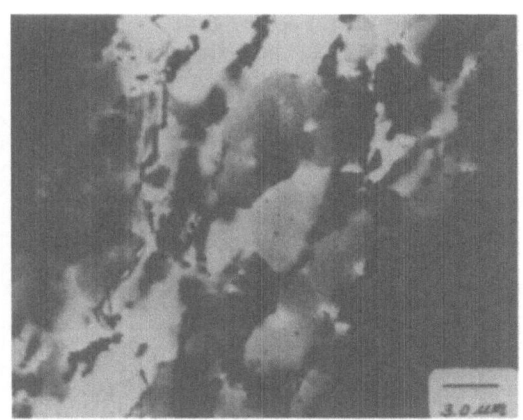

Figure 3: TEM view (BF) of RS-PM Al-Cu-Li (X2020) showing oxide stringers and fine grain size.

Figure 4: Optical view, longitudinal, of RS-PM X2020 alloy showing oxide stringers and localized intercrystalline cracks due to 2.5% cold stretching after solution heat treatment.

Since the aluminium oxides are not reducible or easily removed, benefits would be greater by minimizing or avoiding oxide formation and entrapment. Most recently this has been accomplished by liquid dynamic compaction (LDC), or spray atomization and collection, working with partially solidified atomized droplets which are splat quenched against a metallic substrate (9,10). The resultant deposited product is typically $96 \pm 2\%$ dense, with a fully-recrystallized, fine-grained (15 - 20 μm) structure, with finely scattered near-micron porosity (10). Solidification cooling rates are indicated to be about 10^3°C sec^{-1}, with slower solid state cooling. Even relatively small amounts of hot work (forging, extrusion, rolling) at 400°C resulted in final grain sizes of 2 to 3 μm.

Table 7 illustrates the improvements achievable with an LDC–generated structure for the 7075 + 1% Ni + 0.8% Zr alloy compared to that for conventionally-processed RS-PM material. The improvement in strength without loss of ductility is outstanding.

TABLE 7. Room temperature tensile data for aluminium alloy 7075 + 1% Ni + 0.8% Zr variously processed

Condition and Process	UTS ksi	0.2% ksi	Elong. %	Extrusion Ratio(Area)
RS-PM				
SHT 490°C, cold stretch 3.8%, Age 120°C, 24 h	101	88	7.8	30:1
SHT 490°C, cold stretch 2.5%, Age 120°C, 24 h	95	94	7.3	30:1
Cold work 20%, SHT 490°C, Age 130°C, 24 h	86	80	9.3	30:1
Large Diam Compact, SHT 490°C, Age 120°C, 24 h	105	98	9.0	200:1
Liquid Dyn. Compaction*, SHT 490°C, Age 120°C, 24 h	114	105	6.0	16:1
Liquid Dyn. Compaction*, SHT 490°C, Age 20°C, 24 h + 120°C, 24 h	118	108	8.6	16:1

*LDC-Spray atomization and collection; 96% dense body: hot extruded.

3. SUPERALLOYS

Superalloys, frequently in the form of precision cast products, by virtue of their highly alloyed, complex multiphase structures, with severe segregation and coarse cast structures, are naturally suited for improvements in toughness , ductility, formability, fatigue performance, etc. Only the very high temperature creep rupture properties are worsened. Yet even here the attempted modifications of composition, processing, and structure have not been profound. Table 8 lists typical compositions of two of the well-known superalloys; one is the Ni-base superalloy IN-100, a classical highly—alloyed $\gamma - \gamma'$ product containing up to 65 v/o of the hardening γ'-phase. As a casting the alloy is a coarse-grained, segregated, coarse-phased product which is not hot workable. This is one of the very best alloys for precision-cast blades for the gas turbine industry with operating capability to 950°C. The second alloy, Mar-M509, is one of the strongest cast cobalt-based alloys, hardened by 0.60 - 0.65% carbon. Coarse-grained (3 - 4 mm grain size), with coarse $Cr_{23}C_6$, M_6C and TaC carbide precipitates, this alloy is also not hot workable, and is relatively brittle.

By removing the Ta in Mar-M509 and substituting Hf in an amount required to combine with all of the carbon, leaving a small excess of Hf, and using rapid solidification techniques, one finds that the powders or twin roller foils retain about 90% of the Hf and C in supersaturated solid solution (11) (see Table 8). On heating to about 800°C, fine HfC precipitates are formed to the exclusion of $Cr_{23}C_6$ and M_6C. This product, on hot extrusion at 1150°C, was so ductile that the carbon was increased to 1.2% C and stoichiometrically balanced with Hf to form HfC. Again, on hot extrusion, a strong ductile product was produced which contained only the HfC phase as precipitate (11).

TABLE 8. Composition of IN-100, Mar-M 509 and hafnium-modified M 509 (in weight percent)

Alloy	Co	Ni	Cr	Mo	W	Al	Ti	Hf	Ta	B	Zr	C
In-100	15	bal	10	3	--	5.5	4.5	--	--	0.014	0.06	0.18
Mar-M 509	bal	10	23	--	7.5	--	0.2	--	2.7	0.02	--	0.62
Mod. M 509A	bal	9	22	8	--	--	--	6.7	--	--	--	0.68
Mod. M 509B	bal	8.3	20	7.1	--	--	--	17.6	--	--	--	1.01

Table 9 summarizes the room temperature and the 982°C stress-rupture properties for precision-cast Mar-M 509 and IN-100, and for RS foils of the Hf modified Mar-M 509 and the IN-100 alloys.

TABLE 9. Room temperature tensile and 982°C stress-rupture properties of cast Mar-M 509 and IN-100 and for RS-modified Mar-M 509 and IN-100.

A. Tensile Properties at 20°C

Alloy	Condition	Grain Size μm	0.2% YS ksi	MPa	UTS ksi	MPa	Elong. %
IN 100	Prec. Cast	3000	136	950	143	997	4
IN 100	RS-PM*	5-8	173	1194	241	1663	20
Mar-M 509	Prec. Cast	4000	80	552	115	793	2-5
Mar-M 509	RS-foils*	5-6	130	897	192	1324	14
Mod. 509A	RS-foils*	5	190	1310	250	1725	12
Mod. 509B	RS-foils*	5	230	1587	280	1932	9

B. Stress rupture at 982°C

		G.S., μm	Stress for 100 h Life	Elong., %
IN 100	Prec. Cast	3000	22,000 psi	10
IN 100	RS PM*	5-8	2,500	43
Mar-M 509	Prec. Cast	4000	17,000	6
Mod. Mar-M 509**	RS-foils*	5-6	8,000	10

* Hot extruded with area reductions of 16 to 20:1.
** HfC modified Mar-M 509.

At room temperature (Table 9A) the tensile properties of the hot extruded, rapidly-solidified, fine-grained IN-100 and Mar-M 509 alloys give very much higher strength than their precision cast counterparts. This, combined with a 3 to 4 fold improvement in ductility over the cast, coarse-grained structures, results in strong, tough alloys. These alloys readily responded to hot working of the canned RS-particulates by extrusion at temperatures several hundred degrees lower than would nominally have been used for ingot structures.

Of greater interest, the RS HfC-strengthened cobalt-base alloys show greater strength values than those achieved by the classical $\gamma - \gamma'$ age-hardened nickel-base IN-100 alloy. This is a result which was not and would not have been predicted from alloy theory or prior-precision-cast or ingot technology. The A and B modifications of Mar-M 509 (see Table 8) have about 12 and 24 vol.pct., respectively, of HfC precipiates in the 0.05 to 0.2 µm size range. For these strength levels, the ductility values are outstanding.

Stress rupture properties are summarized in Table 9B for tests at 982°C for the same alloys reported in Table 9A. Here the extremely important role of grain size becomes apparent, namely in that the cast coarse-grained (3 mm) IN-100 alloy supports a stress for an 100 hour rupture life at 982°C which is almost 10 times greater than for the fine-grained (5 to 8 µm) RS-PM alloy, with a difference of 20 to 1 in ductility values at fracture in favor of the fine-grained RS-PM alloy. The Mar-M 509 alloy shows similar differences in strength and ductility between the cast and the RS-PM alloys, with the important additional difference that RS-PM Modified Mar-M 509 alloy containing hafnium carbide is more stable at 982°C than is the $\gamma - \gamma'$ IN-100 alloy, and, in spite of its equally fine-grain size, the modified Mar-M 509 alloy is stronger than the RS-PM IN-100 alloy (for an 100 h rupture life: the stresses are 8000 psi versus 2500 psi).

Totally unexpected for such a high carbon cobalt-base alloy was the high ductility of the HfC-modified Mar-M 509 alloy, due to its highly refined grain size and microstructure. Both of the RS alloys show distinct superplastic tendencies even at a temperature as low as 982°C, based on the duplex phase structures and very fine grain sizes (less than 10 µm). There will be more discussion below regarding these superplastic trends.

4. AUSTENITIC STAINLESS STEELS (MODIFIED 316 GRADE)

An opinion, often expressed, is that rapid solidification technology is primarily suitable for complex, highly-alloyed materials because of known refinements in grain size and general microstructure. The simple austenitic stainless steels such as the 18Cr - 8Ni, 17Cr - 12Ni - 2.5Mo, 20Cr - 10Ni and others are single-phase alloys with little hardening potential expected through cold work, and therefore don t stand to benefit importantly from potential structural refinements. Basically this is true, but there is no valid reason why these alloys should forever remain simple, single-phase austenitic structures. Perhaps the cheapest, most practical alloying element is carbon (and/or nitrogen); however, either the wrong carbides form on the grain boundaries, usually as coarse precipitates and result in embrittlement and poor corrosion resistance, or the more refractory carbides are too coarse to be of benefit to strength. Through rapid solidification, selected, fine refractory carbides can be formed, not preferentially located on grain boundaries, and therefore capable of enhancing strength, without embrittlement, and without negative effects on corrosion.

Type 316 stainless steel (17Cr-12Ni-2.5Mo) will serve as the example in the current instance, modified by additions of titanium, and balanced against the carbon content so as to form TiC, with a small excess of titanium (12-15). Quantities of carbon and titanium (nominal amounts), with appropriate alloy designations are:

PA-1	0.05% C and 0.30% Ti
PA-2	0.10% C and 0.60% Ti
PA-3	0.15% C and 0.90% Ti
PA-4	0.20% C and 1.20% Ti
PA-5	0.25% C and 1.50% Ti

Twin-roller RS foils, less than 100 μm thick, of random short lengths and widths, were produced at solidification cooling rates of about $10^5 \, Ks^{-1}$. Chopped foils were packed into mild steel cans, welded shut with an inserted evacuation tube, pumped to a vacuum of less than 10 μm, sealed, and hot extruded at 1100°C at an extrusion ratio of about 25:1 (area reduction).

Titanium carbo-nitrides in ingot-grade stainless steels are typically 2 to 10 μm on edge, of regular geometry, and do not influence the strength or ductility of the alloy, but may lead to minor surface cracking in thin strip and sheet products. Figure 5 shows the as-extruded microstructure for the PA-2 alloy. The average TiC particle size is 150 Å; the carbides are uniformly distributed in the fine-grained structure, which has a grain size of about 5 μm, in contrast to that of the ingot products where the grain size is more typically about 80 μm.

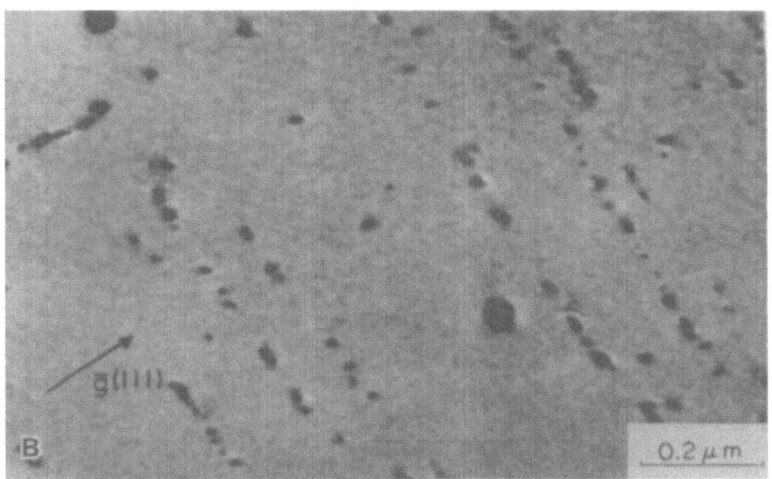

Figure 5: Austenitic 316 + Ti stainless steel extruded from RS-PM material. Average particle size of precipitated TiC particles is 150 Å.

On the basis of the ultrafine TiC dispersion, there is a distinct improvement in strength over ingot material with increasing content of TiC (from PA-1 to PA-5), associated with a TiC dispersion hardening mechanism. Since the yield and tensile strength values of austenitic stainless steels are very low (see Table 10), the 33% increase in yield strength is highly welcome. Note the small decrease in ductility for this increase in strength (14).

The production of the fine carbide dispersion, plus the presence of fine chromium oxides formed during the handling and processing of these alloys, resulted in a very fine distribution of relatively stable dispersoids in sufficient quantities to serve as effective pinning sites for grain boundaries as well as subgrain boundaries. The PA series of RS alloys, after about 62% cold rolling, will undergo recrystallization at 850 °C and be fully annealed at about 900 °C. At an annealing temperature of about 650 °C to 700 °C, after a prior cold reduction of 62%, the alloys undergo recovery after about 0.5 to 1 h, with major readjustments in dislocation arrays into very fine high-angle subgrain boundaries. Whereas the ingot alloy of equivalent composition to PA-1 undergoes partial recrystallization at 650 °C to

222

700 °C, the RS PA-1 and PA-2 alloys form a uniform, fine subgrain structure, the excellent properties of which are tabulated in Table 11 (14).

TABLE 10. Room temperature tensile properties of the cold worked and annealed 316 + Ti alloys PA-1 to PA-3.

Alloy*	Treatment	Y.S. 0.2%, ksi	UTS ksi	Elong. %
IM PA-1	Annealed, 1050°C	37	80	62
IM PA-1	62% cw; 850°C 0.5 h	45	88	53
PA-1 R.S.	62% cw; 850°C 0.5 h	51	97	49
PA-2 R.S.	62% cw; 850°C 0.5 h	56	105	44
PA-3 R.S.	62% cw; 850°C 0.5 h	60	109	44

*IM (ingot product); RS (twin-roller rapidly-solidified foils).

TABLE 11. Room temperature tensile properties of cold-worked and of cold-worked and recovered (at 650°C) PA-1 and PA-2 alloys.

Alloy*	Treatment	Y.S. 0.2%, ksi	UTS ksi	Elong. %
IM PA-1	Annealed plus 20% cw	121	125	16.7
PA-1 RS	62% cw; recover at 650°C	133	145	22.8
PA-2 RS	62% cw; recover at 650°C	136	148	21.3

* IM is ingot material

These studies have shown that even minor alloying and structural modifications of the simple austenitic stainless steels can result in significant improvements in mechanical properties (without degradation of the corrosion properties or loss of creep resistance at 500 to 650°C (12-15). More extensive compositional and structure changes appear warranted and attractive.

5. SUPERPLASTICITY
We discuss here rather than an alloy system, which was done in the first three sections of this chapter a behavior pattern. Specifically the behavior pattern of interest is superplasticity, a mode of deformation above about 0.5 T_m (the melting temperature in degrees absolute), wherein fine-grained (less than about 10 to 15 μm), two phased (multiphased) alloys undergo extensive deformation by grain boundary sliding (16), readily recovering from local work-hardening effects due to the low strain rate deformation. Necking is not a discrete phenomenon because of the recovery which takes place during deformation, leading not only to very large reductions of area but even larger elongation values to fracture.
Critical structural features are the two or multiphased alloy structures (only one phase needs to be a ductile solid solution) necessary to minimize grain growth or coarsening during hot deformation, and a fine grain size

(typically less than about 15 microns). The finer the grain size the greater the extent of grain boundary sliding. Two limits prevail in terms of
the grain size: grain sizes coarser than about 20 microns undergo much more
limited grain boundary sliding than that achieved by finer grain sizes; and
grain sizes finer than about one or two microns are relatively unstable
during slow, hot deformation, and tend to coarsen during deformation,
resulting in decreasing ductility.

One very important attribute of rapid solidification is the fine grain
size which tends to form as a result of the lower than usual temperatures
used for consolidation as compared to those used for hot working ingot
based alloys. The fine dendritic structures of RS alloys also tend to
promote a finer grain size in RS alloys. One might say that it is necessary
to deliberately, improperly, hot-consolidate RS particulates to avoid the
natural tendency to produce the desired fine grain sizes necessary for
superplastic behavior.

Figure 6 shows a typical very fine grain size for an RS microduplex
(ferrite plus austenite) stainless steel hot extruded at 1050°C to yield a
5 to 10 μm grain size (17). Grain sizes as fine as one micron and even in
the submicron size range are readily produced in a variety of alloys (18);
one such alloy will be discussed below.

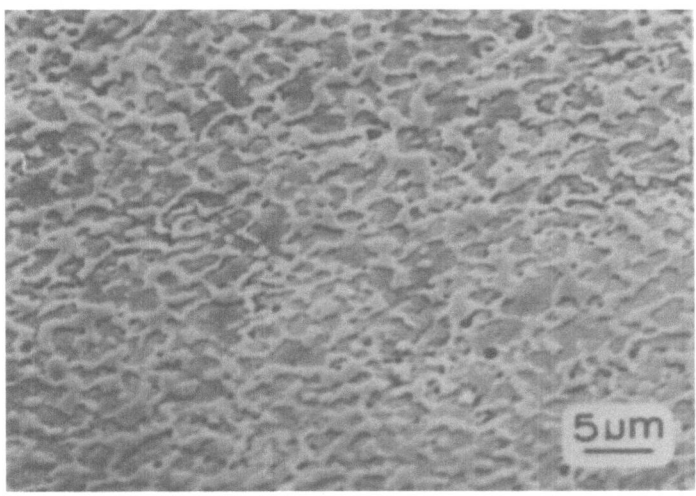

Figure 6: Microduplex stainless steel of 50 % ferrite –
50 % austenite, see Table 12. As extruded and annealed
at 1050°C. Note fine, uniform grain size.

A series of microduplex ferrite-plus-austenite stainless steels were rapidly solidified and hot extruded to produce grain sizes in the 5 to 10 μm
size range. The compositions and volume contents of ferrite and austenite
are shown in Table 12 for these alloys (17).

224

TABLE 12. Composition, volume percent and grain sizes of ferrite and austenite in RS-microduplex stainless steels deformed superplastically.

Alloy No.	Grain Size	Ratio α to γ	C	N	Mn	Cr	Ni	Cu	Si	Mo	P	Nb
A	∿ 2 μm	50-50	0.035	0.145	0.53	23.42	5.02	1.03	0.62	1.49	0.006	--
B	1.5-2.5μm	30-70	0.039	0.152	0.54	22.93	7.04	1.03	0.50	1.49	0.006	--
C	∿ 6 μm	70-30	0.086	0.214	1.03	27.28	5.57	1.05	0.51	1.49	0.132	0.11

Figure 7 is a plot of elongation at fracture versus the log of the strain rate for the 50α - 50γ alloy, the most stable of the three alloys, tested at 950, 1000, and 1070°C (17). More or less typically the highest ductility values are at the lowest strain rates, rather impractical rates of deformation for commercial applications; however at 10^{-3} sec^{-1}, a more practical deformation rate, ductilities are an attractive 250 to 300 %, depending on the test temperature. Of interest, the 70α - 30γ and the 30 α - 70γ alloys show lower but similar ductility behavior, with the 30α - 70γ alloy more ductile than the 70α - 30γ alloy. No grain coarsening is observed in these tests because the grain sizes were on the coarser side.

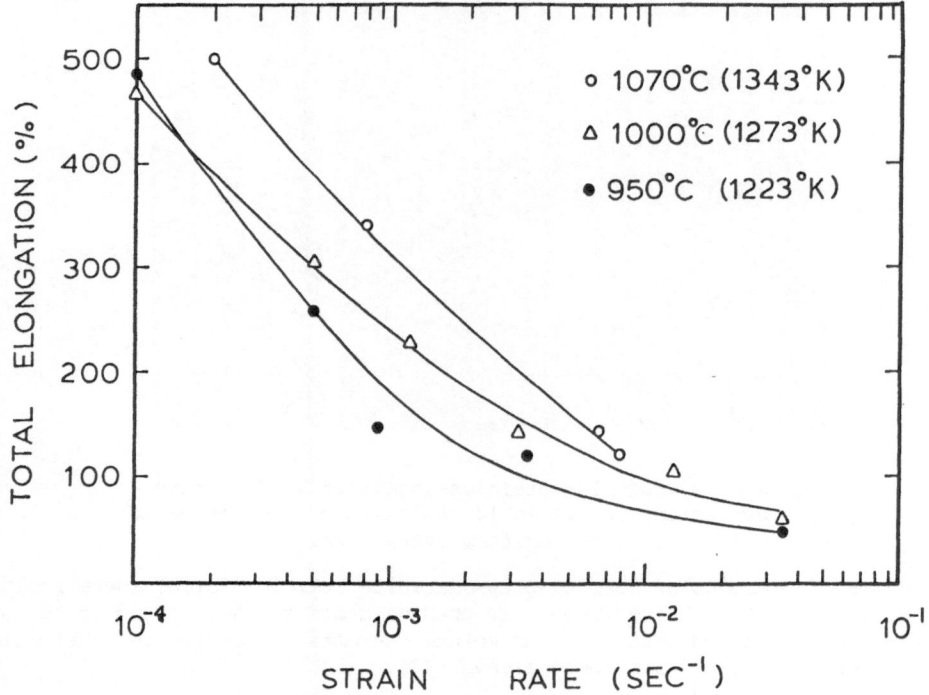

Figure 7: Elongation versus log strain rate for microduplex (ferrite and austenite) RS-splat stainless steel showing superplastic behavior at test temperatures of 950 to 1070°C. Structure is 50α - 50γ.

A series of RS–PM aluminium alloys which were prepared for purposes of achieving improved high room temperature tensile properties were tested for superplastic behavior because they were multiphased alloys with grain sizes finer than about 5 μm. Figures 8 and 9 are plots of ductility versus log strain rate for alloy 7091, a 7075 aluminium alloy modified with 0.4% Co for grain stability, and for an Al – 6.7% Mg – 1.6% Li alloy, respectively (18). Highly different ductility-strain rate patterns are shown by these two alloys both in terms of the temperature response and the strain rate effect. The 7091 alloy shows optimum ductility at 400°C, with decreasing elongation values at 450°C and severe loss at 500°C due to grain coarsening during deformation. Optimum ductility, in contrast to that for the stainless steel, Figure 6, is at the high strain rate of 10^{-2} sec^{-1}, an attractive commercial deformation rate.

In contrast, the Al-Mg-Li alloy shows maximum superplastic response at 325°C, with lower values at 300 and 350°C. Again, high elongation values are achieved at the high strain rates of 10^{-2} sec^{-1} (18).

These three plots are representative of the excellent superplastic response of RS alloys which have a duplex or multiphase structure and have a grain size finer than about 10 to 15 μm; such structures are expected to be found in many alloy systems.

Of unusual interest are the results of superplasticity studies of submicron grain size structures, specifically 0.2, 0.6, and 1.3 μm. These very fine grain sizes were achieved by producing an Fe – 25Cr – 20Ni – 12B (all atom percent) glassy alloy by melt spinning, chopping the ribbon into a flaked "powder", crystallizing the glass at 800, 900, and 1100 °C, and extruding at the same temperatures to produce the indicated grain sizes. The crystallized alloy has an austenitic matrix with about 30 vol. pct. of boride phase, with the boride particle size somewhat finer than the austenitic grain size.

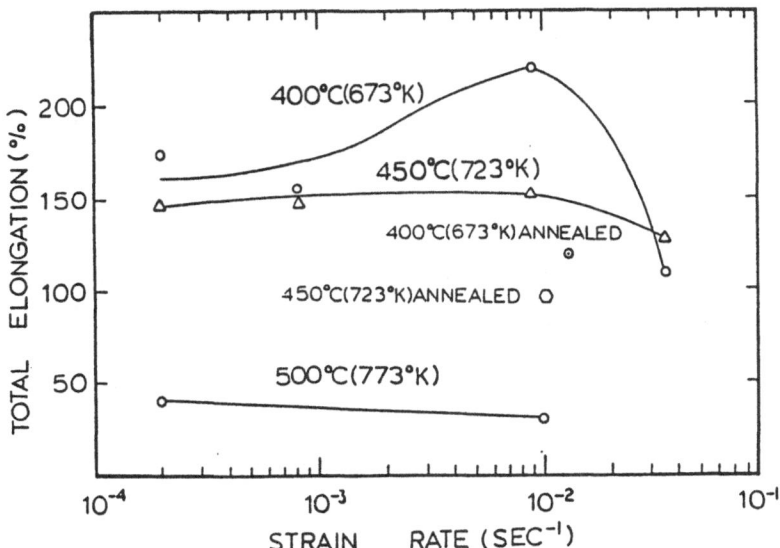

Figure 8: The effect of strain rate on the ductility of RS alloy X7091 at temperatures of 400, 450, and 500°C (673K, 723, and 773°K). Tests are for as-extruded alloy unless noted otherwise.

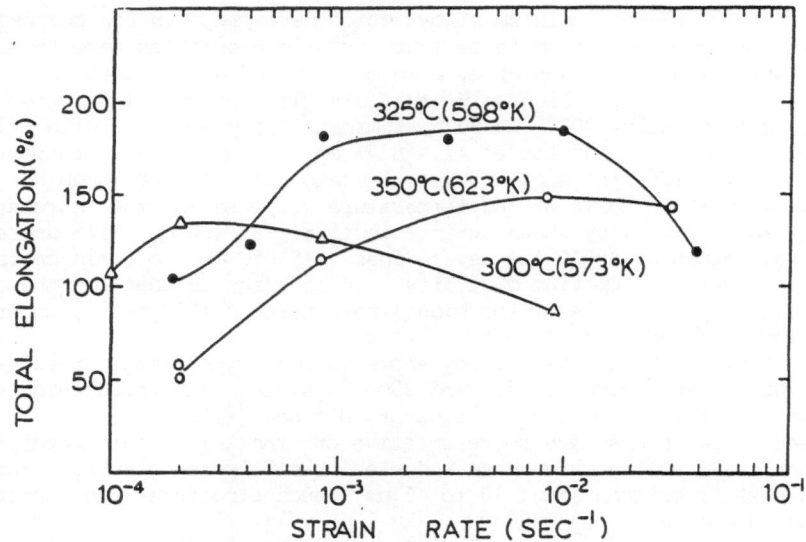

Figure 9: The effect of strain rate on the ductility of RS alloy Al-6.7 Mg- 1.6 Li at temperatures of 300, 325, and 350°C (573, 598, and 623°K).

Figure 10 is a plot of elongation versus the strain rate (log scale) for tests at 1000°C. A comparison is made with microduplex (50% α + 50% γ) stainless steel of grain sizes 20 μm and 5 to 10 μm (see also Figure 7). Rather typically, the 20 μm grain size product is not superplastic at 1000°C, and the finer 5 to 10 μm grain size achieves its highest ductilities at the lowest strain rates.

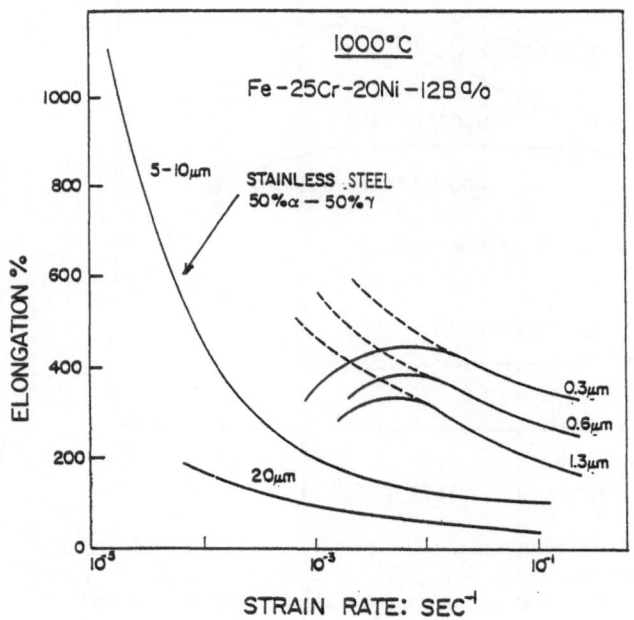

Figure 10: Superplastic behavior of crystallized, ultra-fine grained Fe - 25Cr - 20Ni -12B (atom pct.) prepared from the glassy state. Comparisons made with microduplex austenitic stainless steel of 50 ferrite - 50 austente, with 20 μm and 5-10 μm grain sizes. Maximum ductility for submicron grains sizes is at $10^{-2}sec^{-1}$.

By contrast the submicron grain sizes of the Fe-Cr-Ni-B alloy attain ductilities approaching 500 % at the very high strain rates of 10^{-2} an 10^{-1} sec^{-1}, rates which are extremely interesting for commercial purposes. The dashed curves show where the ductility values would have been if these ultra-fine grain sizes had remained stable; unfortunately, the driving force for grain coarsening is extemely large at 1000 °C. Other, more stable fine dispersions than the $(Fe,Cr)_2B$ might in turn stabilize the fine grain size and thereby permit still higher elongation values.

6. SUMMARY

The results reported in this presentation demonstrate rather clearly that major refinements in overall structure control are possible through rapid solidification, with attendant benefits in strength, ductility,toughness, formability and other properties. Improvements in many combinations of properties are possible in an unlimited number of alloy types and alloy systems, including many of the simpler alloys. A few pitfalls exist. One of the more severe problems is that of incorporating the oxide films which form on the RS particulates of reactive metals and alloys. Such oxides tend to form stringers which are detrimental to transverse mechanical properties, especially fracture toughness values. But even with such alloys, alternate RS techniques exist which will permit consolidation of essentially oxide-free structures.

REFERENCES

1. Proceedings of the International Conference on Metastable Metallic Alloys, Brela, Yugoslavia, Fizika,Vol.2, Supplement 2, 1970.
2. Lyle JP and Cebulak WS: Met. Trans. A, Vol. 6A: 685 (1975).
3. Webster WW: Structure and Properties of Rapidly Solidified X7091 Aluminium Alloy, Massachusetts Institute of Technology, M.S. Thesis, Sept. 1982.
4. Wang W and Grant NJ: Aluminium-Lithium-Alloys II, Conference Proceedings, The Met. Soc. of AIME, Sanders TH, Jr. and Starke AE,Jr. eds., Warrendale, PA, p 447 (1984).
5. Sankaran K and Grant NJ: Mater. Sci. and Eng., Vol. 44: 213 (1980).
6. Fridlyander IM, Ambartsumyan SM, Shiryaeva NV and Gavidullin: Metal-Science and Heat Treatment, Vol. 3: 211 (1978).
7. Kang S and Grant NJ: Aluminium-Lithium alloys, Conference Proceedings, The Met. Soc. of AIME, Sanders TH, Jr. and Starke EA, Jr., eds., Warrendale, PA, p.469 (1984).
8. Domalavage PK, Grant NJ, and Gefen Y: Met. Trans. A., Vol 14A: 1599 (1983).
9. Brooks, RG, Leatham AG, Coombs JS, and Moore C: Metallurgia and Metal Forming, Vol. 9: p.1, Apr. 1977.
10. Lavernia EJ: Liquid Dynamic Compaction of a Rapidly Solidified 7075 Aluminium Alloy Modified with 1% Ni and 0.8% Zr, Massachusetts Institute of Technology, M.S. Thesis, 1984.
11. Domalavage PK and Grant NJ: Metal Powder Report, Vol.38 (No. 10): 555 (1983).
12. Megusar J, Arnberg L, Vander Sande JB and Grant NJ: J. Nucl. Mater., Vols. 103&104: 1103 (1981).
13. Testart E, Megusar J, Arnberg L , and Grant NJ: J. Nucl. Mater., Vols. 103&104: 961 (1981).
14. Megusar J, Arnberg L, Vander Sande JB, and Grant NJ: J. Nucl. Mater., Vol. 99 (Nos. 2&3): 190, Sept. 1981.

15. Megusar J, Harling OK, and Grant NJ: J. Nucl.Mater., Vols.103&104: 961 (1981).
16. Paton NE and Hamilton CH, eds.: Proceedings of a Symposium on the Superplastic Forming of Structural Alloys, AIME, Warrendale, PA, 1982.
17. Zhang Y, Dabkowski F, and Grant NJ: Mater. Sci. and Eng., Vol. 65: 265 (1984).
18. Zhang Y and Grant NJ: Mater. Sci. and Eng., Vol. 68: 119 (1984).

ABSTRACTED DISCUSSION OF THE PAPER BY N.J. GRANT

Participants: J. Ågren, R.W. Cahn, R.E. Maringer and J.V. Wood

The main concern of the discussion was the superplastic behavior of rapidly solidified powders. First, RS aluminium powder technology was discussed. The question was whether or not superplastic shaping would introduce cavities. Due to the fact that in this particular case these powders had been atomized in an argon atmosphere there was no residual gas in pores and the argon was not considered detrimental. Only if local strain rates changed too much would cavitation in certain cases be observable.

Of more concern was the superplastic behavior of RS steel powders. Here the question arose as to what extent two phase mixtures of α- and γ-phase would introduce problems due to their different plastic deformabilities. The grain sizes, it was contended, were so small that no discernable deformation differences were evident. In fact, an added boride phase (FeCr$_2$B), being 30 % smaller than the austenite grains, turned out to be a very stable precipitate. It could not be disrupted even at annealing temperatures up to 1300°C. The grain sizes in question lay between 2 and 0.2 μm. Even smaller grain sizes could readily be achieved.

B. POSTER CONTRIBUTIONS

THERMODYNAMIC ASPECTS OF GLASS FORMATION IN Ni-Zr ALLOYS

N. CLAVAGUERA
Fisica del Estado Solido, Facultad de Fisica
08028 Barecelona, Spain

There is growing evidence that chemical interaction between the components of an alloy plays a significant role on its glass forming ability. One of the most thermodynamically simple ways of treating the attractive interactions between unlike atoms, given a specific chemical short range order, is by means of the so-called associated complexes. Thus, assuming the existance of NiZr and NiZr$_2$ associated complexes we look the Ni-Zr mixtures as strongly associated solutions (1), and we obtain the liquid-solid boundaries.

In our model we use four parameters:

a) the equilibrium constants $K_1 = X_{Ni} X_{Zr} / X_{NiZr}$ and $K_2 = X_{Ni} X_{Zr}^2 / X_{NiZr2}$, x_j being the molar fraction of the j-chemical species (i.e. Ni, Zr, NiZr, NiZr$_2$) and

b) the interaction parameters α and β between the pairs Zr-NiZr$_2$ and NiZr$_2$-NiZr respectively.

Experimental results and the phase diagram calculated are shown in figure 1. The following values were used: $K_1 = 10^{-2}$; $K = 10^{-1}$; $\alpha = -3.4$ kcal/mol K, $\beta = -1.5$ kcal/mol K.

It is seen that a good agreement for the Zr-NiZr$_2$ subsystem is obtained although the NiZr$_2$ - NiZr shows a departure from the theoretical model, the deep eutectic being non-reproducible.

However, it seems evident that, due to the values for K_1 and K_2 used, the chemical ordering is quite important in the neighborhood of the compounds NiZr$_2$ and NiZr. To this ordering effect which minimizes the internal energy the effect of mixing effects should be added which, on the other hand, maximizes the entropy.   This is found by taking into account the Gibbs free energy difference, $G^l - G^s$, between the liquid mixture and the equilibrium crystalline phases. The results obtained as a function of composition at two temperatures (800 and 1000 K) are shown in figure 2. As expected the $G^l - G^s$ function has its minimum value not far from the eutectic compositions. There, the glass forming ability comes to a maximum.

(1) M.T. Clavaguera -Mora and N. Clavaguera, J. Phys. Chem. Solids, 43, (1982), 963.

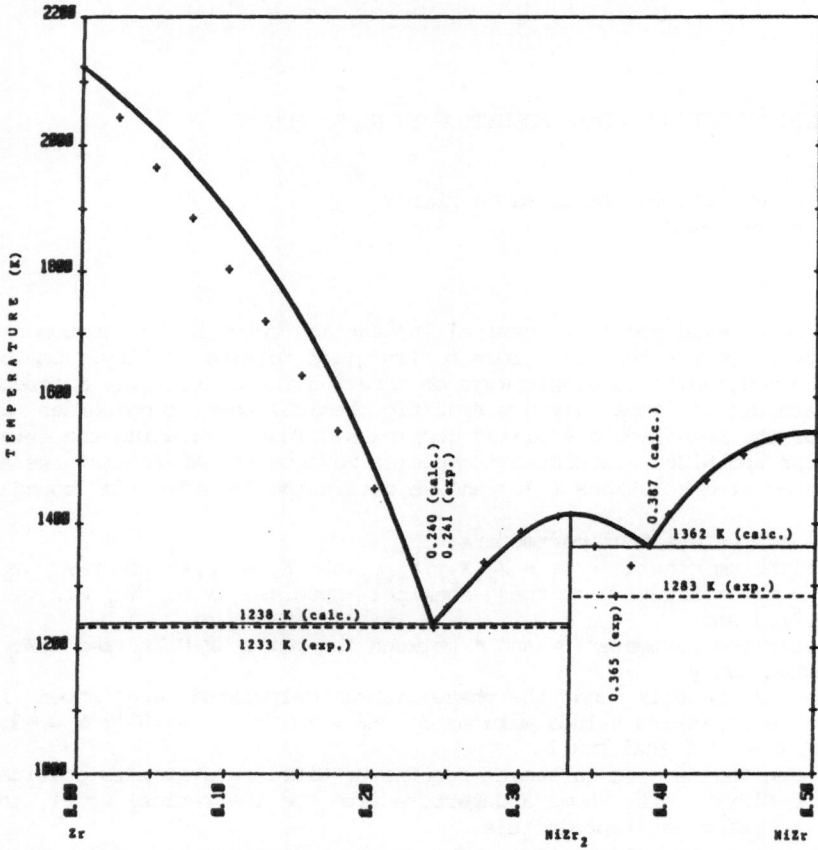

Figure 1: Calculated phase diagram and experimental points.

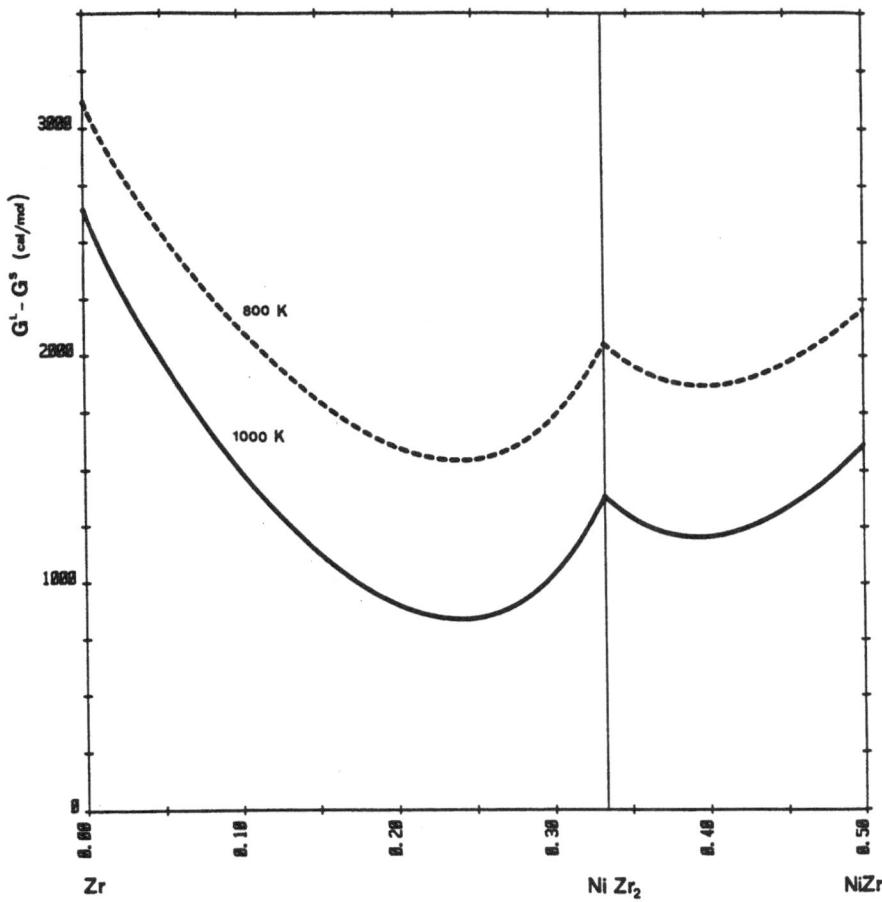

Figure 2: Calculated Gibbs free energy difference, $G^L - G^S$
for temperatures of 800 and 1000 K.

THERMODYNAMICS OF CHEMICAL ORDERING AND GLASS-TRANSITION IN LIQUID Mg-Sn ALLOYS

B. JÖNSSON and J. ÅGREN
Division of Physical Metallurgy
The Royal Institute of Technology
S-100 44 Stockholm, Sweden

Thermochemical data and phase diagram information on the Mg-Sn system have been assessed in terms of thermodynamic models of the individual phases. Special attention has been paid to the effect of chemical order in the liquid. The thermodynamic description obtained has been used to calculate the phase diagram, T_0-curves, isoentropic temperatures and nucleation temperatures. The C_p of the pure liquid substances was extrapolated into the supercooled region by postulating isoentropic temperature at a third of the melting point. An alternative evaluation based on a different kind of extrapolation gave basically the same agreement with the experimental data. This indicates that it is not possible to extrapolate the high temperature properties to large undercoolings without making additional assumptions about the behaviour at very large undercoolings.

STRUCTURAL CRITERIA FOR GLASS FORMING ABILITY: APPLICATION TO ALUMINIUM-BASED ALLOYS.

J. M. DUBOIS, G. LE CAER and K. DEHGHAN
Laboratoire de Metallurgie, Parc de Saurupt
54042 Nancy Cedex, France

The glass forming ability (GFA) of metallic alloys is usually accounted for in terms of thermodynamics and kinetics parameters. Structural parameters do not explicitly enter such approaches though the atomic size ratio r_X/r_M in binary $M_{1-x}X_x$ alloys has obviously some link with GFA as was recently shown by EGAMI and WASEDA (1).

Their study of the minimum concentration of the glass forming range in more than 60 binary alloys indeed proves that large volume differences between atomic species are correlated with high GFA in these systems. A fairly good agreement was obtained by EGAMI and WASEDA (1) between this experimental correlation and the minimum concentration for which the solid solution of X atoms substituting for M atoms becomes unstable because of internal stresses. When the X concentration overcomes this value, the amorphous state may thus offer an attractive alternative for the system, especially if no stable crystalline phase exists at the same stoichiometry.

However, this model does not account for the GFA in several classes of glass formers. This is mainly true in iron - metalloid alloys: Fe-B is a good glass former, Fe-C is a poor glass former whereas Fe-N has not yet been obtained amorphous ($r_N < r_C < r_B < r_{Fe}$).

In order to clarify this point, we may assume that the smaller atom X does not substitute for the larger atom M but dissolves interstitially in the (basically metallic) M network. In this case, internal stresses may develop if the X species is not small enough to fit into the octahedral sites of the M lattice which, because it is metallic, tends towards the highest packing fraction. Such stresses may also destabilise the interstitial solid solution as an alternative is to distort the octahedral interstices into triangular prisms which on the one hand offer more room to the small atoms and on the other hand preserve the close packed fraction of the metal atoms M (2).

Due to the relatively large solute concentration, such a transformation involves a certain degree of correlation between triangular prismatic units as M atoms may belong to two or more units. Several such correlations, which i- account for the stoichiometry and ii- satisfy to the steric constraints of space filling, are known in the crystalline state (2). A satisfactory description of the amorphous structures may thus be obtained i- by limiting the distance over which the correlation applies along a given direction to typically a few nanometres and ii- by restricting the inter-units connections to be invariant for a local rotation in order to avoid grain boundary formation (3).

The central point to our present purpose is that several distinct types of structural correlations may fulfil the above requirements. Given the nature and the concentration of the different species, there are indeed different ways to pack together triangular prisms, the most simple called Simple Chemical Twinning and Fourling (2) being the structure building operations acting in the Al_3Ni and Al_2Cu compounds, respectively.

As such operations are known to act in the crystalline counterparts of several readily glass forming systems (e.g. simple chemical twinning in Ni_3B or Pd_3Si, fourling in Co_3P or Fe_2B), some predictions of the composition of glass forming alloys in a given system may be infered from the knowledge of the corresponding structural operations. Moreover, alloying of different solute species (in strong interaction with the solvent element) may involve distinct structural operations, thus significantly increasing the configurational entropy of the system and consequently enhancing its GFA.

In order to test the validity of these ideas, several aluminium-based alloys containing both Ni and Cu were quenched from the liquid state by using a conventional melt-spinning device. Following the above arguments, this specific choice was made because the Al_3Ni and Al_2Cu compounds are isotypic with Ni_3B and Fe_2B, respectively, the crystalline counterparts of well known glass forming alloys.

Figure 1 shows that a $Al_{81}Cu_{12}Ni_7$ alloy my be retained in a partially amorphous state. Adding a small amount of Mo, which presumably favors the kinetic avoidance of crystallization as in iron-based glasses, leads to a fully amorphous sample of composition $Al_{80}Cu_{10}Ni_8Mo_2$ (figure b).

Though this approach does not include kinetics parameters (which, however, determine the success or the failure of a quench with a given machine), it has been successful enough to define the composition of amorphous ribbons in aluminium-based alloys (at the time of our first report (4), such alloys were not reported to give bulk glassy samples). More details about this study are published elsewhere (5).

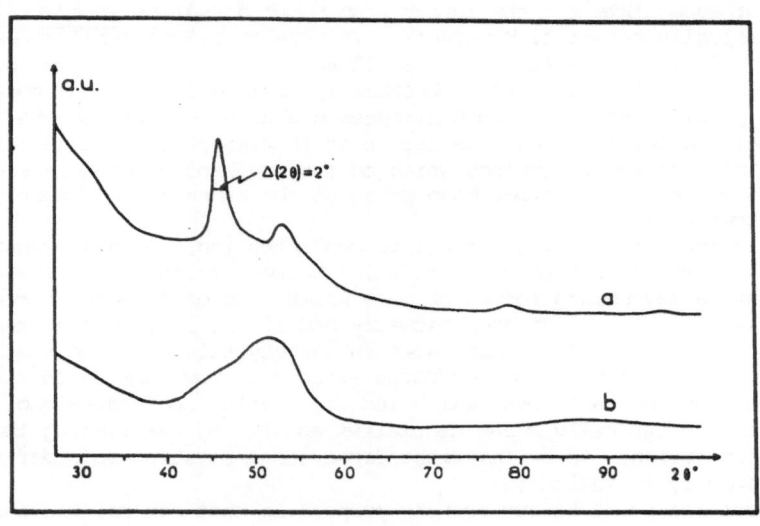

Figure 1 : X-ray diffraction patterns (λ = 1.789 Å) of as-quenched alloys:
a: $Al_{81}Cu_{12}Ni_7$; b: $Al_{80}Cu_{10}Ni_8Mo_2$

REFERENCES

1. T. Egami and Y. Waseda , J.Non Cryst. Sol. 64, (1984), 113
2. B.G. Hyde, S. Andersson, M. Bakker, C.M. Plug, M. O'Keeffe, Prog. Sol.
    State Chem. 12, (1979), 273
3. J. M. Dubois, P.H. Gaskell, G. Le Caer, RQ5 Conf. S. Steeb and H.
    Warlimont ed. in the press
4. J. M. Dubois and G. Le Caer, C. R. A. S., pli cacheté du 07-06-1982
    in the press
5. J. M. Dubois, G. Le Caer, K. Dehghan, RQ5 Conf. S. Steeb and H. Warli-
    mont ed. in the press.

EFFECT OF THE COOLING RATE ON THE AS-CAST FRACTION CRYSTALLIZED

W. HUG AND P.R. SAHM
Foundry-Institute, Aachen Institute of Technology
D-5100 Aachen, F. R. Germany

The objective of this work was to investigate the fraction of as-cast crystals in an otherwise glassy matrix as function of cooling rate and solidification time in rapidly quenched $Fe_{40}Ni_{40}B_{20}$ alloys. The ribbons were produced by planar flow casting, figure 1, with a continuous variation

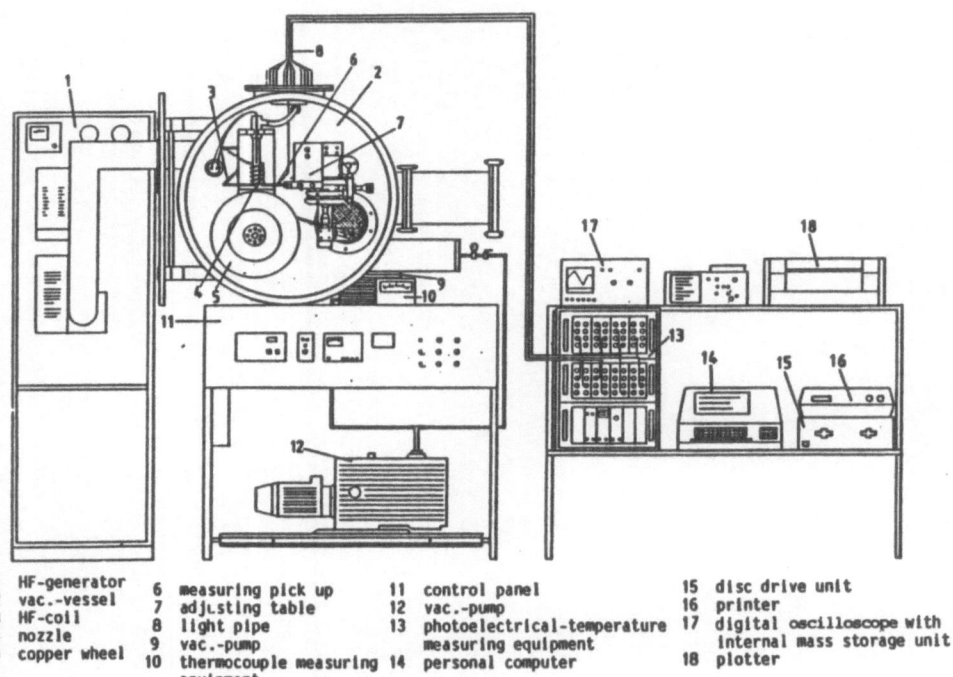

1 HF-generator	6 measuring pick up	11 control panel	15 disc drive unit
2 vac.-vessel	7 adjusting table	12 vac.-pump	16 printer
3 HF-coil	8 light pipe	13 photoelectrical-temperature	17 digital oscilloscope with
4 nozzle	9 vac.-pump	measuring equipment	internal mass storage unit
5 copper wheel	10 thermocouple measuring equipment	14 personal computer	18 plotter

Figure 1: Planar flow casting unit and photoelectrical temperature monitor.

of thickness in the range between 20 and 250 µm. This was accomplished by decelerating the rotating disc during the casting process. The solidification process which, occuring in only few milliseconds, was monitored by a photoelectric temperature measuring device developed at the Foundry-Institute.

Figure 2 illustrates a series of typical solidification curves of $Fe_{40}Ni_{40}B_{20}$ ribbons produced by planar flow cast-quenching on a copper wheel. Cast-in crystals of $(FeNi)_{23}B_6$ are observed by light microscopy and TEM, figure 3. First crystals appear at ribbon thicknesses of approximately 60 µm. Semicrystalline structure then is found in the range between 60 and 110 µm, with corresponding cooling rates of $1,8 \cdot 10^5$ K/s, figure 4. The main aspiration of these investigations is to correlate the fraction crystallized and nucleation frequency with growth velocity and the solidification time.

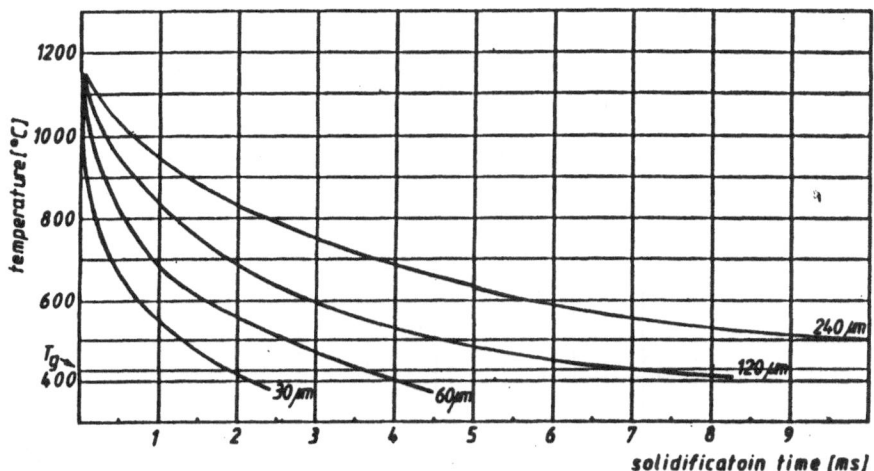

Figure 2: Typical cooling curves of $Fe_{40}Ni_{40}B_{20}$ ribbons produced by planar flow casting on a copper wheel.

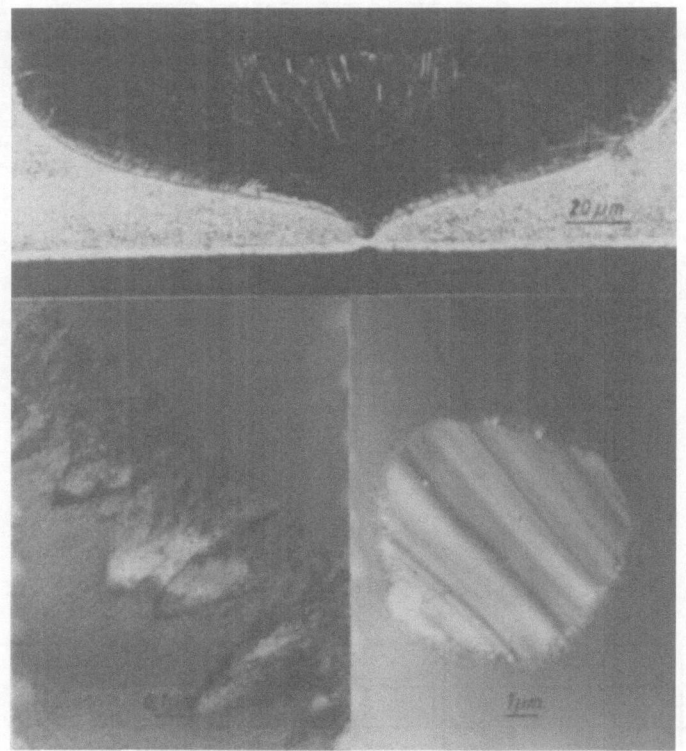

Figure 3: $(FeNi)_{23}B_6$ –crystal in semi-amorphous
ribbons.

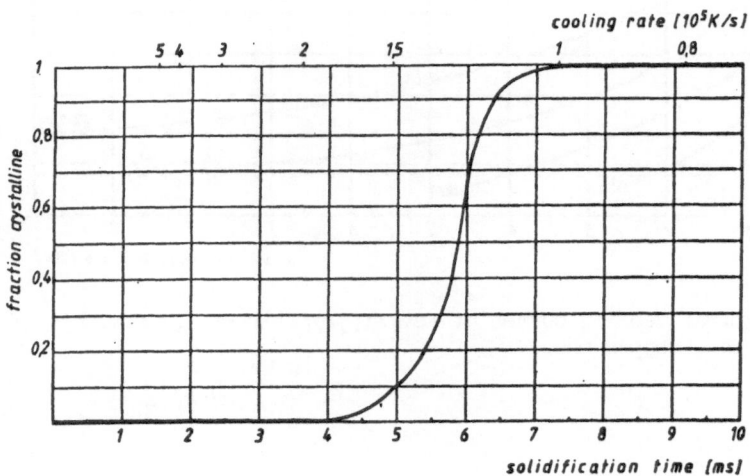

Figure 4: Crystalline fraction of $Fe_{40}Ni_{40}B_{20}$
ribbons as function of solidification time and
cooling rate.

SOLIDIFICATION OF AMORPHOUS $Pd_{77,5}Si_{16,5}Cu_6$ IN A DROP TUBE

L. KALLIEN AND P.R. SAHM
Foundry-Institute, Aachen Institute of Technology
D-5100 Aachen, F. R. Germany

All liquids undergo the glass transition if they are sufficiently under-cooled and crystallization does not intervene. The cooling rates necessary to prevent crystallization, range from a fraction of a degree per second for silicate glasses to 10^6 K/s and more for certain metallic alloys. In order to avoid heterogeneous nucleation for decreasing critical cooling rates it helps if contact of the melt with a surface can be avoided. One method towards this end is to use a low gravity containerless method. Figure 1 represents the drop tube set-up.

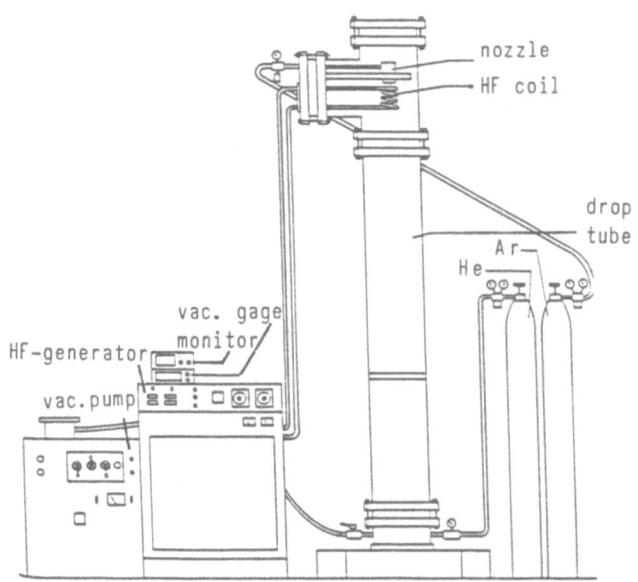

Figure 1: Drop tube for containerless solidifica-tion producing a low gravity condition.

By pressing the melt through a quartz capillary droplets are produced in the range of 100 to 1500 μm. DSC-analysis showed that the $Pd_{77,5}Si_{16,5}Cu_6$-samples were amorphous to crystalline in the range between 300 and 1300 μm, figure 2. The cooling rates were estimated assuming Newtonian cooling.

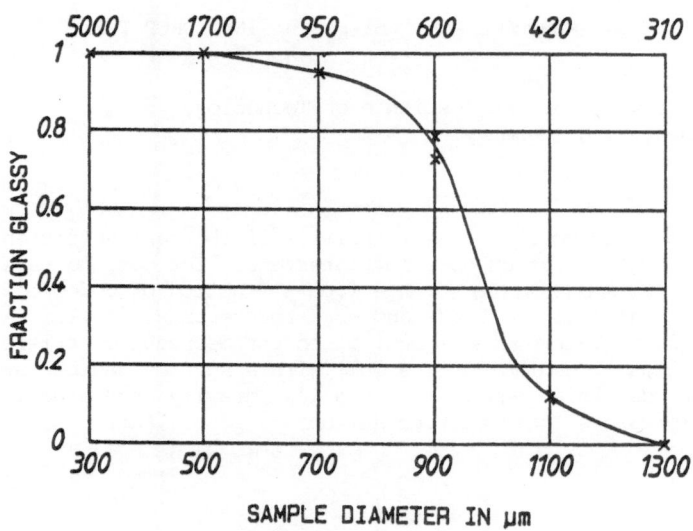

Figure 2: Fraction glassy as a function of droplet diameter and cooling rate.

Figure 3a shows amorphous $Pd_{77,5}Si_{16,5}Cu_6$-samples with a diameter of 300 μm and less. Pd_3Si-dendrites in amorphous matrix are presented in figures 3b (particle diameter: 700μm) and 3c (particle diameter: 900 μm).

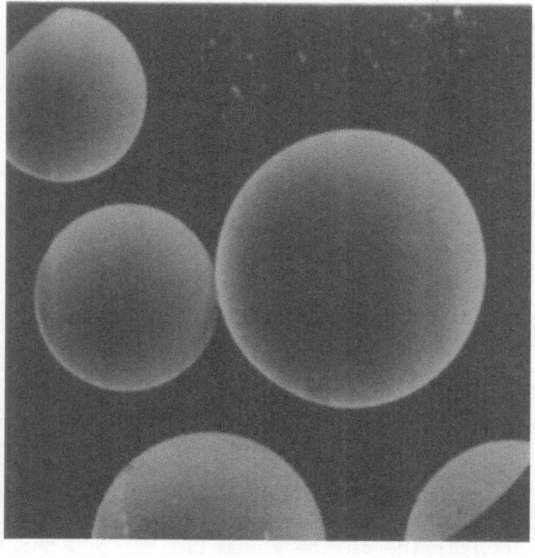

Figure 3a: PdCuSi
dia 300 μm.

Figure 3b: PdCuSi
dia 700 μm.

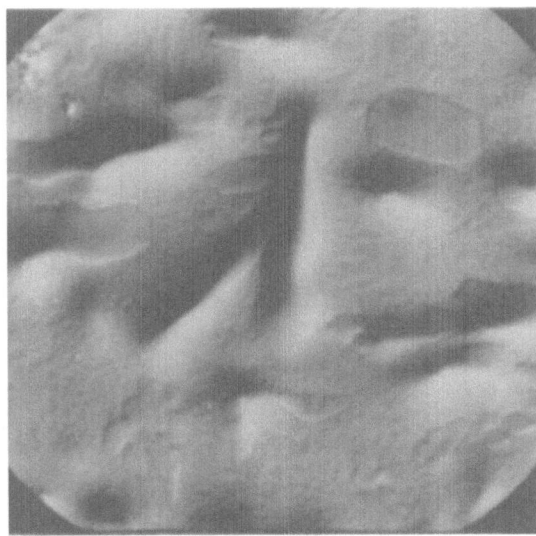

Figure 3c: PdCuSi
dia 900 μm.

THE FORMATION KINETICS OF PdCuSi- AND PdNiP-METALLIC GLASSES

C.S. KIMINAMI* AND P.R. SAHM
Foundry-Institute, Aachen Institute of Technology
D-5100 Aachen, F. R. Germany

The critical cooling rate ($\dot{T}_c$) required for glassy solidification of $Pd_{79}Cu_4Si_{17}$, $Pd_{77,5}Cu_6Si_{16,5}$ and $Pd_{40}Ni_{40}P_{20}$ alloys were measured by monitoring the critical velocity for the formation of amorphous structure during directional solidification, figure 1. The measured critical velocities (v_c) and the critical cooling rates ($\dot{T}_c$) for the three alloys are given in table 1. These results were compared with the $\dot{T}_c$ values computed from CCT-diagram using the Uhlmann treatment considering only homogeneous nucleation. Influences of activation energy of nucleation (ΔG^*) and Gibbs volume free energy difference between solid and liquid (ΔG_v) on the computed $\dot{T}_c$ were analysed. The computed $\dot{T}_c$ are presented in figure 2. Computed $\dot{T}_c$-values for both PdCuSi alloys are in satisfactory agreement with experimental data when the Thompson-Spaepen ΔG_v-approximation and $\Delta G^* = 65$ kT at 0.2 ΔT_r are used. For $Pd_{40}Ni_{40}P_{20}$ alloy none of the computed $\dot{T}_c$ agreed with experimental results in the ranges analysed.

Alloy	$Pd_{79}Cu_4Si_{17}$	$Pd_{77,5}Cu_6Si_{16,5}$		$Pd_{40}Ni_{40}P_{20}$
Tube	alumina	alumina	quartz	alumina
V_c[mm s^{-1}]	8,22	2,64	3,65	0.40
$\dot{T}_c$ K s^{-1}	411 $\pm$ 40	221 $\pm$ 10	253 $\pm$ 25	35,0 $\pm$ 0,5

Table 1: The critical velocities (v_c) and the critical cooling rate (T_c) for formation of metallic glasses measured by rapid directional solidification.

---

* on leave of absence from Federal University of Paraiba, CCT-DEM, Brazil

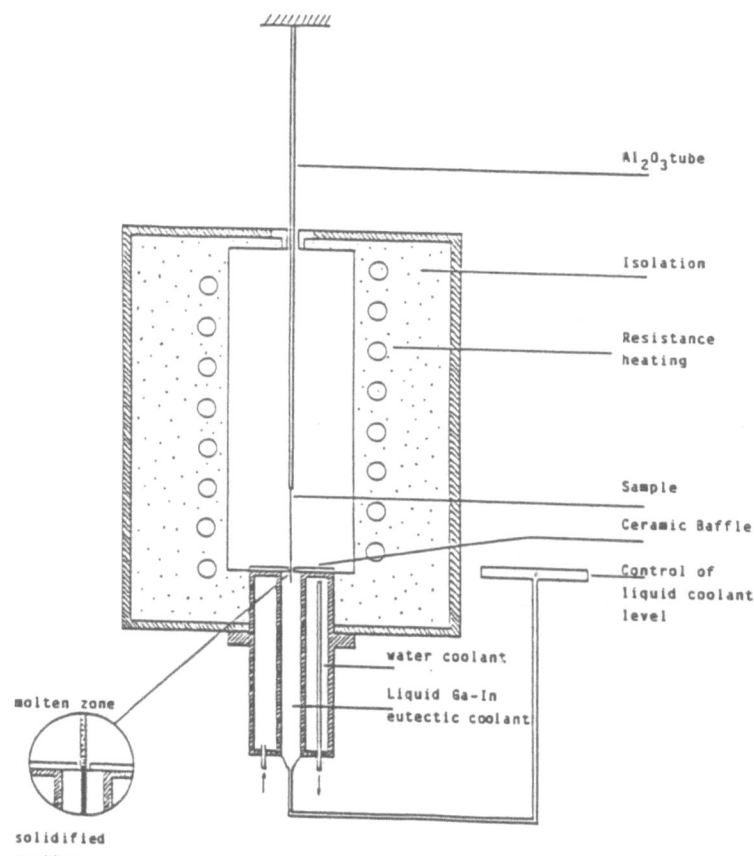

Figure 1: High cooling rates are not only achieved by rapid directional solidification velocities but also by utilizing small diameter samples.

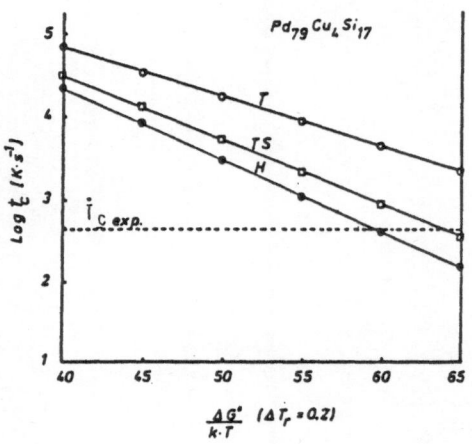

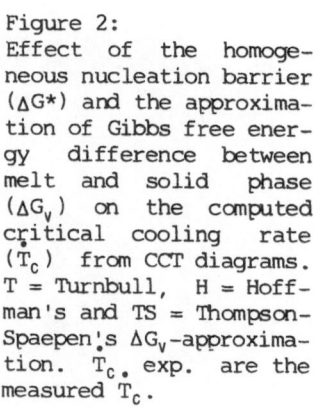

Figure 2:
Effect of the homogeneous nucleation barrier (ΔG^*) and the approximation of Gibbs free energy difference between melt and solid phase (ΔG_v) on the computed critical cooling rate (T_c) from CCT diagrams. T = Turnbull, H = Hoffman's and TS = Thompson-Spaepen's ΔG_v-approximation. T_c exp. are the measured T_c.

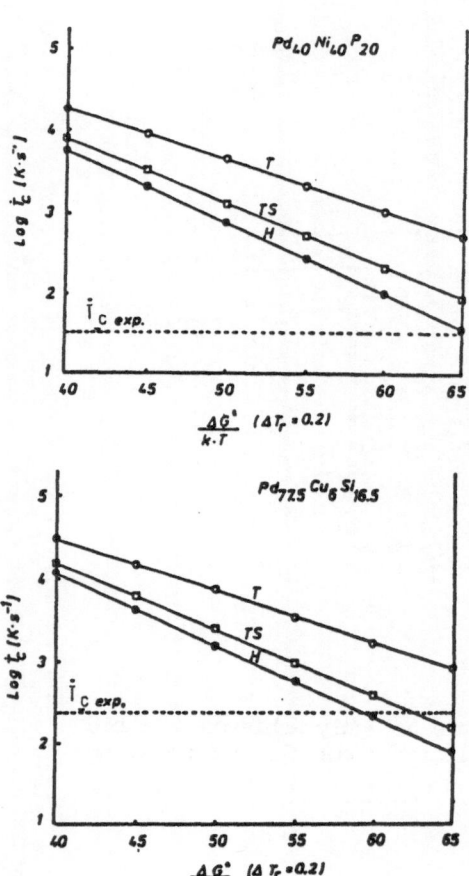

CRYSTALLIZATION BEHAVIOUR OF Ni-Nb AMORPHOUS ALLOYS: COMPARISON OF CALORI-
METRIC RESULTS AND HOMOGENEOUS NUCLEATION - GROWTH MODELLING

F. MORET and S. PAÏDASSI
CENG - DMG - 85 X - 38041 GRENOBLE Cedex
France

Nickel-niobium amorphous ribbons have been made by planar flow casting in
vacuum (figure 1). Light and SEM micrographs of annealed ribbons showed
that crystallization occured in the bulk and not from the surface giving a
fine microcrystallized structure (10^{19} gains/m^3).

The crystallization behaviour has been studied in a Perkin-Elmer DSC2
calorimeter with baseline and heating rate dependent temperature lag corrections (1) (table 1). Peak shapes were calculated from numerical integra-
tions of JMA transformation rate equation using the activation energies and
the kinetic coefficients derived from the KISSINGER plots of the peak
temperatures.

$Ni_{60}Nb_{40}$ experimental peak shapes are wider than calculated ones because
of the kind of reaction involved (figure 2) (primary (2)).

For $Ni_{50}Nb_{50}$, alloy x-ray diffractometry on annealed DSC samples re-
veales the equilibrium phase (NiNb intermetallic compound). The reaction
involved is polymorphous and experimental and calculated peak shapes are in
good agreement (figure 3).

Homogeneous nucleation and growth calculations based on KELTON and GREER
(3), (4) modelling applied in the non-isothermal case (quench and DSC an-
neal) show that the larger amount of nuclei is produced 40 K above the peak
temperature. This result justifies the good agreement of JMA coefficient
n = 3 based calculations with experimental peak shapes (figure 3).

TABLE 1.

composition	Ni Nb 60 40	Ni Nb 50 50
thickness	32 µm	29 µm
DTA from 750 to 1200K	2 peaks	1 peak
first peak = temperature (20 k/min) enthalpy activation energy	 935 K 1200 J/mol 620 kJ/mol	 998 K 1850 J/mol 405 kJ/mol

REFERENCES

1. A.L. Greer, Acta Met. 30, (1983), 172-176.
2. L.E. Collins, N.J. Grant, J.B.Van der Sand, Mat.Sci. 18,(1983), 804-814.
3. K.F. Kelton, A.L. Greer, C.V. Thompson, J. Chem. Phys. 79 (12),
   (1983), 6261-6276.
4. K.F. Kelton, A.L. Greer, R.Q.5 proceedings.

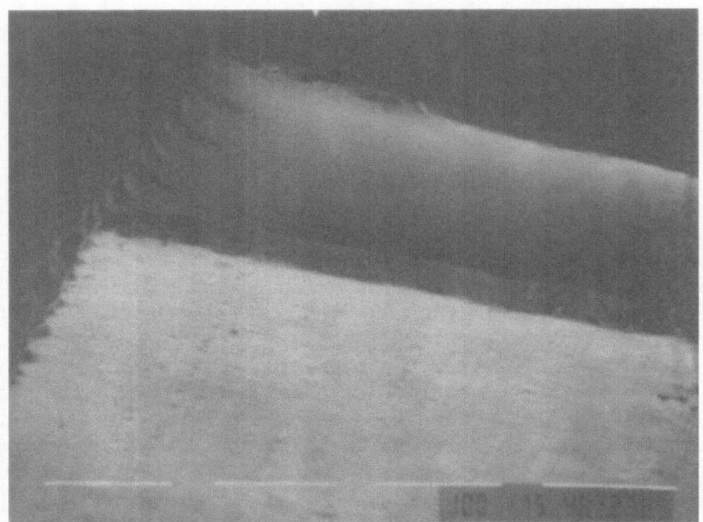

Figure 1

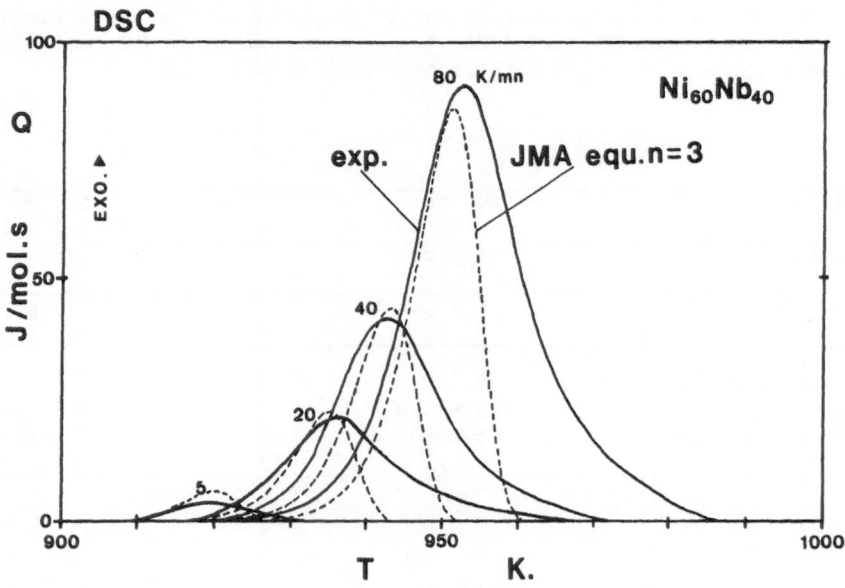

Figure 2

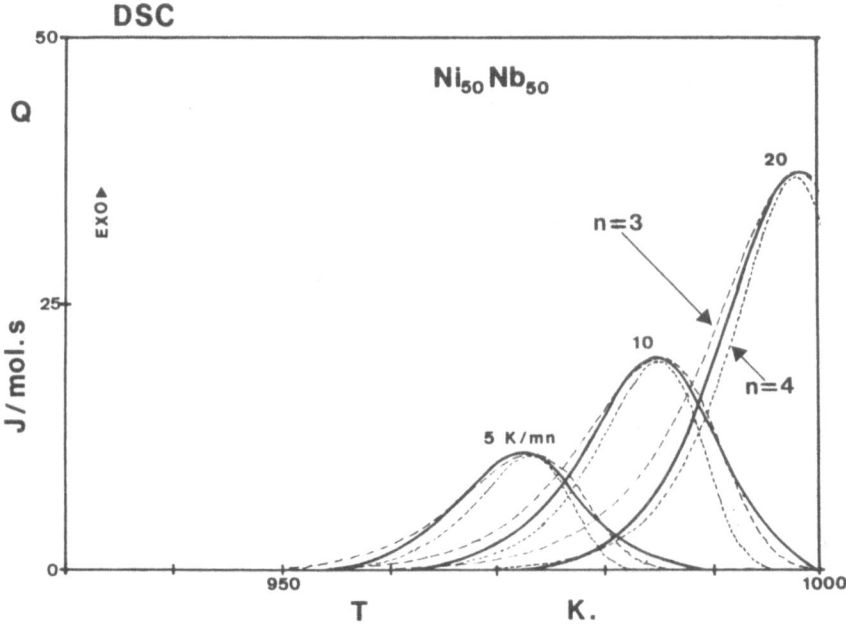

Figure 3

NUCLEATION AND CRYSTALLIZATION KINETICS IN $GeSe_2$-Sb_2Te_3 SYSTEMS

S. SURIÑACH, M.D. BARO and M.T. CLAVAGUERA-MORA
Facultat Ciencies, Universitat Autonoma Bellaterra
08028 Barcelona, Spain

We present results on the formation, thermal stability and crystallization of $(GeSe_2)_{100-y}$ $(Sb_2Te_3)_y$ glasses prepared by quenching of the melt. Previous studies have shown that this system is quasi-binary. The phase diagram (figure 1) shows a limited solubility of $GeSe_2$ in Sb_2Te_3; the solid solution limit being 43 mol% Sb_2Te_3 at the eutectic temperature 758 K. The eutectic is located at 24 mol% Sb_2Te_3.

Water quenching 0.5 g samples of a molten alloy inside quartz ampoules of 5 mm internal diameter and 1 mm thickness, produces a cooling rate at the eutectic temperature of the order of 10^3 K/min. The glass-forming region by water quenching goes from 5 to 30 mol% Sb_2Te_3. The temperatures of the structural transformations are plotted in figure 1. Glasses in the range 5 to 20 mol% Sb_2Te_3 show on heating a single crystallization peak, whereas for higher Sb_2Te_3 content they show two crystallization peaks. The activation energy for each stage of crystallization has been evaluated by use of the peak method. An estimate of the free-exponential A of the kinetic equation $(d\alpha/dt)_T = A f(\alpha) \exp(-E/RT)$ has also been obtained. (Here α is the fraction of material crystallized at temperature T). Both the activation energy and ln A increase as $GeSe_2$ is replaced by Sb_2Te_3 (figure 2).

For compositions lying in the $GeSe_2$ primary crystallization region (Sb_2Te_3 content less than 24mol %) crystals appear preferentially at the surface of the glass. For greater Sb_2Te_3 content nucleation develops typically at random in the bulk; the nucleation density, measured at fixed crystallized fractions, increases with Sb_2Te_3 content. A lamellar eutectic crystallization seems to occur in all the glasses studied.

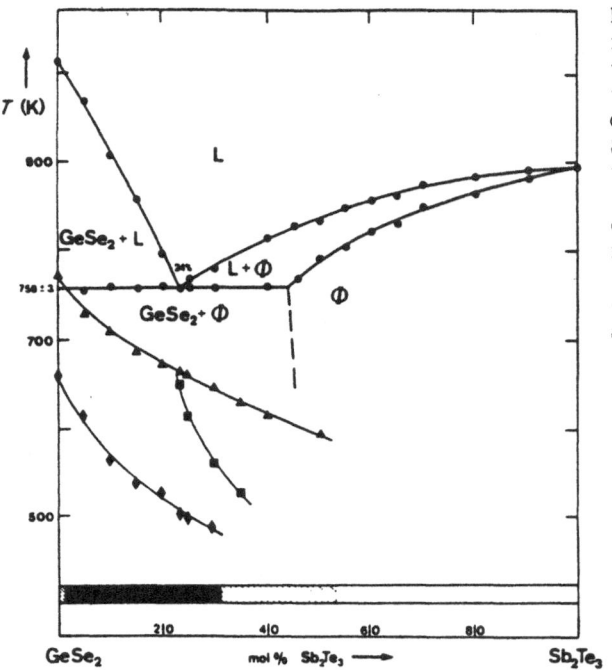

Figure 1: Glass-forming region, temperatures of the relevant structural transformations and phase diagram of the GeSe₂-Sb₂Te₃ quasi-binary system. ■ glassy, ▨ partially crystalline and □ crystalline alloys. Temperatures of: (♦) glass-transition, (■,▲) peaks of crystallization, (●) solidus and liquidus. Heating rate: 20 K/min⁻¹.

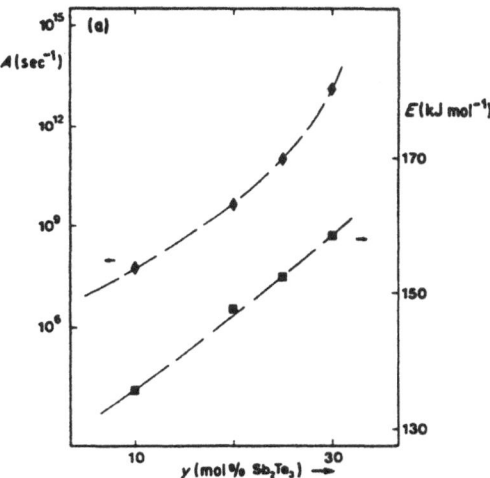

Figure 2: Composition dependence of the kinetic parameters: A and E.

DIRECT THERMAL MEASUREMENT OF ADIABATIC DENDRITIC COARSENING

M.E. GLICKSMAN
Materials Engineering Department
Rensselaer Polytechnic Institute
Troy, NY 12181
USA

This poster portrays the microstructural changes occuring in pure systems freezing from the undercooled melt. We show that changes occur in dendritic arm spacings concurrently with the solidification process. The changes involve local rearrangements, including local melting and coarsening of side arms, plus initiation of chaotic structures bearing little relationship to the well-known periodic microstructures associated with dendritic solidification. All the observed changes involve an increasing average microstructural size scale, although locally, decreasing size scales may be seen. The tendency to increase size scales has been modelled as diffusional coarsening, a process akin to Ostwald ripening, except that the diffusing quantity is enthalpy, not solute. We show how to estimate the local microcurrents of heat due to spatial gradients of interfacial curvature, which through the Gibbs-Thomson effect result in thermal gradients. Simple scaling arguments are shown to deduce that the average microstructural length scale will grow as time to the cube-root power. An outline is provided that describes some novel experiments which monitor the progress of recalescence subsequent to rapid dendritic freezing. We show that under adiabatic solidification conditions the recalescence temperature asymptotically approaches the bulk melting point, too. Immediately, after rapid solidification of a melt from the nucleation temperature, T_N , a volume fraction of solid, X_S, is established and maintained constant by virtue of the adiabatic constraint. Under such conditions, we monitor the resistance, R, of a platinum thermometer with a resolution of 10^{-4} K. We invariably observe a transient temperature rise, caused merely by the warming of the thermometer, followed by a long steady-state temperature rise which we associate with the microstructural coarsening process. The steady-state temperature rise occurs as time to the cube-root power. Explicit predictions are shown for calculating the temperature rise as a function of materials constants such as solid-liquid surface energy, γ ; molar volume of the solid, Ω ; the molar entropy change, Δs; the molar latent heat, ΔH; the molar heat capacity of the melt, C_p; and the thermal diffusivity, X . In addition, a rate constant, A, which only depends on the volume fraction of solid, must be included to calculate the progress of microstructural coarsening subsequent to rapid solidification. Data on the monitoring of the recalescence are shown for three model systems: succinonitrile (SCN), ice/water (H_2O), and ethylene carbonate (ETC). These systems freeze as BCC, 4-fold dendrites (SCN), hexagonal flat "snow-flake" dendrites (H_2O), and as branchless needles (ETC). The data presented clearly show that the steady-state recalescence curves for these diverse materials show similar kinetics, viz. the offset from the bulk melting point, $T = T_\infty - T$, decays as time to the negative of the cube-root power. Work is continuing to explore this method as a technique for measuring solid-liquid interfacial energy, as well as a technique for estimating the true rate of microstructural coarsening.

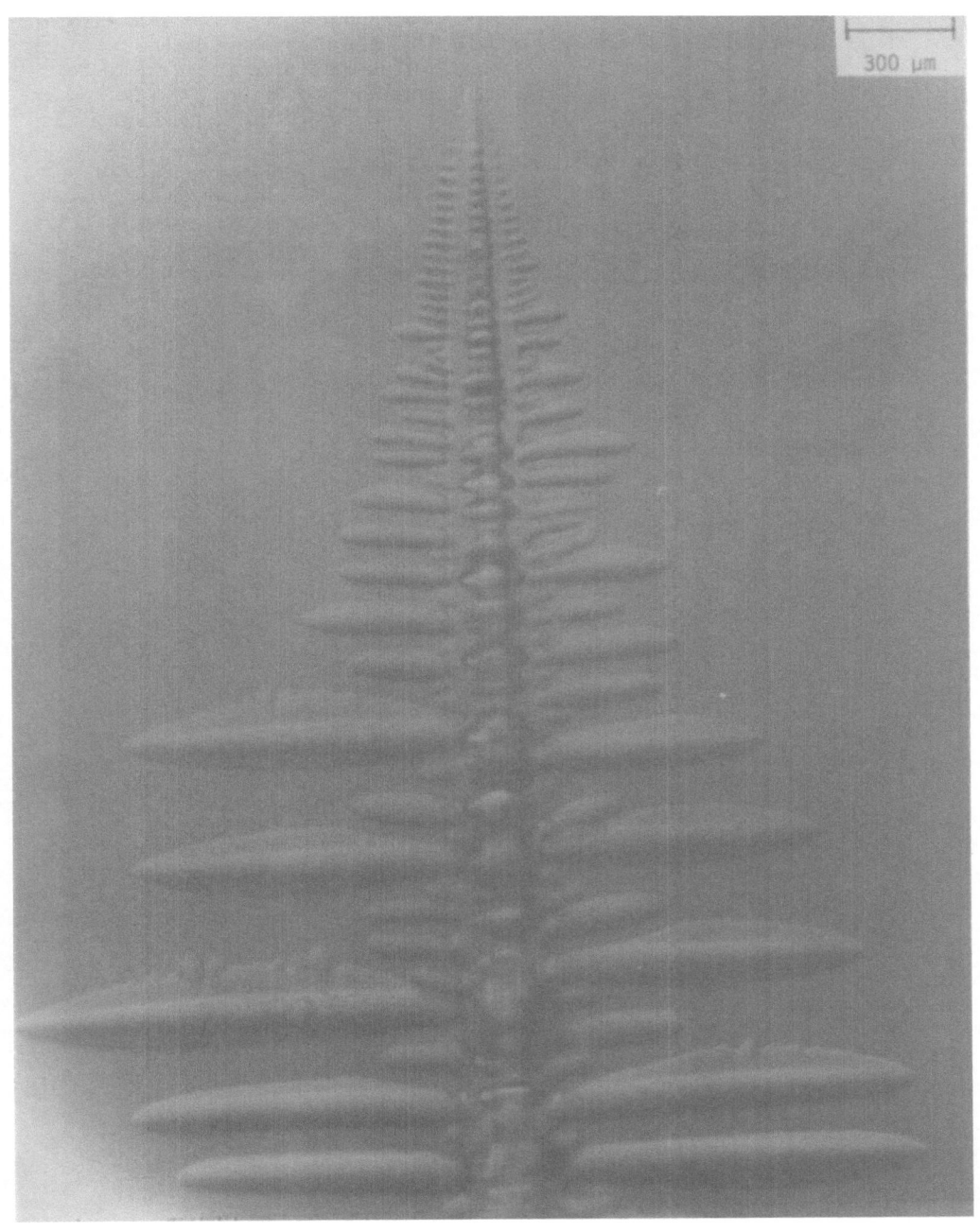

Figure 1: In situ micrograph of a cyclohexanol dendrite growing in the undercooled melt. Note how the secondary and tertiary arms change in the older parts of the microstructure. Eventually, a chaotic pattern develops with an increasing size scale, giving the structure a "pebbled" appearance rather than the customary periodic form of dendritic structures in a less coarsened condition.

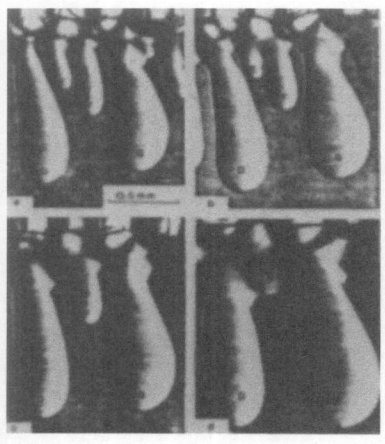

Figure 2: Local microstructural coarsening in succinonitrile. The side arms E and F have slightly lower melting points due to their small size relative to nearby large side arms D and G. Secondary side arm spacing doubles in figure 2c, and quadruples in figure 2d relative to the arm spacings in figures 2a and 2b. Such local events are akin to Oswald ripening, wherein microstructural size scales simultaneously increase and decrease leading to an increasing average size scale.

MICROSTRUCTURE OF MELT-SPUN Al-Mo ALLOYS

E.A. GUEST, J.N. PRATT and M.H. LORETTO
Dept. of Metallurgy and Materials
University of Birmingham
Birm., B 152 TT, U.K.

The melt spinning technique has been used to prepare rapidly solidified Al-Mo alloys over a range of compositions containing up to about 20 at.% Mo, and their microstrucure and thermal stability have been investigated using analytical electron microscopy.

It has been found that the solid solubility of Mo in Al can be increased from the equilibrium value of 0.07 at.% Mo to about 2.4 at.% Mo. Alloys with molybdenum contents too great to form complete solid solutions were found to contain two new metastable and one new stable phase. The two new metastable phases were a diamond cubic and a hexagonal phase. The diamond cubic phase was found to have a composition within the range 7.6 to 11.2 at.% Mo and space group Fd3m with a = 14.5 Å. The hexagonal phase was found to have a composition between 12.3 and 16.4 at.% Mo; its space group was determined to be either P6/mm or P6$_3$/mmc, with a = 4.5 Å and c = 2.7 Å. The new stable phase was rhombohedral in structure and was present in rapidly quenched then annealed alloys containing about 10.7 at.% Mo. Its composition lay between 14.7 and 18.5 at.% Mo and its space group was R$\overline{3}$c, with a = 14.5 Å and α = 86°.

Annealing experiments showed that, at temperatures in excess of 500°C, the supersaturated solid solutions decomposed by grain boundary precipitation of Al$_{12}$Mo. Increasing the supersaturation caused decomposition to occur more rapidly and at lower temperatures. The two new metastable phases both decomposed to give Al$_{12}$Mo.

MICROSTRUCTURE OF PARTIALLY CRYSTALLINE METALLIC GLASSES

U. KÖSTER
Dept. Chem. Eng., University Dortmund
D-4600 Dortmund 50, F.R. Germany

Metallic glasses exhibit a combination of remarkable properties, which are directly derived from their glassy state. However, similar to glass ceramic or pyroceram some partially crystalline metallic glasses (R. Cahn, 1983: pyromets) possess improved properties, e.g. higher ductility (1), less magnetic core loss for high frequency applications (2), or stronger flux pinning interactions in superconducting metallic glasses (3).

Three different ways are known for designing partially crystalline microstructures:

1) Rapid solidification (4) by melt spinning or planar flow casting may result in fully amorphous, partial crystalline or micro-crystalline material. Due to local differences in the cooling rate the microstructure exhibits a strong gradient throughout the thickness of the ribbon: The microstructure near the contact side may be fully amorphous, but may change with increasing distance to partially crystalline and microcrystalline.

2) Partial amorphization by solid state reaction (5),(6) can lead to artificial microstructures with amorphous layers (see figure 1), fibers or dots; their distribution can be controlled by building up the multilayered structure.

3) Partial crystallization of metallic glasses (7): The microstructure can be controlled by an appropriate choice of solidification and/or crystallization parameters:

Solidification parameters, e.g., melt temperature, surface speed of the wheel are known to influence the number of quenched-in nucleation sites; crystallization below T_g occurs by heterogeneous, transient-type nucleation controlled by diffusivity and the number of quenched-in nucleation sites.

Annealing treatments: Above T_g crystallization proceeds by homogeneous, transient-type nucleation controlled by viscosity of the undercooled melt. Therefore, pulse heating above T_g for example can increase nucleation rates by orders of magnitude.

External factors, e.g., stress, which is known to increase volume diffusion and therefore to accelerate nucleation as well as volume diffusion controlled growth.

Nucleation agents, e.g., addition of 2 % Au or Cu in $Fe_{80}B_{20}$ has been found to provide ample nucleation sites for α-Fe primary crystallization.

REFERENCES
1. H.-G. Hillenbrand, E. Hornbogen, U. Köster, Metall 36 (1982), 1059
2. A. Datta, N.J. DeCristofaro, L.A. Davis, Proc. Rapidly Quenched Metals 4 (Sendai 1981), p. 1007
3. W.L. Johnson, in: "Glassy Metals I", eds. H.-J. Güntherodt, H. Beck Topics in Applied Physics, Vol. 46 (Springer Verlag 1981), p. 191

4. U. Köster, U. Herold, D. Krause, in: "Rapidly Solidified Amorphous and Crystalline Alloys", MRS-Boston 1981, eds. B.H. Kear, B.C. Giessen, M. Cohen, Elsevier Publ. Co. 1982), p. 179
5. R. Schwarz, W.L. Johnson, Phys. Rev. Lett. $\underline{51}$ (1983), 415
6. H. Schröder, K. Samwer, U. Köster, Phys. Rev. Lett. $\underline{54}$ (1985), 197
7. U. Köster, Z. Metallkunde $\underline{75}$ (1984), 691

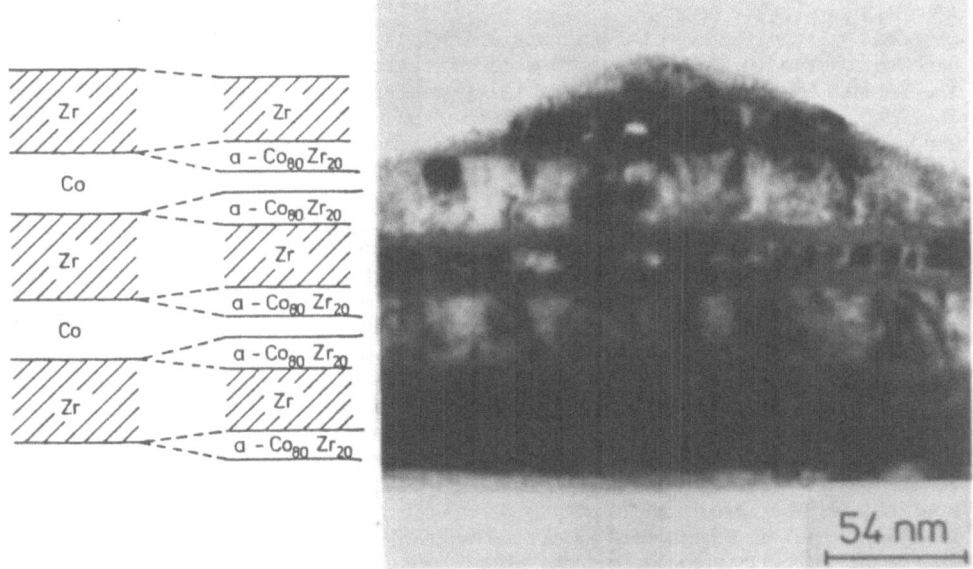

Figure 1: Amorphization by solid state reaction in Co/Zr layered structures (20nm Co/34nm Zr: 2h at 210°C).

THEORY OF MICROSTRUCTURAL DEVELOPMENT DURING RAPID SOLIDIFICATION

R. TRIVEDI (1) and W. KURZ (2)

(1) Iowa State University, 100 Metallurgy Building, Ames, IA 50011, USA
(2) Ecole Polytechnique Federale de Lausanne, CH-1007 Lausanne, Switzerland

ABSTRACT
   Some of the fundamental factors which play important roles in determining microstructures under rapid solidification condition are described. It is shown that the interface stability analysis needs to be extended to include the possibility of high thermal as well as high solute undercoolings. Furthermore, appropriate descriptions of solute and thermal fields for high speed transformations need to be developed.
   An extension of the linear perturbation analysis is proposed which takes into account the high thermal undercooling condition. It is found that above a certain critical velocity (or undercooling), a planar interface will always be stable with respect to any arbitrary perturbations in its shape. Thus, an absolute velocity is predicted for pure and alloy undercooled melts. A general stability criterion is developed for a planar interface stability and this criterion is valid for small as well as large growth rates.
   The results of the stability analysis are applied to study the microstructural changes which occur as a function of undercooling. For this prediction we consider the length scale of the microstructure to correspond to the marginally stable wavelength for a planar interface stability. The application of this criterion to dendritic growth phenomena shows that a limiting condition is achieved at the unit dimensionless undercooling value where the velocity approaches the absolute velocity value, and the radius of dendrite tip approaches infinity. Furthermore, the function VR^2, which is constant at low undercooling conditions, is found to increase sharply as the undercooling is increased.
   The theoretical model is extended to understand microstructural changes which occur under constrained growth conditions. At high growth rates, appropriate stability criterion is coupled with the solute trapping effect to predict microstructural features as a function of velocity. Furthermore, at high growth rates, the undercooling can become sufficiently large so that the diffusion coefficient variation with temperature must also be taken into account. In addition, any changes in phase diagram characteristics with temperature should also be examined. The effects of all these variables on interface curvature and on solute compostion in solid have been examined, and their influence on microstructural characterization under rapid solidification conditions have been evaluated. The model is applied to the Ag-Co system for which detailed experimental microstructural analysis has been published in the literature. A reasonable agreement between the theory and the experimental data is found.
   The theoretical model is also extended to predict eutectic characteristics under rapid solidification conditions. It is found that a coupled eutectic growth can occur only below a certain critical velocity. The effect of phase diagram characteristics on this critical velocity is examined.

PLANAR INTERFACE STABILITY ANALYSIS

Planar interface unstable when

$$m(G_c)_{eff} - (G)_{eff} > (\gamma/\Delta S)\omega^2$$

$$(G_c)_{eff} = (G_c) \cdot \xi_c$$

$$(G)_{eff} = K_S G_S \xi_S + K_L G_L \xi_L$$

The function ξ_S, ξ_L, ξ_c represent the effect of high growth rates.

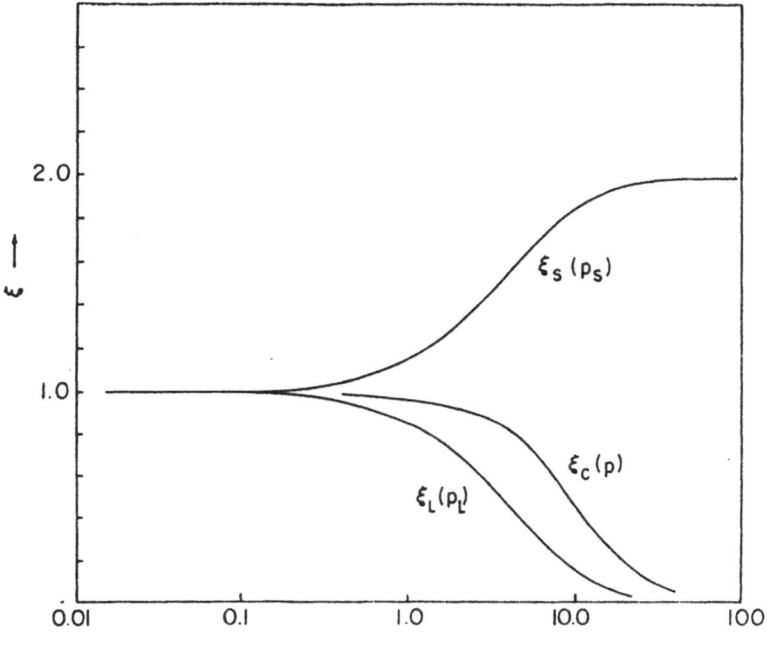

$$P_L = \frac{v\lambda}{2a_L}, \quad P_S = \frac{v\lambda}{2a_S}, \quad P = \frac{v}{2D}$$

a_L, $a_S \rightarrow$ thermal diffusivities
$D \qquad \rightarrow$ diffusion coefficient
$\lambda \qquad \rightarrow$ marginally stable wavelength

CONCLUSIONS

I.   ABSOLUTE STABILITY
     1. Pure thermal case
        $(V_{abs}) = a_L \Delta S \Delta H / \gamma c_L$

     2. Pure solute case
        $(V_{abs}) = D \Delta T_0 \Delta S / \gamma k$

     3. Undercooled alloy melt
        $(V_{abs}) = (V_{abs})_T + (V_{abs})_S$

II. Effective gradients of temperature and concentration differ signi-
     ficantly from actual gradients under rapid solidification conditions.

III. The constitutional supercooling criterion is not valid under rapid
     solidification conditions.

     CSC valid only when $(G_c)_{eff} = G_c$ and $(G)_{eff} = G$. i.e., $V \rightarrow O$, and when
     interface energy effects are negligible.

Application → dendrite growth

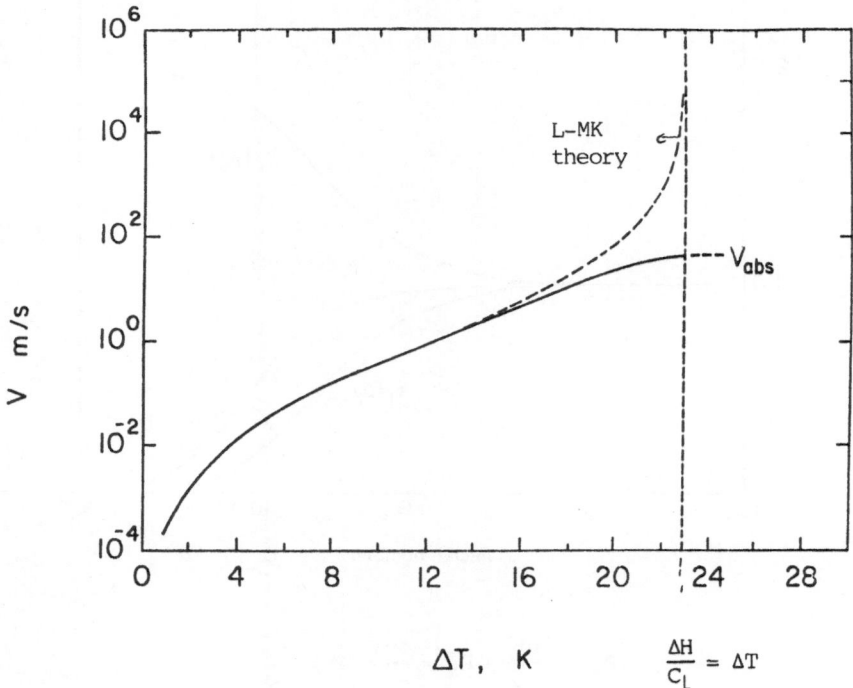

Growth rate vs. undercooling for pure succinonitrile dendrite growth from undercooled melt. L-MK → Langer and
Müller-Krumbhaar theory.

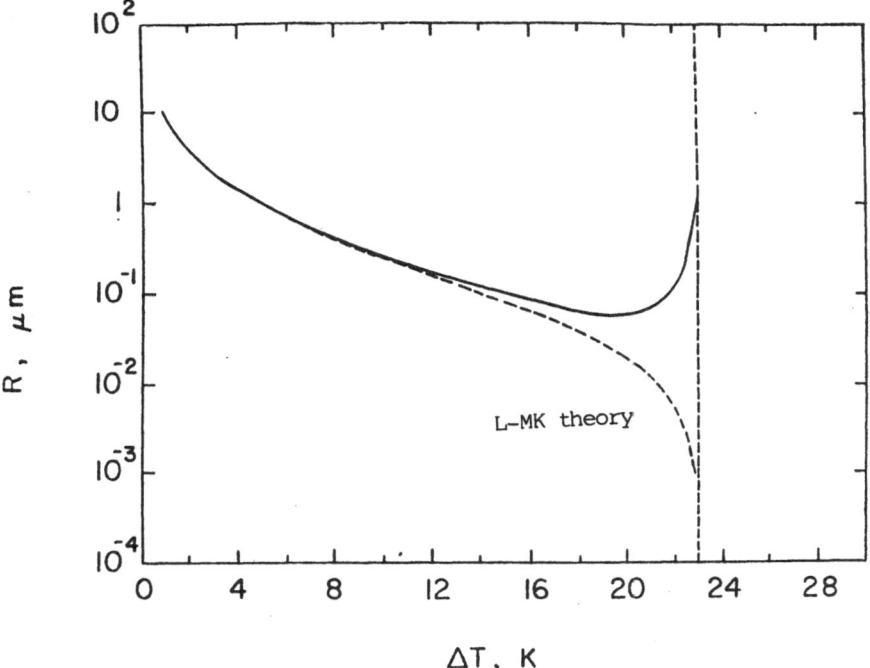

Dendrite tip radius vs. undercooling.

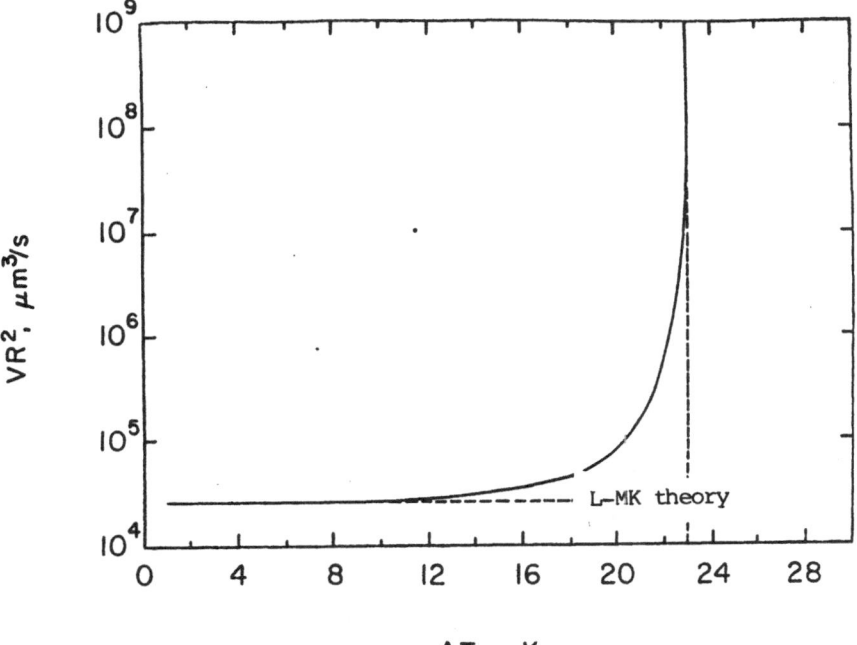

VR^2 vs. undercooling for pure succinonitrile growth.

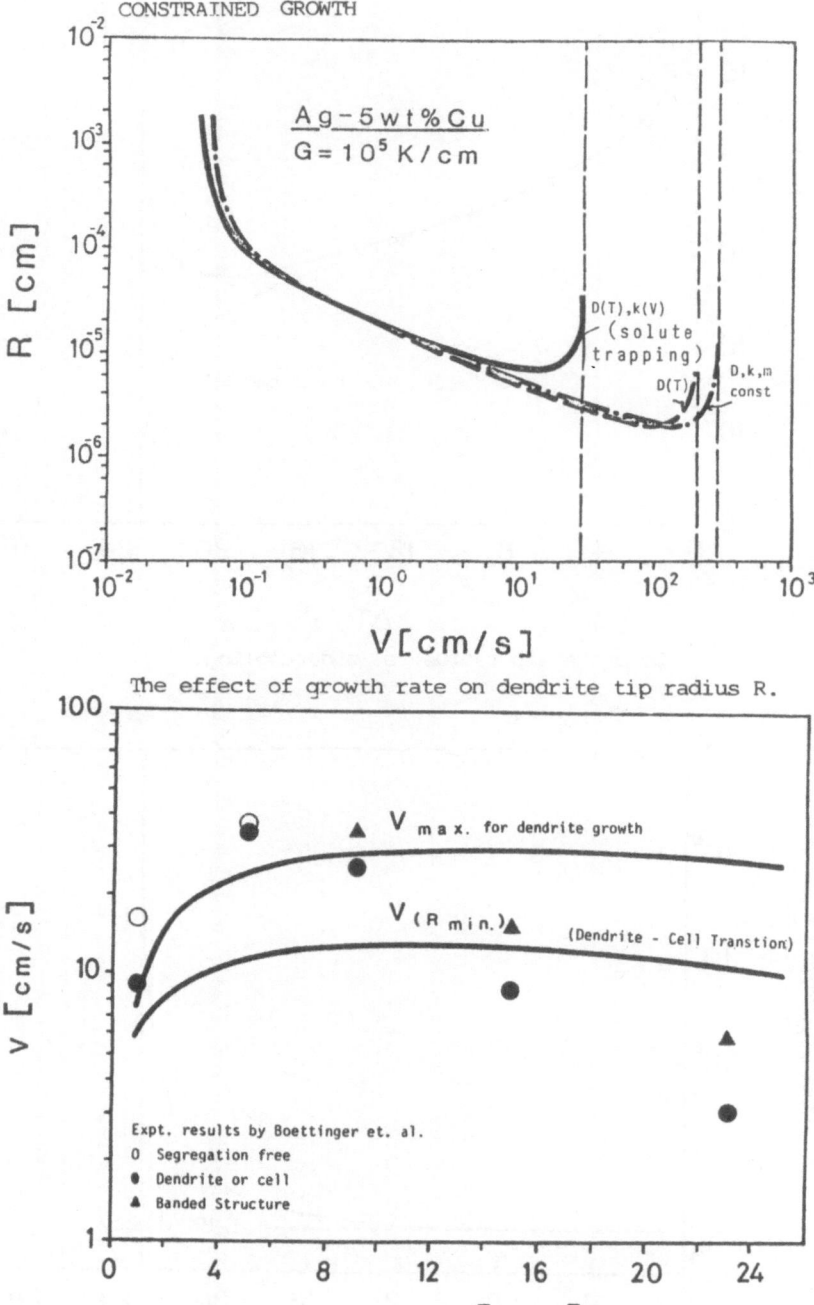

The effect of growth rate on dendrite tip radius R.

Predicted microstructural changes as a function of composition.

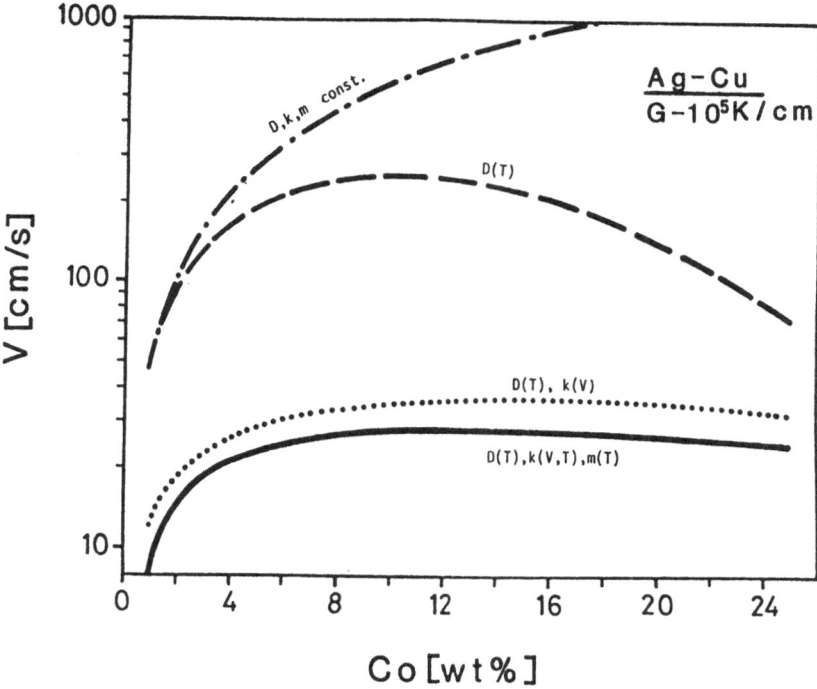

The effects of various factors which become important
at high growth rates.

EUTECTIC GROWTH

THEORY

I.   Jackson-Hunt Model:
     Not applicable to rapid soldification since they assume:
     (a) $V \ll 2Dn\pi/\lambda$
     (b) undercooling is small.

II. These assumptions are relaxed in the present model. The result can be
     written to be equivalent to the Jackson-Hunt result. The major dif-
     ference is in the function P which was constant for a given volume
     fraction in the J-H treatment. The present model shows P to be a
     function of volume fraction (f), equilibrium partition coefficient (k)
     and velocity (V).

III. P(f,k,V) is found to depend on the nature of the phase diagram. Two
     limiting phase diagrams are first considered. (i) Liquidus and solidus
     are parallel, and (ii) $k_\alpha = k_\beta = 0$. The effect of k is then studied
     for a phase diagram in which $k_\alpha = k_\beta$.

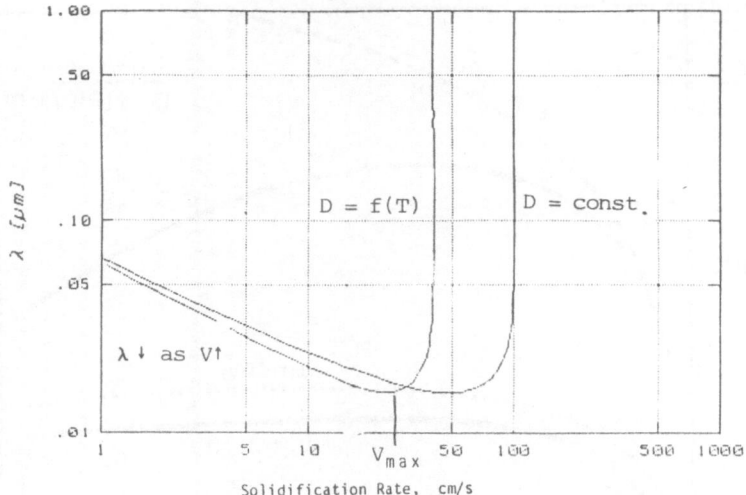

λ vs. V for eutectic system with parallel solidus and liquidus lines, V_{max} is predicted.

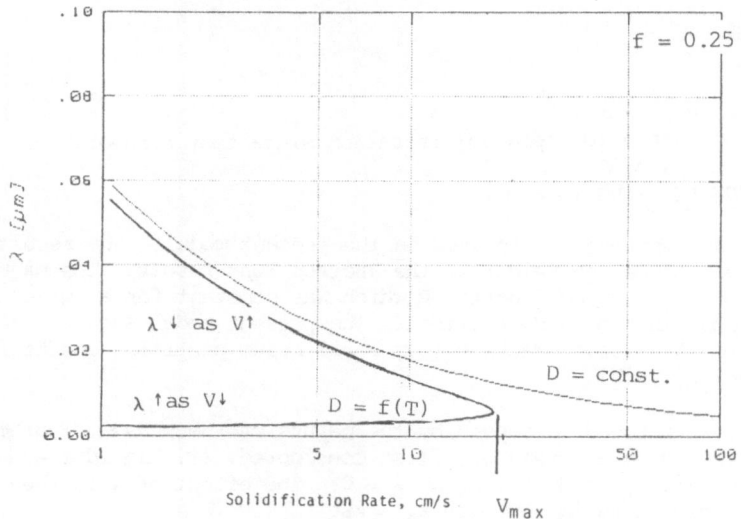

λ vs. V for eutectic system with $k = 0$
V_{max} is predicted.

CONCLUSIONS

1. A significant variation in the Jackson-Hunt function P is found at high velocities. For a given volume fraction, P increases with peclet number when $k = 0$, whereas it decreases when $k \to 1$.

2. The temperature dependence of the diffusion coefficient is very important for small k values.

3. A unique λ vs. V relationship is obtained for large k values. However, λ is a double valued function when k is small.

4. For small k, two branches of λ vs. V curve are observed. The upper branch, for solidification, shows that λ decreases as V increases. The lower branch, for amorphous to crystalline transition shows that λ increases as V increases.

5. Irrespective of the phase diagram, a limiting value of V is found above which an eutectic structure cannot be formed. This is due to either (i) the variation in the diffusion coefficient with temperature, or (ii) the limit of maximum undercooling at the interface which is imposed by the phase diagram.

RS ALUMINIUM - PROCESSING AND EVALUATION

M.J. COUPER and R.F. SINGER
Brown Boveri Research Center
CH-5405 Baden, Switzerland

The poster presented attempts to highlight the main advances that have been made recently in producing rapidly solidified aluminium powders and evaluation of the microstructure of these powders.

A refinement in powder size and hence increase in average cooling rate has been achieved using rapid-cooling gas-atomization (RCGA). Close-coupled nozzle designs are used and operation (or rather optimization) of the nozzles is achieved by judicious choice of nozzle insert geometry, nozzle insert position relative to the gas inlet and gas inlet pressure. These parameters are characterized by the static pressure within the melt flow tube under simulated atomizing conditions. Powder sizes of around 50 wt % <20 µm are achieved. Cooling rates are calculated from convective heat transfer equations.

The microstructure of the powders has been characterized in two alloy systems: X2618 and Al-Fe-Ce. Measured cell sizes are related to the powder diameter. Fine powders ($\leq$15 µm) show a variety of structures indicative of a variation in degree of under-cooling (figure 1a) and a clear transition from "massive solidification" to dendritic solute partitioning due to recalescence can often be seen. In Al-Fe-Ce powders the micro-cellular (or micro-eutectic) structure found in Al-Fe binary alloys is seen, with a more gradual change in cell spacing during recalescence as solidification procedes through the droplets. (figure 1 b)

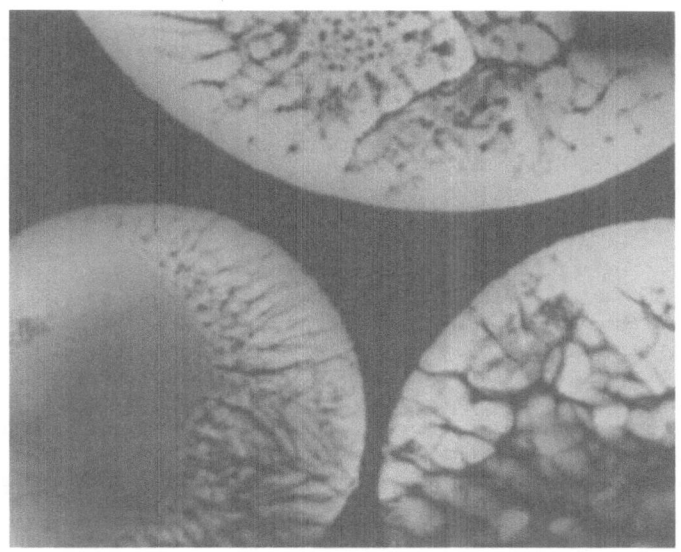

Figure 1a: X2618 Powders

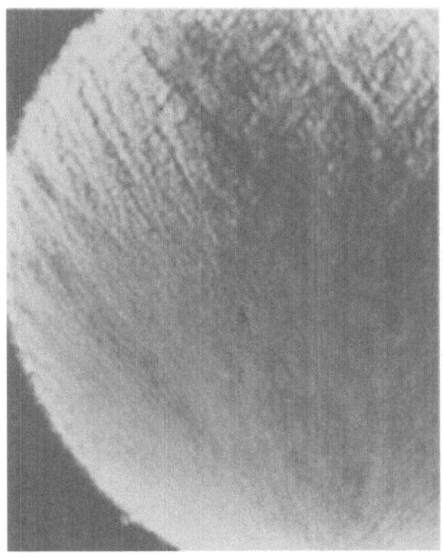

Figure 1b: Al-Fe-Ce Powder

COMPUTER CONTROLLED PYROMETRIC MEASUREMENT OF THE COOLING RATE DURING MELT SPINNING

G. FROMMEYER and E. VOGT
Max-Planck-Institut für Eisenforschung GmbH
D-4000 Düsseldorf 1, F.R. Germany

A computer controlled meltspinning unit allows collection of 12,000 data during each run of the following process parameters:
- melt temperature (measured by a Pt 18 thermocouple)
- argon gas pressure in the crucible
- wheel surface temperature (type "J" sliding thermocouples)
- nozzle - wheel gap distance (measured by modulated IR-light. The gap is adjusted by a computer controlled step motor with a 1 μm step width)
- ribbon surface temperature at up to 5 measuring points, tracing the same point of the ribbon. (Measurement with fast response IR-pyrometers. The cooling rate is derived from the time- temperature differences between the measuring points.)

Typical process parameters for a Fe 6.2 wt % Si alloy:
- melt weight 50 - 150 g, melt superheat 200° C
- ribbon width 12 mm
- argon gas pressure 15 - 25 kPa
- nozzle - wheel gap distance 150 - 200 μm
- wheel temperature 150 - 200°C (mild steel wheel).

Figure 1:
The meltspinning unit at the Max-Planck-Institut für Eisenforschung ,
Düsseldorf.

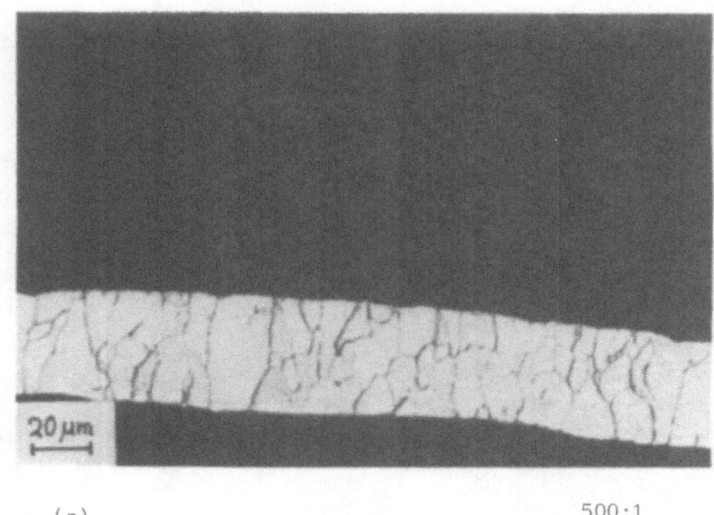

(a)   500:1

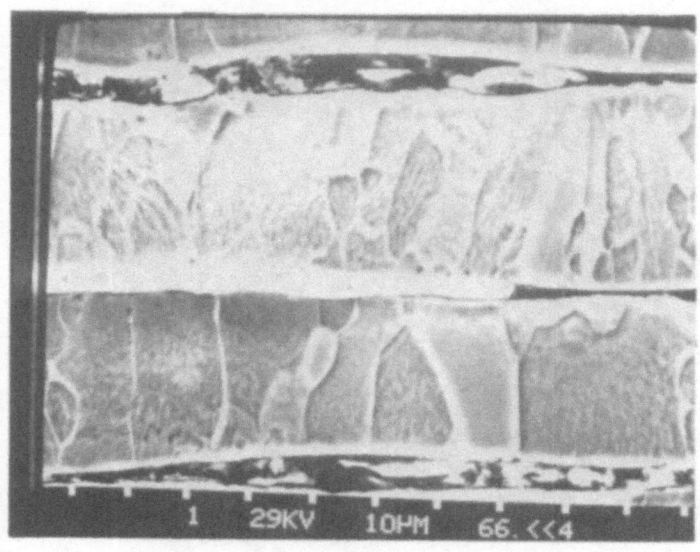

(b)   1000:1

(b)   1000:1

Figure 2:
As solidified structure of an Fe 6.2 wt% Si meltspun ribbon, ribbon thickness 35 µm, mean cooling rate 2.1 x 10^5 K/s.

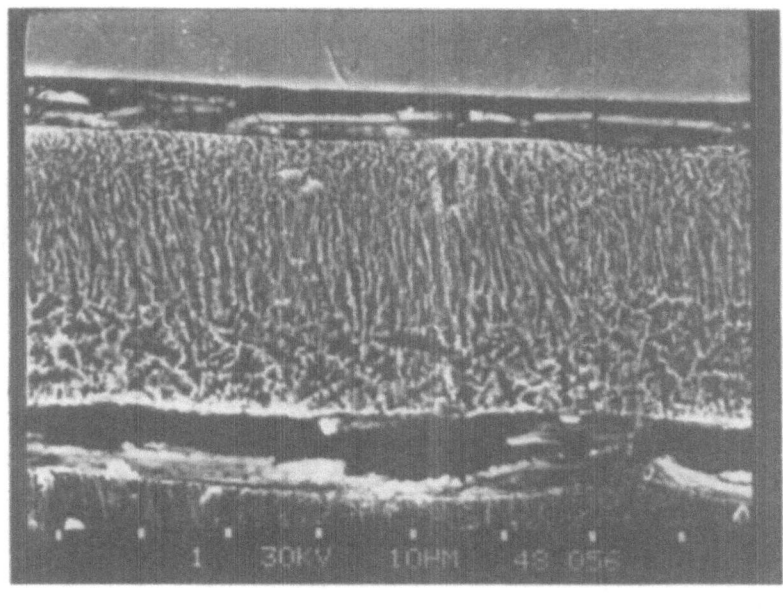

(a)   1500:1

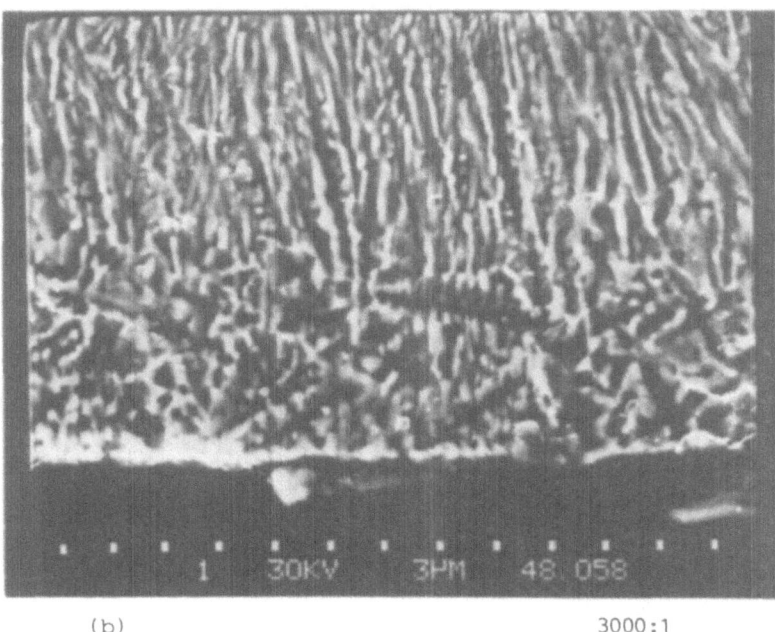

(b)   3000:1

Figure 3:
As solidified structure of an Fe 3.2 wt% C meltspun ribbon, ribbon thickness 30 μm.

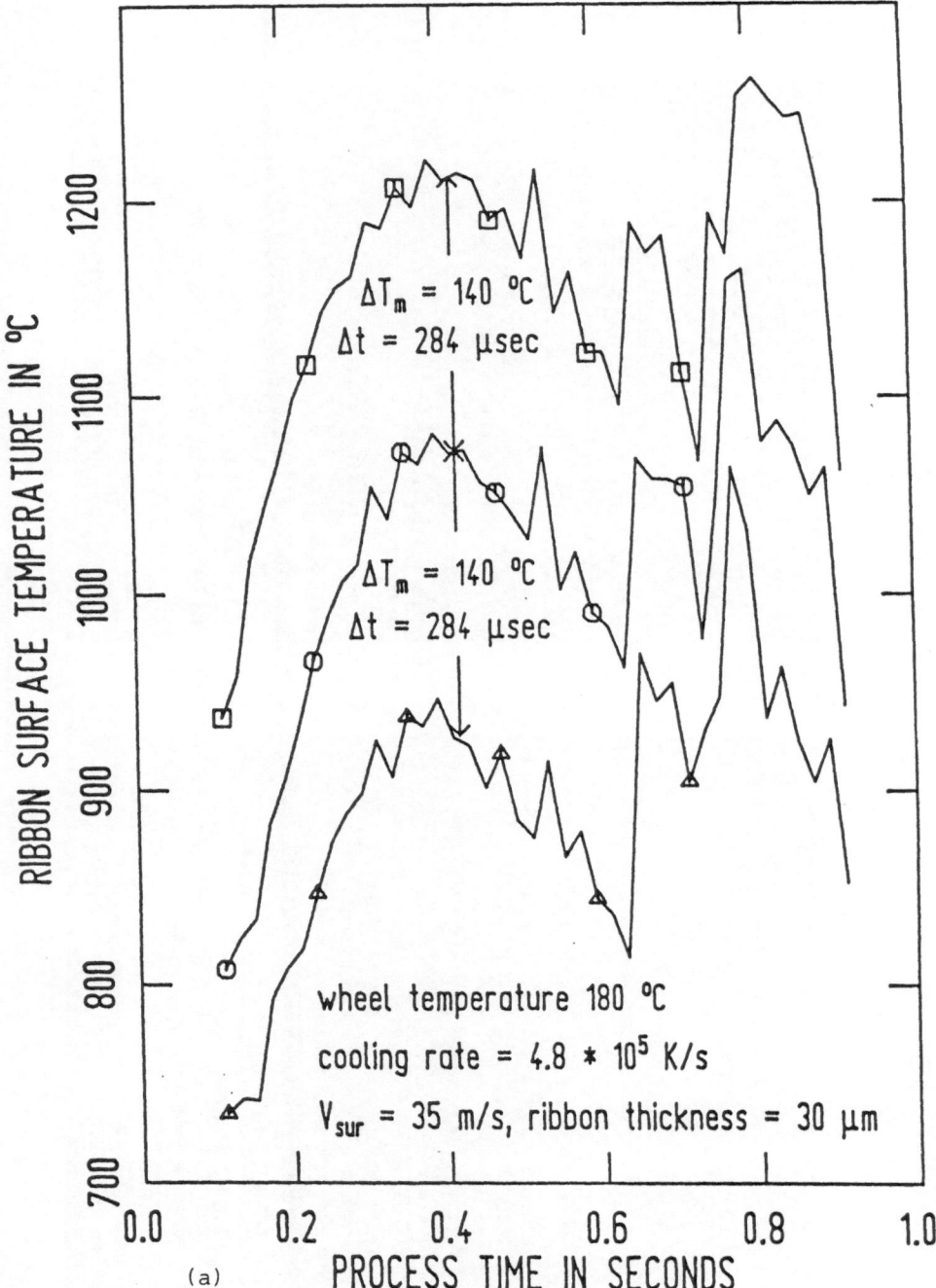

Figure 4:
The two plots show measurements of the ribbon surface temperature vs. process time for 12 mm wide meltspun ribbons of Fe-6.2 wt% Si. The ribbon surface temperature is measured with three IR-pyrometers positiones 25, 35 and 45 mm away from the nozzle. After about 0.2 - 0.4 seconds the process

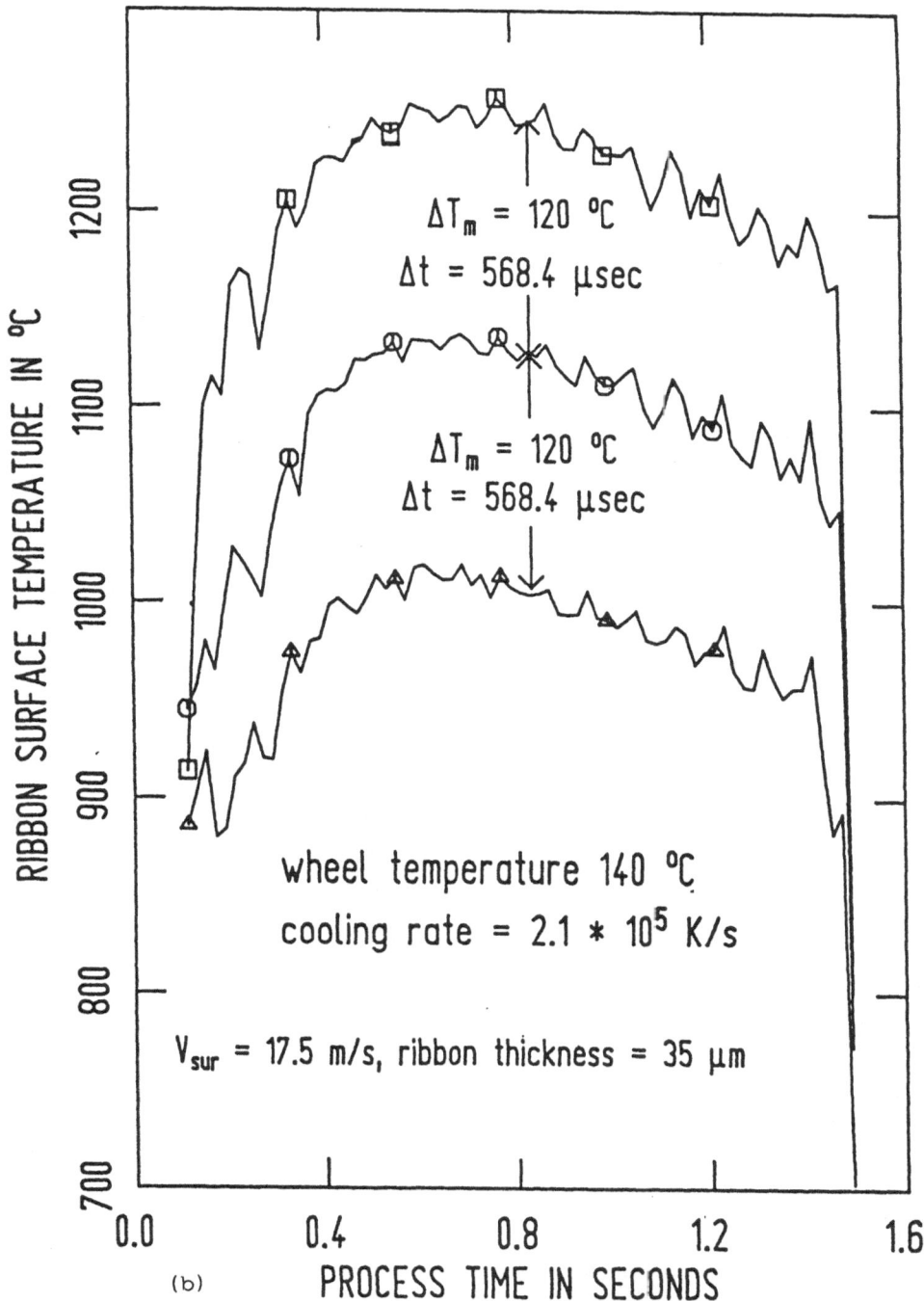

(b)

is fully developed and the cooling rates between the pyrometers are almost the same. An increase in wheel surface temperature produces better contact between ribbon and wheel and a marked increase in cooling rate and contact length.

THE OPTIMIZATION OF AMORPHOUS MATERIALS FOR TOROIDAL TRANSFORMER CORES

K. MÄKELÄ, T. TIAINEN and P. KETTUNEN
Tampere University of Technology
Institute of Materials Science
SF-33101 Tampere, Finland

INTRODUCTION
    The influence of process parameters on the magnetic properties of melt spun amorphous ribbons has been described in many reports (1) - (4). The magnetic properties depend, e.g., on cooling rate, ribbon thickness, surface defects etc. which all can be controlled by proper selection of process parameters.
    The magnetic properties of as-cast amorphous ribbons can be greatly improved by heat treatment carried out in a magnetic field. This annealing is a necessity for achieving the full advantage of the good magnetic properties of certain amorphous metals.
    The magnetic properties of the toroidal transformer cores made from amorphous metals depend also on the preparation of the toroid. The winding force and the possible use of insulative layers between windings affect greatly the magnetic properties of the toroid.
    In this study the effect of process parameters and the preparation method of toroids on the magnetic properties of small amorphous iron-based transformer cores is presented and discussed in the light of some typical examples.

EXPERIMENTAL PROCEDURE
    Several Fe-based alloys were first prepared by melting Armco iron and ferroproducts using alumina crucibles in an argon atmosphere. From these premelted alloys amorphous ribbons with a width of 12,5 mm and a thickness varying from 16 to 52 μm were prepared by the planar flow casting method.
    In the experimental works the influence of several process parameters, magnetic field annealing and the preparation of toroids on the magnetic properties of small transformer cores was studied. The studied cores had an outer diameter of 32 mm and inner diameter of 20 mm. The magnetic properties were measured at frequencies varying from 50 Hz to 10 kHz. In the following, some typical results obtained with the alloy $Fe_{81}B_{13}Si_4C_2$ at the measuring frequency of 50 Hz are presented and discussed.

RESULTS AND DISCUSSION
    The magnetization and iron loss curves for different ribbon thicknesses obtained by varying the speed of the quenching roll surface are shown in figure 1 for cores annealed in magnetic field (2 h at 365°C, magnetic field of 750 A/mm). Magnetizing is easy and iron losses are smallest at low magnetic induction obtainable with ribbon thickness of 38 μm whereas at higher magnetic induction the optimum ribbon thickness is about 33 μm.
    The effect of gas manifold (keeping the solidifying ribbon on the cooling surface for a controlled time) on the magnetization curves as well as on iron losses of the studied alloy is shown in figure 2 for the annealed as well as for the non-annealed state. The gas manifold helps to maintain the cooling conditions at the wanted level which in its turn influences also the magnetic properties of the ribbon.

The influence of the winding force on packing density and iron losses of toroids made from an alloy of $Fe_{40}Ni_{40}B_{20}$ is shown in figure 3 for as-cast ribbons. Increasing winding force increases the packing density of toroids; on later ribbons a packing density of 85 % was achieved at a winding force of 1 N. Simultaneously, the increase in winding force increases the iron losses due to the increased internal stress level of the ribbon.

CONCLUSIONS

The magnetic properties of amorphous ribbons can be greatly influenced by a proper preparation technique of the ribbon itself, by magnetic field annealing as well as by the preparation of toroidal amorphous cores. In the studied Fe-based alloys, the annealing treatment of the ribbons in a magnetic field appeared to be necessary for the transformer applications. The improvement of magnetic properties obtained by heat treatment was quite significant.

ACKNOWLEDGMENT

The financial aid received from the Technical Development Center of Finland (TEKES) is gratefully acknowledged. The authors want also to thank Miss Tiina Aaltonen for carrying out the magnetic measurements.

REFERENCES
1. F.E. Luborsky, H.H. Liebermann and J. L. Walter, Metallic Glasses: Science and Technology, edited by C. Hargitai, I. Bakonyi and T. Kemeny, Vol. 1, 203, Hungary, 1981.
2. S. Takayama and T. Oi, J. Appl. Phys., 50, (1979), 4962.
3. M. Matsura, M. Kikuchi, M. Yaki and K. Suzuki, Jap. J. Appl Phys., 19, (1980), 1781.
4. F. Okazaki, C. Kaido, Y. Matsuo and K. Honma, J. Magn. Magn., Mat., 41, (1984), 1942.

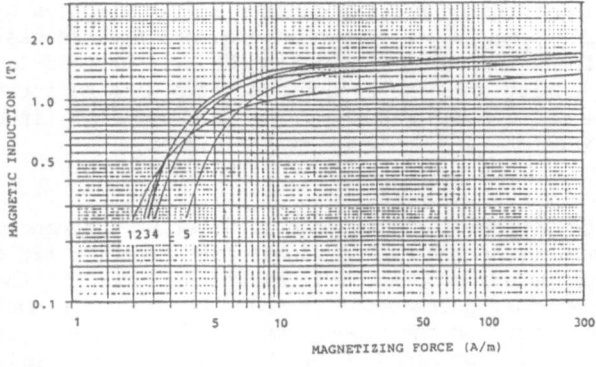

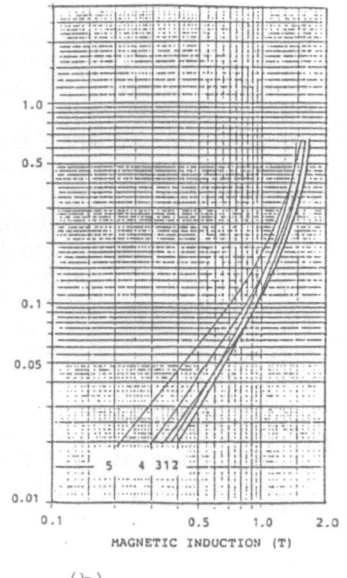

Figure 1: Magnetization curves (1a) and iron losses (1b) of magnetic field annealed toroids. Ribbons for toroids prepared at following surface speeds:
1. 17,5 m/s, thickness 44 μm
2. 20,0 m/s, thickness 38 μm
3. 22,5 m/s, thickness 33 μm
4. 25,0 m/s, thickness 26 μm
5. 27,5 m/s, thickness 23 μm

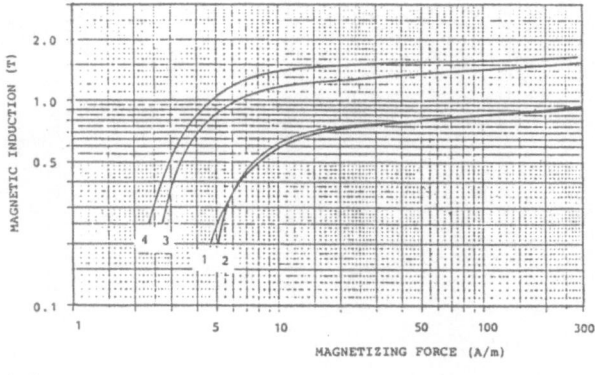

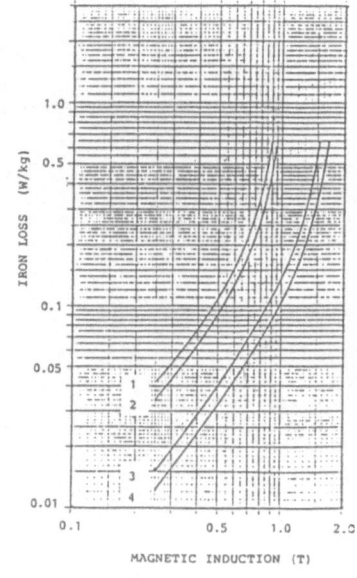

Figure 2: The effect of the use of gas manifold on the magnetization curves (2a) and iron losses (2b).
1. as cast, without gas manifold
2. as cast, with gas manifold
3. annealed in magnetic field, without gas manifold
4. annealed in magnetic field, with gas manifold.

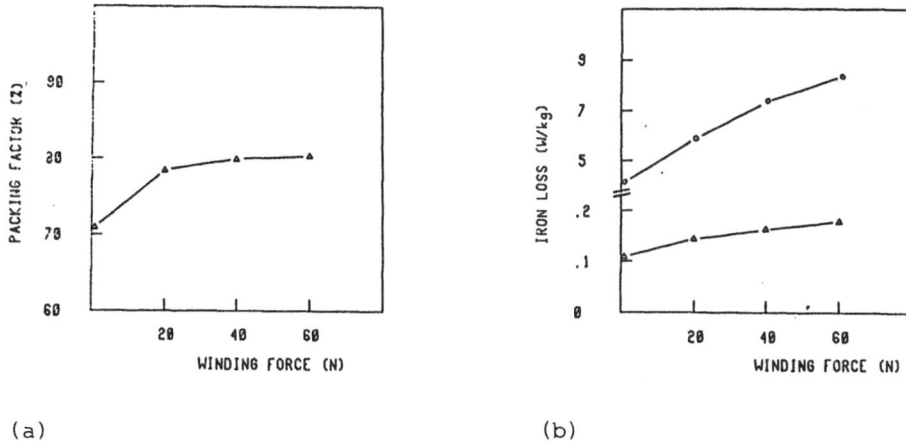

(a)   (b)

Figure 3: The effect of the winding force in as cast Fe+Ni-base alloy: a) on the packing factor, b) on the iron losses Δ = 50 cps, O = 1000 cps. The width of the ribbon is 11 mm and the thickness is 30 μm.

C. CONCLUDING STATEMENT

GENERAL OBSERVATIONS AND IMPRESSIONS

M. E. GLICKSMAN
Materials Engineering Dept.
Rensselaer Polytechnic Institute
Troy, N.Y. 12811, USA

An important aspect of the "NATO Workshop on the Science and Technology of the Undercooled Melt", one which sets this workshop in an unique position, is the focus upon undercooling as the key parameter in the rapid solidification process. Lectures and discussions centered on the considerable recent progress in exploring the bounds of the physical processes which are possible in highly undercooled melts. This attention to basics should provide practical dividends by yielding guidance to technologists in improving both processes for rapid solidification as well as strategies for optimal alloy designs that will benefit from such processing. Indeed, it was shown and discussed that a considerable range of rapid solidification microstructures can now be rationalized from rather basic solidification principles. The complexities of metastability and non-equilibrium processes were also addressed in providing more satisfying explanations of structures observed thus far by rapid freezing. It still remains to establish "hard" limits on undercooling and nucleation as well as incorporating the hydrodynamic state of the melt, especially in processes such as planar flow casting and melt spinning. These are obvious areas for future research. The WORKSHOP also clearly showed that rapid solidification is only one link - albeit an important one - in the overall chain of technological application and end-use. Open questions remain on the strategies for consolidation and subsequent achievement of engineering properties. Often subsequent processing can undo the improved microstructural state, although proper process strategy and alloy design can both add improvements which are not mutually exclusive. Lower working temperatures during subsequent processing are generally needed to preserve the microstructural improvement, although lower working temperatures where possible are desirable in themselves, e.g. for achieving finer grain sizes independent of the as-solidified grain size. Discussions of electrical and magnetic properties also broadened the viewpoint within the WORKSHOP, and showed interesting insights provided by solid-state physics and chemistry. The views of physicists, chemists, metallurgists, and technologiysts were actively shared by the participants, giving the strong impression that progress was significant on all the disciplinary fronts of rapid solidification. Indeed, the impending maturity of the fundamental aspects of this field augurs well for solid progress and important developments in the future.

INDEX